PREALGEBRA
AND ALGEBRA

CONTENTS

11 ALGEBRAIC EXPRESSIONS 134

12 LINEAR EQUATIONS 154

13 WORD PROBLEMS 174

14 GRAPHING STRAIGHT LINES 197

15 SYSTEMS OF LINEAR EQUATIONS 222

16 INEQUALITIES 236

17 FACTORING 248

18 FRACTIONS CONTAINING VARIABLES 269

19 FORMULA AND EQUATION MANIPULATION 304

20 PROPERTIES OF EXPONENTS 313

PREFACE

This book is intended to be used for any basic mathematics course in which the students are expected to learn elementary algebra. A presentation of essential arithmetic is provided for review or reference. In any case, students are led gently into a study of standard elementary algebra topics. The text is suitable for three types of courses:

1. Prealgebra and algebra
2. Elementary algebra
3. Arithmetic and algebra

FEATURES OF THE TEXT

The book retains many features that have proven successful in previous editions:

1. Theory and discussions have been kept brief and are followed by numerous examples.
2. Because I have found that an extra step or example can make the difference between understanding and confusion, students are led step by step through basic concepts.
3. Many exercises are included to ensure mastery of the material. Chapter review exercises are also included.
4. Answers to more than half of the exercises are provided.
5. Important ideas are boxed for easy reference.
6. Step-by-step procedures are given and boxed where appropriate.

NEW IN THIS EDITION

Several changes have been made to improve the text:

1. The use of *two colors* makes drawings clearer, reference easier, and the text more attractive and easier to use.

2. *Cumulative* review exercises appear in Chapters 11 through 22. They are intended to help students reinforce and remember concepts studied earlier in the course.

3. *Many more word problems* appear throughout the text. *The emphasis is on realism.* In particular, Chapter 13 ("Word Problems") has been rewritten to include more examples and many more exercises.

4. The *Example/Solution format* makes the examples easier to locate and use.

5. Exercise sets have been examined, refined, and expanded to better suit the needs of both students and instructor.

6. Many clarifying steps, statements, and examples have been added.

7. The appendix topic "ratio and proportion" has been merged into the main text for easier use by those who choose to include it.

8. Section 10.5 on *notation* has been simplified.

9. The presentation of *slope* has been moved forward to Chapter 14. However, the topic of slope (Sections 14.3 and 14.4) can be omitted or taught later without any loss of continuity.

10. Chapters 19 and 20 have been interchanged. As a result, Section 18.7 ("Equations with Fractions") leads nicely into 19.2, which contains many equations with fractions. Also, fractional exponents (20.5) now leads gently into Chapter 21, "Radicals and Roots."

11. Chapter 19 on *formula manipulation* now consists of two sections. As a result, the first material (19.1) can be taught earlier, say, as an extension or application of Chapter 12.

TITLE CHANGE

The text *Arithmetic and Algebra* was published originally in 1973 as the first book to offer chapters of arithmetic review (for use in class or as a reference) followed by a complete presentation of elementary algebra. More recently the marketplace has been filled with huge books offering *developmental* approaches to arithmetic via thousands of exercises. Those books assume the students know virtually nothing about basic arithmetic and must learn everything from the beginning. By contrast, I assume the student has forgotten some things and perhaps never really understood certain concepts, especially fractions and percent.

The approach is brief, to-the-point, and stresses understanding. My intent in these early chapters is to present the essence of arithmetic, to prepare students sufficiently so that they have a reasonable chance for success in algebra. This is why I now choose to call the approach *prealgebra* and the text *Prealgebra and Algebra.*

SUPPLEMENTS

Special Instructor's Edition

The publisher has created a special edition of the text for use by instructors. It includes answers to *all* exercises, course outlines, and teaching suggestions.

Test Item File

There is a Test Item File consisting of many questions (and answers) for each chapter. The file can be used to create review questions, quizzes, or chapter tests.

Computerized Testing System

A Computerized Testing System is available. It generates individual tests from a bank of over 700 items.

ACKNOWLEDGMENTS

I am grateful for the assistance of the following reviewers of *Prealgebra and Algebra* who have helped me to improve the text.

Previous Editions

Margaret G. Aldrich, Montgomery College; Robert Alwin, St. Petersburg Junior College; Dorys J. Barban, Montgomery College; Marty Campbell, College of the Virgin Islands at St. Croix; James W. Gorham, Jr., Harris-Stowe College; Anita Kitchens, Appalachian State University; Janice McFatter, Gulf Coast Community College; Gerhart Moore, Southeastern Louisiana University; Carla B. Oviatt, Montgomery College; Dudley R. Pitt, Northwestern State College of Louisiana; Patricia D. Roecklein, Montgomery College; Patricia H. Rubenstein, Montgomery College; Ronald E. Ruemmler, Middlesex County College; George Schultz, St. Petersburg Junior College; Robin Symonds, Indiana University at Kokomo.

Fourth Edition

Ann S. Bretscher, *University of Georgia*

Pat C. Cook, *Weatherford College*

Joanne Crowley, *Wayne College*

Irwin Goldfine, *Truman College*

Henry Harmeling, *North Shore Community College*

Martin W. Johnson, *Montgomery County Community College*

David L. Newton, *ITT Technical Institute*

John A. Oppelt, *Bellarmine College*

Patricia D. Roecklein, *Montgomery College*

Karen J. Schroeder, *Bentley College*

Arlene Sherburne, *Montgomery College*

Robin G. Symonds, *Indiana University at Kokomo*

Herbert C. Voltz, *Slippery Rock University*

Richard Walker, *Mansfield University*

James P. Winner, *College of William & Mary*

DEDICATION

This book is dedicated to my wife Sylvia and her mother Sylvania Prince.

Daniel D. Benice
Rockville, Maryland

PREALGEBRA
AND ALGEBRA

1

REVIEW OF
BASIC OPERATIONS

To succeed in your study of algebra, you must have a solid foundation in arithmetic. Why? Because most of the properties and manipulations used in algebra are the same as those used in arithmetic. Algebra is a generalization and extension of arithmetic. You will use a great deal of arithmetic when you study algebra. So if your arithmetic background is weak, you will not be able to master algebra.

The assumption is made that arithmetic is not totally foreign to you. In other words, we assume that you have seen fractions and decimals before, and that you have added, subtracted, multiplied, and divided whole numbers before. On the other hand, the assumption is made that many things have been forgotten, confused, or never really learned.

1.1 NUMBER STRUCTURE

We begin our study of arithmetic with a look at the structure of numbers and the basic operations used to combine numbers. Much of our attention will be on the **whole numbers**—that is, 0, 1, 2, 3, 4, 5, 6, 7, 8, 9, 10, 11, 12, 13, 14, 15, and so on. At the very beginning we will be using the numbers called **digits:** 0, 1, 2, 3, 4, 5, 6, 7, 8, 9.

When you read the number 572, you don't think five-seven-two but rather five hundred seventy-two. In other words, each digit (5, 7, and 2) has an additional meaning that corresponds to where it appears in the number, that is, its *place value*.

$$572 = 500 + 70 + 2$$
$$752 = 700 + 50 + 2$$
$$275 = 200 + 70 + 5$$

A digit placed in the rightmost column of a number contributes that many *ones* to the value of the number. In 572 there are 2 ones.

A digit placed in the next column to the left contributes that many *tens* to the number. In 572 there are 7 tens.

A digit placed in the next column over contributes that many *hundreds* to the number. In 572 there are 5 hundreds.

Thus

$$572 = (5 \times 100) + (7 \times 10) + (2 \times 1)$$
$$894 = (8 \times 100) + (9 \times 10) + (4 \times 1)$$

Similarly,

$$5360 = (5 \times 1000) + (3 \times 100) + (6 \times 10) + (0 \times 1)$$
$$47{,}815 = (4 \times 10{,}000) + (7 \times 1000) + (8 \times 100) + (1 \times 10) + (5 \times 1)$$

Here is a chart showing place values.

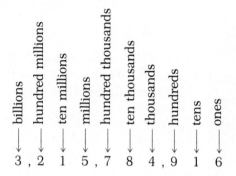

The *word equivalent* of any whole number can be read from the number. Here are a few examples.

$$894 = \text{eight hundred ninety-four}$$
$$47{,}815 = \text{forty-seven thousand eight hundred fifteen}$$
$$3{,}050{,}907 = \text{three million fifty thousand nine hundred seven}$$

Notice the use of commas to separate at every three digits, counting from the right.

EXERCISES 1.1

Answers to starred exercises are given in the back of the book.

1. Write each number as the sum of ones, tens, hundreds, and so on, as in the example below.

Example: 1563 = (1 × 1000) + (5 × 100) + (6 × 10) + (3 × 1)

*(a) 143	(b) 159	*(c) 50
*(d) 1603	*(e) 777	(f) 490
*(g) 600	*(h) 19,734	*(i) 34,851
(j) 65,127	*(k) 648,713	(l) 952,006
*(m) 4,980,617	(n) 1,187,543	

2. Write the word equivalent of each whole number.

*(a) 754
(d) 8694
*(g) 563,432
(j) 7,734,015
*(m) 654,456,546
(p) 9,017,658,432

(b) 879
*(e) 54,617
(h) 782,996
*(k) 78,800,167
(n) 780,078,009

*(c) 1341
(f) 71,955
*(i) 1,897,400
(l) 44,608,500
*(o) 8,098,764,123

*3. Which of the following numbers are digits?

(a) 9 (b) 0 (c) 15 (d) 3 (e) 2
(f) 10 (g) 11 (h) 167 (i) 23 (j) 32

*4. Which of the following numbers are whole numbers?

(a) 6 (b) 16 (c) 160 (d) $4\frac{1}{2}$
(e) $\frac{1}{4}$ (f) 6.4 (g) .78 (h) 10,000,000

1.2 ADDITION

Hopefully you will not need the basic addition table presented next. But if you don't already know it, memorize it. The entry 8 is circled to demonstrate how to read the table: $3 + 5 = 8$.

+	0	1	2	3	4	↓ 5	6	7	8	9
0	0	1	2	3	4	5	6	7	8	9
1	1	2	3	4	5	6	7	8	9	10
2	2	3	4	5	6	7	8	9	10	11
→ 3	3	4	5	6	7	(8)	9	10	11	12
4	4	5	6	7	8	9	10	11	12	13
5	5	6	7	8	9	10	11	12	13	14
6	6	7	8	9	10	11	12	13	14	15
7	7	8	9	10	11	12	13	14	15	16
8	8	9	10	11	12	13	14	15	16	17
9	9	10	11	12	13	14	15	16	17	18

You should be able to add 59 and 25 and get 84 as the sum.

$$
\begin{array}{r}
59 \\
\underline{25} \\
84
\end{array}
$$

Your reasoning may be as follows.

1. $9 + 5 = 14$.

2. Put down 4 and carry 1.

3. $5 + 2 = 7$ plus 1 (carried) gives 8.

But do you *understand* what you are doing? Does the procedure "put down 4 and carry 1" make sense to you or is it something that you memorized in grade school without any real understanding? Let's look at the problem differently. The number 59 has 5 tens and 9 ones; 25 contains 2 tens and 5 ones. Thus

$$
\begin{array}{c}
59 \\
25
\end{array}
\longrightarrow
\begin{array}{cc}
tens & ones \\
5 & 9 \\
2 & 5 \\
\hline
\end{array}
$$

Add the ones to get

$$
\begin{array}{cc}
tens & ones \\
5 & 9 \\
2 & 5 \\
\hline
& 14
\end{array}
$$

Since 14 ones are the same as 1 ten and 4 ones, we can rewrite the problem as

$$
\begin{array}{cc}
tens & ones \\
^{1}\;5 & 9 \\
2 & 5 \\
\hline
& 4
\end{array}
$$

where the boldface 1 indicates the "1 ten" part of the 14 ones. This is the "carry." The 4 is the "4 ones" part of the 14 ones. The final sum is obtained by adding the 5, 2, and carried 1. Thus

$$
\begin{array}{c}
^{1} \\
59 \\
25 \\
\hline
84
\end{array}
$$

The process is similar for addition of numbers having more than two digits; for example,

$$
\begin{array}{c}
395 \\
280 \\
155 \\
\hline
\end{array}
\longrightarrow
\begin{array}{c}
^{1} \\
395 \\
280 \\
155 \\
\hline
0
\end{array}
\longrightarrow
\begin{array}{c}
^{2\,1} \\
395 \\
280 \\
155 \\
\hline
30
\end{array}
\longrightarrow
\begin{array}{c}
^{2\,1} \\
395 \\
280 \\
155 \\
\hline
830
\end{array} \;\checkmark
$$

Addition may involve decimal numbers as well as whole numbers. *We will postpone until Chapter 6 any attempt to explain the meaning of decimal numbers. But here we will present the mechanics of adding decimal numbers, since we know that you have used decimals before.*

When the numbers being added have decimals, arrange the numbers so that all the decimal points are aligned. (This will align all the place values, so you will be adding tenths to tenths, and so on.) Then add as in previous examples. To add 342.6, 17.93, and 4.65, arrange the numbers and add as

$$\begin{array}{r} 342.6 \\ 17.93 \\ \underline{4.65} \\ 365.18 \quad \checkmark \end{array}$$

A whole number, such as 173, can be written as the decimal number 173. or 173.0. Thus to add the numbers 154.7, 173, and 23.02, set up the problem as

$$\begin{array}{r} 154.7 \\ 173.0 \\ \underline{23.02} \end{array}$$

and then add. You may prefer to supply enough zeros after the last decimal digit to make the numbers themselves align on the right side, as

$$\begin{array}{r} 154.70 \\ 173.00 \\ \underline{23.02} \end{array}$$

EXERCISES 1.2

*1. Perform each basic addition.

(a) $\begin{array}{r} 5 \\ \underline{4} \end{array}$
(b) $\begin{array}{r} 0 \\ \underline{6} \end{array}$
(c) $\begin{array}{r} 6 \\ \underline{2} \end{array}$
(d) $\begin{array}{r} 7 \\ \underline{6} \end{array}$
(e) $\begin{array}{r} 8 \\ \underline{0} \end{array}$
(f) $\begin{array}{r} 5 \\ \underline{9} \end{array}$

(g) $\begin{array}{r} 9 \\ \underline{9} \end{array}$
(h) $\begin{array}{r} 7 \\ \underline{8} \end{array}$
(i) $\begin{array}{r} 1 \\ \underline{8} \end{array}$
(j) $\begin{array}{r} 9 \\ \underline{1} \end{array}$
(k) $\begin{array}{r} 8 \\ \underline{3} \end{array}$
(l) $\begin{array}{r} 6 \\ \underline{7} \end{array}$

(m) $\begin{array}{r} 6 \\ \underline{4} \end{array}$
(n) $\begin{array}{r} 8 \\ \underline{8} \end{array}$
(o) $\begin{array}{r} 3 \\ \underline{7} \end{array}$
(p) $\begin{array}{r} 5 \\ \underline{8} \end{array}$
(q) $\begin{array}{r} 6 \\ \underline{6} \end{array}$
(r) $\begin{array}{r} 7 \\ \underline{7} \end{array}$

2. Perform each addition.

*(a) $\begin{array}{r} 57 \\ \underline{12} \end{array}$
*(b) $\begin{array}{r} 36 \\ \underline{53} \end{array}$
*(c) $\begin{array}{r} 49 \\ \underline{26} \end{array}$
(d) $\begin{array}{r} 58 \\ \underline{37} \end{array}$

(e) $\begin{array}{r} 67 \\ \underline{79} \end{array}$
*(f) $\begin{array}{r} 99 \\ \underline{54} \end{array}$
(g) $\begin{array}{r} 428 \\ \underline{351} \end{array}$
*(h) $\begin{array}{r} 785 \\ \underline{517} \end{array}$

*(i) $\begin{array}{r} 670 \\ \underline{920} \end{array}$
(j) $\begin{array}{r} 906 \\ \underline{809} \end{array}$
*(k) $\begin{array}{r} 6520 \\ \underline{1349} \end{array}$
(l) $\begin{array}{r} 4503 \\ \underline{2917} \end{array}$

*(m) $\begin{array}{r} 9743 \\ \underline{1059} \end{array}$
*(n) $\begin{array}{r} 94 \\ 37 \\ \underline{15} \end{array}$
*(o) $\begin{array}{r} 691 \\ 473 \\ \underline{120} \end{array}$
(p) $\begin{array}{r} 157 \\ 845 \\ \underline{419} \end{array}$

(q)	452	*(r)	9503	(s)	9913	*(t)	1508
	107		7914		4527		1974
	59		1506		1602		6461
							3915

*(u)	5218	(v)	4397	*(w)	59406	*(x)	78532
	4729		1562		18733		94164
	1001		2468		29464		60878
	7362		1619				

(y) 46521
    94374
    16857

3. Perform each addition.

*(a)	76.2	*(b)	1.89	(c)	54.107	*(d)	145.09
	81.9		7.68		77.099		23.992

*(e)	876.9	(f)	78.981	*(g)	86.74	(h)	185.09
	752.8		47.097		17.91		329.19
					23.85		762.29

(i)	5692.13	*(j)	7888.99
	9987.98		2111.01

1.3 SUBTRACTION

Consider the subtraction problem

$$\begin{array}{r} 5947 \\ \underline{1362} \end{array}$$

Just as in addition, you proceed from column to column: first subtract the digits in the ones column, then in the tens column, and so forth. To begin, 2 from 7 is 5.

$$\begin{array}{r} 5947 \\ \underline{1362} \\ 5 \end{array}$$

Next, try to subtract 6 from 4. But wait! 6 is larger than 4; so the subtraction cannot be performed at this point. Instead, "borrow" 1 from the column to the left of the 4. And since numbers placed in that column are worth ten times as much, the borrowed 1 is like a borrowed 10. Thus 10 is added to the 4, and the 9 is reduced to an 8. The effect is

$$\begin{array}{r} {}^{10} \\ 5847 \\ \underline{1362} \\ 5 \end{array}$$

As you can see, we didn't really *borrow* anything. Instead, we changed the form of 94 to 84 + 10; that is, 90 + 4 became 80 + 14.
 Now subtract 6 from 14 (10 + 4).

$$
\begin{array}{r}
\overset{10}{5847} \\
1362 \\
\hline
85
\end{array}
$$

Next, subtract 3 from 8, and subtract 1 from 5.

$$
\begin{array}{r}
\overset{10}{5847} \\
1362 \\
\hline
4585 \quad \checkmark
\end{array}
$$

We can check our result by an addition process. If 5947 − 1362 = 4585, then 1362 + 4585 = 5947.

$$
\textit{Check:} \quad
\begin{array}{r}
1362 \\
4585 \\
\hline
5947 \quad \checkmark
\end{array}
$$

 Note, however, that we usually do this addition mentally by looking at the completed subtraction problem.

$$
\left.
\begin{array}{r}
5947 \\
1362 \\
\hline
4585
\end{array}
\right\} \text{ Add these two mentally to get 5947.}
$$

 As another example, consider the subtraction below.

$$
\begin{array}{r}
4783 \\
2459 \\
\hline
\end{array}
\longrightarrow
\begin{array}{r}
\overset{10}{4773} \\
2459 \\
\hline
4
\end{array}
\longrightarrow
\begin{array}{r}
\overset{10}{4773} \\
2459 \\
\hline
24
\end{array}
\longrightarrow
\begin{array}{r}
\overset{10}{4773} \\
2459 \\
\hline
324
\end{array}
\longrightarrow
\begin{array}{r}
\overset{10}{4773} \\
2459 \\
\hline
2324 \quad \checkmark
\end{array}
$$

$$
\textit{Check:} \quad
\left.
\begin{array}{r}
4783 \\
2459 \\
\hline
2324
\end{array}
\right\} \quad 2459 + 2324 = 4783
$$

 Subtraction may involve decimal numbers. As in addition, align the numbers according to their decimal points. Then subtract as before. Here are two examples.

$$
\begin{array}{r}
65.37 \\
\underline{19.42} \\
45.95
\end{array}
$$

$$
\begin{array}{r}
139.07 \\
\underline{85.4}
\end{array}
\quad \longrightarrow \quad
\begin{array}{r}
139.07 \\
\underline{85.40} \\
53.67 \;\; \checkmark
\end{array}
$$

EXERCISES 1.3

*1. Perform each basic subtraction.

(a) 7	(b) 9	(c) 6	(d) 8	(e) 16	(f) 18
$\underline{\;\;2}$	$\underline{\;\;8}$	$\underline{\;\;0}$	$\underline{\;\;6}$	$\underline{\;\;9}$	$\underline{\;\;8}$
(g) 23	(h) 24	(i) 35	(j) 26	(k) 37	(l) 26
$\underline{\;\;9}$	$\underline{\;\;7}$	$\underline{\;\;8}$	$\underline{\;17}$	$\underline{\;18}$	$\underline{\;18}$

2. Subtract and check.

*(a) 53	(b) 74	*(c) 94	(d) 253
$\underline{29}$	$\underline{38}$	$\underline{65}$	$\underline{134}$
*(e) 540	(f) 619	*(g) 2538	(h) 1108
$\underline{265}$	$\underline{387}$	$\underline{1939}$	$\underline{794}$
*(i) 9420	*(j) 16046	*(k) 59213	(l) 35984
$\underline{3428}$	$\underline{9477}$	$\underline{49896}$	$\underline{26415}$
*(m) 94603	(n) 84716	*(o) 666666	
$\underline{82583}$	$\underline{75908}$	$\underline{99999}$	

3. Subtract and check.

*(a) 29.7	*(b) 87.6	*(c) 92.37	(d) 132.1
$\underline{15.2}$	$\underline{23.7}$	$\underline{45.43}$	$\underline{85.1}$
*(e) 8.765	(f) 56.90	*(g) 65.00	(h) 19.3
$\underline{1.766}$	$\underline{47.82}$	$\underline{17.32}$	$\underline{8.07}$
*(i) 212.66	(j) 654.783		
$\underline{132.76}$	$\underline{453.974}$		

1.4 MULTIPLICATION

Multiplication can be viewed as a shortcut for repeated addition. For instance, $5 \times 7 =$ five times seven = five sevens (added) $= 7 + 7 + 7 + 7 + 7$. You should have learned multiplication tables like the one shown here. If you do not know this table, memorize it. The entry 42 is circled to demonstrate how to read the table: $6 \times 7 = 42$.

								↓		
×	0	1	2	3	4	5	6	7	8	9
0	0	0	0	0	0	0	0	0	0	0
1	0	1	2	3	4	5	6	7	8	9
2	0	2	4	6	8	10	12	14	16	18
3	0	3	6	9	12	15	18	21	24	27
4	0	4	8	12	16	20	24	28	32	36
5	0	5	10	15	20	25	30	35	40	45
→ 6	0	6	12	18	24	30	36	(42)	48	54
7	0	7	14	21	28	35	42	49	56	63
8	0	8	16	24	32	40	48	56	64	72
9	0	9	18	27	36	45	54	63	72	81

You should know how to apply the multiplication tables to more complicated problems, such as

$$
\begin{array}{r}
27 \\
45 \\
\hline
\end{array}
$$

giving

$$
\begin{array}{r}
27 \\
45 \\
\hline
135 \\
108 \\
\hline
1215 \\
\end{array}
$$

Some understanding of this mechanical operation can be gained by noting the following.

$$45 \times 27 = 5 \times 27 + 40 \times 27 \qquad \text{(Consider 45 as 40 and 5.)}$$

and

$$
\begin{array}{l}
5 \times 27 = 27 + 27 + 27 + 27 + 27 = \quad 135 \\
40 \times 27 = \underbrace{27 + 27 + \quad \cdots \quad + 27}_{40 \quad 27\text{'s}} = \underline{1080} \\
 1215 \quad \checkmark
\end{array}
$$

The sum of *5* 27's and *40* 27's is 1215. This is the same as *45* 27's or 45 *times* 27.

Here is another important multiplication rule.

> If the multiplication involves numbers with decimal points, then the result will have after its decimal point the total number of digits found after the decimal points of the numbers being multiplied.

For example,

$$
\begin{array}{r}
14.2 \\
21.3 \\
\hline
42\ 6 \\
142 \\
284 \\
\hline
302.46
\end{array}
$$

14.2 ⟵ *one* digit after decimal
21.3 ⟵ *one* digit after decimal

302.46 ⟵ *two* (*one* + *one*) digits after decimal

and

$$
\begin{array}{r}
1.212 \\
.5 \\
\hline
.6060
\end{array}
$$

1.212 ⟵ *three* digits after decimal
.5 ⟵ *one* digit after decimal

.6060 ⟵ *four* (*three* + *one*) digits after decimal

Although we will not explain here why this process works, note that in the last example the result should be half (.5) of 1.212, or .606. And indeed, it is the result that we obtained by using the rule for decimals. An explanation of why this decimal multiplication method works is given after multiplication of fractions is presented in Chapter 6.

Here is the rule for multiplication involving zero.

> *Multiplication of any number by zero produces zero as a result.*

$$
\begin{aligned}
0 \times 3 &= 0 \\
0 \times 56 &= 0 \\
0 \times 0 &= 0 \\
0 \times \tfrac{1}{2} &= 0 \\
0 \times 7.4 &= 0
\end{aligned}
$$

If a multiplication involves a number containing a digit that is zero, it can be handled as

$$
\begin{array}{r}
513 \\
204 \\
\hline
2052 \\
000 \\
1026 \\
\hline
104652
\end{array}
$$

000 since
$0 \times 3 = 0$
$0 \times 1 = 0$
$0 \times 5 = 0$

Usually, however, we do not put in the whole row of zeros (as done above); instead, we simply use one zero and place the next row of the product on the same line, as

$$
\begin{array}{r}
513 \\
204 \\
\hline
2052 \\
10260 \\
\hline
104652
\end{array}
$$

Using a zero serves to indent the next row of digits (1026). It has the same effect as using the entire row of zeros.

EXERCISES 1.4

*1. Perform each basic multiplication.

(a)	8	(b)	9	(c)	7	(d)	8	(e)	0	(f)	4
	9		9		5		1		7		7

(g)	3	(h)	1	(i)	4	(j)	6	(k)	6	(l)	5
	6		0		3		2		8		9

2. Multiply.

*(a)	193	*(b)	165	*(c)	69	*(d)	75
	9		8		52		158

(e)	287	(f)	473	*(g)	4906	(h)	6754
	194		509		743		963

*(i)	6578	(j)	1692	*(k)	5600	*(l)	6754
	3495		4876		4732		4000

(m)	6700	*(n)	87591	(o)	7000000	*(p)	14783
	9000		60200		7000000		52000

3. Multiply.

*(a)	19.32	*(b)	9.874	(c)	85.14	(d)	78.39
	25.49		7.62		.603		56.10

*(e)	6.1697	(f)	842.93	*(g)	549.017	(h)	56.001
	42.392		618.01		8581.36		10.070

*(i)	805.001	(j)	8888.8
	1.00080		9999.9

1.5 DIVISION

Just as multiplication can be viewed as a shortcut for repeated addition, division can be viewed as a shortcut for repeated subtraction. For instance, if you are dividing 4 into 12, you are answering the question "How many fours are there in 12?" This question can be answered by subtracting fours from 12 until the result is zero and counting the numbers of fours subtracted. Thus

12	original number
4	⟵ subtract the first 4
8	result
4	⟵ subtract the second 4
4	result
4	⟵ subtract the *third* 4
0	result

Since subtracting three 4's from 12 left zero as a result, there are three 4's in 12. So 4 goes into 12 *three* times. This can be expressed by using such notation as

$$4\overline{)12}\;^{\textstyle 3}$$

$$12 \div 4 = 3$$

$$12/4 = 3$$

$$\frac{12}{4} = 3$$

Again (as in multiplication) we learn tables so that we can divide directly and quickly and avoid the actual subtraction and counting process.

In more complicated division problems, tables cannot be used *directly* to determine the *final result*. Several applications are needed. For example, in

$$5\overline{)170}$$

first divide 5 into 17 and get 3. Multiply 3 by 5 to get 15 and subtract the 15 from 17. There is a remainder of 2 at this point.

$$
\begin{array}{r}
3 \\
5\overline{)170} \\
15 \\
\hline
2
\end{array}
$$

The 15 is from 3 × 5.

You then "bring down" the next digit (0) and divide 5 into the newly formed number (20). This result is 4. The final result is 34.

$$
\begin{array}{r}
34 \quad\checkmark \\
5\overline{)170} \\
15 \\
\hline
20 \\
20 \\
\hline
\end{array}
$$

But this is a shortcut. Can you see what is really going on? If you look carefully at

$$5\overline{)170}$$

you can see that 5 × 30 = 150, which is smaller than 170, whereas 5 × 40 = 200, which is larger than 170. So divide 5 into 170 and get 30.

$$
\begin{array}{r}
30 \\
5\overline{)170} \\
150 \\
\hline
20
\end{array}
$$

Now 20 of the 170 is left. Next, divide 5 into that 20 to see how many more fives there are in 170. Since $20 \div 5 = 4$,

$$
\begin{array}{r}
30 + 4 = 34 \quad \checkmark \\
5\overline{)170} \\
\underline{150} \\
20 \\
\underline{20} \\
0
\end{array}
$$

So you can see that division (just as multiplication) can be done without shortcuts; however, meaningful shortcuts are available—and the shortcuts should be used. Consider a more advanced example.

$$34\overline{)75265}$$

$$
\begin{array}{r}
2 \\
34\overline{)75265} \\
68 \\
\hline
72
\end{array}
$$

$$
\begin{array}{r}
22 \\
34\overline{)75265} \\
68 \\
\hline
72 \\
68 \\
\hline
46
\end{array}
$$

$$
\begin{array}{r}
221 \\
34\overline{)75265} \\
68 \\
\hline
72 \\
68 \\
\hline
46 \\
34 \\
\hline
125
\end{array}
$$

$$
\begin{array}{r}
2213 \quad \checkmark \\
34\overline{)75265} \\
68 \\
\hline
72 \\
68 \\
\hline
46 \\
34 \\
\hline
125 \\
\underline{102} \\
23 \quad \checkmark
\end{array}
$$

The result is not exact, as before. There is a **remainder** of 23. The answer can be written as 2213, R: 23 (R for *Remainder*) or as $2213\frac{23}{34}$, where the fraction $\frac{23}{34}$ means that the process could be continued by dividing 34 into 23. But how can we divide 34 into 23, since 34 is larger than 23? Continue the computation by placing a decimal point after 75265 and zeros after the decimal point. This does not change the value of the number 75265.

$$
\begin{array}{r}
2213. \\
34\overline{)75265.0000} \\
\underline{68} \\
72 \\
\underline{68} \\
46 \\
\underline{34} \\
125 \\
\underline{102} \\
23
\end{array}
$$

Let us continue as before.

$$
\begin{array}{r}
2213.6 \\
34\overline{)75265.0000} \\
\underline{68} \\
72 \\
\underline{68} \\
46 \\
\underline{34} \\
125 \\
\underline{102} \\
230 \\
\underline{204} \\
26
\end{array}
\qquad\longrightarrow\qquad
\begin{array}{r}
2213.67 \\
34\overline{)75265.0000} \\
\underline{68} \\
72 \\
\underline{68} \\
46 \\
\underline{34} \\
125 \\
\underline{102} \\
230 \\
\underline{204} \\
260 \\
\underline{238} \\
22
\end{array}
$$

We can continue until the result comes out exact (with no remainder) or until we have enough digits after the decimal point. The result of division does not always come out exact; for example, $\frac{1}{3} = .333333333\ldots$. It is a nonterminating decimal.

Division involving numbers with decimal points is similar except that the decimal points of the numbers involved are shifted, as explained in the next example. The reason that this shifting process works is given after multiplication of fractions is explained in Chapter 6.

If you have a problem like

$$7.42\overline{)269.3516}$$

then shift the decimal two places to the right in both numbers. Doing so will make the divisor a whole number, 742.

$$7.42\overline{)269.3516}$$

So we get

$$742\overline{)26935.16}$$

instead of

$$7.42\overline{)269.3516}$$

We use multiplication to *check* that the result produced by division is correct. That is,

$$\begin{array}{r} 34 \\ 5\overline{)170} \end{array} \text{ if } 34 \times 5 = 170$$

When the division does not produce an exact result, the check is not complete until the remainder is added to the product. To illustrate, consider the division

$$34\overline{)75063}$$

The division:

$$
\begin{array}{r}
2207 \\
34\overline{)75063} \\
68 \\
\hline
70 \\
68 \\
\hline
26 \\
0 \\
\hline
263 \\
238 \\
\hline
25\ R
\end{array}
$$

The check:

$$
\begin{aligned}
(2207 \times 34) + 25 &= 75038 + 25 \\
&= 75063 \ \checkmark
\end{aligned}
$$

EXERCISES 1.5

*1. Perform each basic division. Then check, using multiplication.

(a) $5\overline{)20}$ (b) $6\overline{)54}$ (c) $9\overline{)63}$
(d) $8\overline{)24}$ (e) $4\overline{)28}$ (f) $3\overline{)18}$
(g) $1\overline{)8}$ (h) $2\overline{)18}$ (i) $7\overline{)35}$
(j) $6\overline{)42}$ (k) $5\overline{)45}$ (l) $3\overline{)27}$

2. Divide and check.

*(a) $4\overline{)2844}$ *(b) $2\overline{)16846}$ *(c) $3\overline{)2763}$
(d) $9\overline{)1827}$ (e) $8\overline{)64016}$ *(f) $7\overline{)5915}$
*(g) $21\overline{)1974}$ (h) $15\overline{)5730}$ *(i) $18\overline{)2754}$

3. Divide. Obtain the remainder in each case. Then check.

   *(a) 9)78   (b) 7)59   *(c) 12)1531
    (d) 15)9409   *(e) 23)5827   (f) 31)14762
   *(g) 84)19526   (h) 64)37682   *(i) 57)23155

4. Divide. Continue to two decimal places.

   *(a) 27)8416   *(b) 45)7894   (c) 40)15433
    (d) 102)56781   (e) 153)67854   *(f) 78)99999
   *(g) 205)19864   (h) 234)78695   *(i) 132)53187

5. Divide. Continue until your answer contains five digits.

   *(a) 7.9)259.204   (b) 2.1)54923   *(c) .62)78.65
    (d) 10.3)5.73029   *(e) 2.83)67435   (f) 8.79).015934
   *(g) .06)987   (h) .009)743.64   *(i) 1.09)12.007

1.6 ROUNDING

Suppose that you paid $10 for six house plants and you want to know how much you paid per plant. Naturally you would divide the $10 by 6 in order to obtain the cost per plant. The division:

$$\begin{array}{r} \$\ 1.666666 \\ \hline 6)\$10.000000 \end{array}$$

The division could be continued indefinitely. Actually, it has been carried too far already. Certainly $1.666666 is a strange form for an amount of money. Since dollars and cents are the standard way in which we express money, two digits after the decimal point are sufficient. But should the result be simply $1.66?

When *rounding* a number, the last digit retained should reflect the size of the digits being dropped. Specifically, if a number is being rounded to two decimal places (that is, two digits after the decimal point), then the third digit after the decimal should be examined. If it is 5 or greater (5, 6, 7, 8, or 9), the second digit should be increased by 1. If it is 4 or smaller (0, 1, 2, 3, or 4), the second digit should not be changed. All digits beyond the second digit are dropped. Here are some examples. More will be said about tenths and hundredths in Section 7.1.

Number	Rounded to Two Decimal Places (nearest hundredth)
$1.666666	$1.67
75.287	75.29
1.492	1.49
849.1254	849.13
16.2964	16.30
735.1337	735.13

The procedure for rounding to one decimal place is similar, of course.

Number	Rounded to One Decimal Place (nearest tenth)
9.27	9.3
9.24	9.2
18.35	18.4
1.791	1.8
2.96	3.0

Here are some examples of numbers rounded to the nearest whole number.

Number	Rounded to Nearest Whole Number
81.6	82
81.2	81
9.51	10
9.49	9
86.499	86
89.96	90

When rounding to the nearest ten, hundred, thousand, etc. (that is, to the *left* of the decimal point), digits being "dropped" to the left of the decimal point are replaced by zeros used as placeholders.

Number	Nearest Ten	Comment
863	860	since 3, leave 6 as is
7249	7250	since 9, add 1 to the 4

Number	Nearest Hundred	Comment
863	900	since 6, add 1 to the 8
7249	7200	since 4, leave 2 as is

EXERCISES 1.6

1. Round each number to two decimal places (nearest hundredth).
   *(a) 73.769   *(b) 15.222   (c) 4.128991   *(d) 98.1351
   (e) 1.87499   *(f) 765.919   *(g) 56.895   (h) 159.995

2. Round each number to one decimal place (nearest tenth).
   *(a) 17.83   *(b) 99.68   *(c) 187.15   *(d) 40.049
   (e) 76.555   *(f) 456.789   (g) 29.008   *(h) 45.986

3. Round each number to the nearest whole number.
   *(a) 19.8   *(b) 23.43   (c) 67.9102   (d) 56.5
   (e) 456.7   *(f) 9.999   *(g) 34.4999   (h) 298.18

4. Round each number to three decimal places.
   *(a) 58.6439   *(b) 46.1631   (c) 1.6765   *(d) 98.0017
   (e) 1009.9944   (f) 76.7996   *(g) 9.9998   (h) 1.78632
   *(i) 5.76509   (j) 384.0008

5. Round each number to the nearest ten.
   *(a) 86   *(b) 147   *(c) 541   (d) 894
   *(e) 5675   (f) 8365   *(g) 90   (h) 50

6. Round each number to the nearest hundred.
   *(a) 1770   *(b) 1850   *(c) 5449   (d) 1784
   *(e) 1990   (f) 3956   (g) 2849   *(h) 9979

1.7 DIVISION INVOLVING ZERO

The concept of division is based on multiplication. For example, $20 \div 5 = 4$, since $5 \times 4 = 20$, and 4 is the only number that will "check" this way.

$$5\overline{)20} \quad \text{(quotient 4)} \quad \text{since} \quad 4 \times 5 = 20$$

Similarly,

$$6\overline{)18} \quad \text{(quotient 3)} \quad \text{since} \quad 3 \times 6 = 18$$

and

$$3\overline{)21} \quad \text{(quotient 7)} \quad \text{since} \quad 7 \times 3 = 21$$

Then what about

$$5\overline{)0}$$

The result is 0:

$$5\overline{)0} \quad \text{(quotient 0)} \quad \text{since} \quad 0 \times 5 = 0; \text{ that is, 0 checks!}$$

Similarly,

$$3\overline{)0} \quad \text{(quotient 0)} \quad \text{since} \quad 0 \times 3 = 0$$

$$29\overline{)0} \quad \text{(quotient 0)} \quad \text{since} \quad 0 \times 29 = 0$$

So *if we divide a number into 0, the result is 0.*

Another interesting case is division of a nonzero number by zero; for example,

$$0\overline{)3}$$

The result of this division is the number that when multiplied by zero produces 3. In other words, we are looking for a number (call it $\square$) such that $0 \times \square = 3$. But zero times any number is zero; so there is no number that checks. Consequently, we *cannot divide by zero*; it is an *undefined operation*, an *impossible operation*.

> You cannot divide by zero.
> Division by zero is undefined.

EXERCISES 1.7

1. Each division problem below involves *zero*. Indicate whether the result is zero or whether the division cannot be done.

*(a) $16\overline{)0}$ (b) $3\overline{)0}$ *(c) $1\overline{)0}$ (d) $0\overline{)6}$

*(e) $0\overline{)17}$ (f) $5\overline{)0}$ *(g) $0\overline{)0}$ (h) $9\overline{)0}$

*(i) $0\overline{)13}$ (j) $8\overline{)0}$ *(k) $0\overline{)1}$ (l) $49\overline{)0}$

1.8 TERMINOLOGY

The following examples introduce the terminology often used to identify the different numbers used in addition, subtraction, multiplication, and division.

Addition:

$$\begin{array}{ll} 17 & \longleftarrow \text{addend} \\ \underline{35} & \longleftarrow \text{addend} \\ 52 & \longleftarrow \text{sum} \end{array}$$

Subtraction:

$$\begin{array}{ll} 37 & \longleftarrow \text{minuend} \\ \underline{12} & \longleftarrow \text{subtrahend} \\ 25 & \longleftarrow \text{difference} \end{array}$$

Multiplication:

$$\begin{array}{ll} 16 & \longleftarrow \text{multiplicand} \\ \underline{4} & \longleftarrow \text{multiplier} \\ 64 & \longleftarrow \text{product} \end{array}$$

Division:

$$\begin{array}{ll} 5 & \longleftarrow \text{quotient} \\ 4\overline{)23} & \longleftarrow \text{dividend} \\ \underline{20} & \\ 3 & \longleftarrow \text{remainder} \end{array}$$

divisor

1.9 THE MEANING OF EXPONENT

Look at the examples in the table below to see how *exponents* can be used as a shorthand notation for multiplication. Here we are using a dot to indicate multiplication, instead of using ×. Thus 4 × 5 can be written as 4 · 5. Still other ways of writing it include (4)(5), 4(5), (4)5, and variations of them. We use the word **factor** to mean any of the numbers that are multiplied together to form a product; for example, 3 and 7 are factors of 21.

Multiplication Form	Exponent Form	Base	Exponent
$5 \cdot 5$	5^2	5	2
$3 \cdot 3 \cdot 3 \cdot 3 \cdot 3 \cdot 3 \cdot 3$	3^7	3	7
$4 \cdot 4 \cdot 4$	4^3	4	3
$2 \cdot 2$	2^2	2	2
$x \cdot x \cdot x \cdot x \cdot x$	x^5	x	5

The number that is repeated in the multiplication is called the **base.** The number of times that it appears as a factor of the multiplication is called the **exponent** or **power.** Thus in 5^7 (or $5 \cdot 5 \cdot 5 \cdot 5 \cdot 5 \cdot 5 \cdot 5$), 5 is the base and 7 is the exponent. The base and exponent are given for each example in the chart.

EXAMPLE 1 Write $7 \cdot 7 \cdot 7 \cdot 7 \cdot 7 \cdot 7 \cdot 7 \cdot 7$, using exponents.

Solution The base is 7. It appears 8 times as a factor; so the exponent is 8. Thus

$$7 \cdot 7 \cdot 7 \cdot 7 \cdot 7 \cdot 7 \cdot 7 \cdot 7 = 7^8$$

EXAMPLE 2 Write 5^6 without using exponents.

Solution $5^6 = 5 \cdot 5 \cdot 5 \cdot 5 \cdot 5 \cdot 5$, since the base 5 must appear 6 (exponent) times as a factor.

If a number is raised to the second power, we say that it is **squared.** Thus 5^2 can be read "5 squared" as well as "5 to the second power."

If a number is raised to the third power, we say it is **cubed.** Therefore 5^3 can be read "5 cubed" as well as "5 to the third power."

EXERCISES 1.9

1. Express each by using exponents; for example, $3 \cdot 3 \cdot 3 \cdot 3 \cdot 3 = 3^5$.
   *(a) $6 \cdot 6$ (b) $3 \cdot 3$
   *(c) $2 \cdot 2 \cdot 2$ *(d) $9 \cdot 9 \cdot 9 \cdot 9 \cdot 9 \cdot 9 \cdot 9$

(e) $4 \cdot 4 \cdot 4 \cdot 4 \cdot 4$ *(f) $3 \cdot 3 \cdot 3 \cdot 3 \cdot 3 \cdot 3 \cdot 3 \cdot 3 \cdot 3 \cdot 3$
(g) $12 \cdot 12 \cdot 12 \cdot 12 \cdot 12 \cdot 12$ *(h) $m \cdot m$
*(i) $x \cdot x \cdot x \cdot x \cdot x$ *(j) $(x + y)(x + y)(x + y)$

2. Express as a product of factors; for example, $3^5 = 3 \cdot 3 \cdot 3 \cdot 3 \cdot 3$.
   *(a) 8^2 *(b) 2^3 (c) 5^4 *(d) 6^8
   (e) 3^7 (f) 10^{13} *(g) x^2 (h) y^5
   *(i) $(a + b)^4$

3. Compute the value; for example, $3^4 = 3 \cdot 3 \cdot 3 \cdot 3 = 81$.
   *(a) 2^3 (b) 4^2 *(c) 5^4 (d) 2^5
   (e) 9^2 *(f) 7^3 *(g) 1^8 (h) 15^2
   *(i) 3^5 (j) 6^3 *(k) 0^3 (l) 4^4

CHAPTER 1 REVIEW EXERCISES

1. Perform each addition.
   *(a) 768   (b) 7699   *(c) 67.89
         924   9901   82.04
         659   5643

   (d) 679.06   *(e) 56789
       321.94   12345
       756.71   77777
                         76804

2. Perform each subtraction.
   *(a) 8976   (b) 7644   *(c) 96666
         4563   6545   77676

   *(d) 45.987   (e) 516.765
        32.899   498.886

3. Perform each multiplication.
   *(a) 638   (b) 1007   *(c) 8739
         542   588   4702

   *(d) 56.78   (e) 459.12
        10.80   88.46

4. Perform each division, if possible. Continue to two decimal places.
   *(a) $45\overline{)7389}$ *(b) $67\overline{)0}$ *(c) $0\overline{)892}$
   *(d) $67.1\overline{)987.3}$ (e) $8.92\overline{)999.99}$

5. Round off each number to one decimal place.
   *(a) 5.98   (b) 87.04   *(c) 9.91
   (d) 19.99   *(e) .85216

*6. Find the value of each expression by actual computation.
   (a) 2^4 (b) 0^2 (c) 5^3
   (d) 12^2 (e) 1.3^2

2

ORDER OF OPERATIONS

2.1 ORDER OF OPERATIONS

In Chapter 1 you reviewed the basic arithmetic operations of addition, subtraction, multiplication, division, and exponentiation. Hopefully, you will have no problem carrying out such operations in the future. Unfortunately, however, when a *series* of such operations is encountered, the order in which they must be carried out may not be clear. For example, in the calculation $2 + 3 \cdot 4$, should you first add the 2 and 3 or should you multiply the 3 and 4? Certainly if parentheses were used, there would be no doubt. Thus $(2 + 3) \cdot 4$ specifies that the 2 and 3 should be added and that sum should then be multiplied by 4. The result is 20. On the other hand, $2 + (3 \cdot 4)$ specifies that the 3 and 4 should be multiplied and that result should be added to 2. The result is 14.

But what do you do when there are no parentheses to guide you? There are rules for the order in which operations must be carried out when parentheses do not specify. For instance, multiplication is done before addition. So in the case of $2 + 3 \cdot 4$, the result is 14 rather than 20.

Here is the entire set of rules for determining the order of operations *when there are no parentheses to guide you.*

When there are no parentheses,	
First:	Apply all exponents.
Second:	Perform all multiplications and divisions, proceeding from left to right.
Third:	Perform all additions and subtractions, proceeding from left to right.

In the examples that follow, the rules for order of arithmetic operations (given above) will be applied. In each case the expression is simplified down to a one-number final answer. The directions, however, may be given in a variety of ways, including "Simplify the expression," "Simplify," "Evaluate the expression," "Determine the value of," "Calculate," and "Evaluate."

EXAMPLE 1   Simplify the expression $4 + 9 \cdot 2$.

Solution $4 + 9 \cdot 2 = 4 + 18$ Do multiplication before addition.
                                $= 22 \;\; \checkmark$

EXAMPLE 2   Simplify $2 \cdot 3^2$.

Solution $2 \cdot 3^2 = 2 \cdot 9$ Apply the exponent before multiplying.
                                    $= 18 \;\; \checkmark$

EXAMPLE 3   Evaluate the expression $6 \cdot 5 \div 2 \cdot 3$.

Solution $6 \cdot 5 \div 2 \cdot 3 = 30 \div 2 \cdot 3$ $\begin{cases} \text{since multiplication and division} \\ \text{are performed from left to right} \end{cases}$

                          $= 15 \cdot 3$ $\begin{cases} \text{since multiplication and division} \\ \text{are performed from left to right} \end{cases}$

                          $= 45 \;\; \checkmark$

EXAMPLE 4   Evaluate $15 + \dfrac{5}{5}$.

Solution $15 + \dfrac{5}{5} = 15 + 1$ Do division before addition.
                                      $= 16 \;\; \checkmark$

EXAMPLE 5   Determine the value of $5 \cdot 2^2 + 7$.

Solution $5 \cdot 2^2 + 7 = 5 \cdot 4 + 7$ Do exponentiation before multiplication.
                              $= 20 + 7$ Do multiplication before addition.
                              $= 27 \;\; \checkmark$

EXAMPLE 6   Calculate $10 + 8 \div 2 + 1$.

Solution $10 + 8 \div 2 + 1 = 10 + 4 + 1$ Do division before addition.
                                  $= 15 \;\; \checkmark$

EXAMPLE 7   Simplify $2 + 3^2 + 8 + \dfrac{12}{2}$.

Solution $2 + 3^2 + 8 + \dfrac{12}{2} = 2 + 9 + 8 + \dfrac{12}{2}$ Apply the exponent first.

$$= 2 + 9 + 8 + 6 \qquad \text{Do division before addition.}$$
$$= 25 \quad \checkmark$$

Parentheses and other grouping symbols can be used to change the order of operations when desired. *If parentheses are used, the expression inside them is treated as if it were one number. That is, its value is computed before it is combined with other quantities.*

> When parentheses are present, simplify all expressions inside parentheses.
> *Then* apply the order of operations rules.

EXAMPLE 8   Simplify $(5 + 2)6$.

Solution $(5 + 2)6 = (7)6$ $\begin{cases} \text{because parentheses indicate that} \\ \text{the addition is done first} \end{cases}$

$$= 42 \quad \checkmark$$

EXAMPLE 9   Evaluate $(9 + 3) \div 4 + 2$.

Solution $(9 + 3) \div 4 + 2 = 12 \div 4 + 2$ $\begin{cases} \text{First, evaluate the expression} \\ \text{inside parentheses.} \end{cases}$

$$= 3 + 2 \qquad \text{Do division before addition.}$$
$$= 5 \quad \checkmark$$

In the next example the notation / is used for division, as is often the case in computer programming.

EXAMPLE 10   Simplify $(9 + 3)/(4 + 2)$.

Solution $(9 + 3)/(4 + 2) = 12/6$ $\begin{cases} \text{after first determining the values} \\ \text{of the quantities in parentheses} \end{cases}$

$$= 2 \quad \checkmark$$

Very often division, such as that used in Example 10, is not written on one line. In other words, $(9 + 3)/(4 + 2)$ would appear on two lines, as

$$\frac{9 + 3}{4 + 2}$$

In this latter form, it is obvious that the 9 and 3 should be added and that the 4 and 2 should be added before any division takes place. The long bar of the division serves the same purpose as the parentheses in the one-line form. Thus

$$\frac{9 + 3}{4 + 2} = \frac{12}{6} = 2$$

If the expression of Example 9 is written in this form, it becomes

$$\frac{9 + 3}{4} + 2 = \frac{12}{4} + 2 = 3 + 2 = 5$$

EXAMPLE 11   Simplify $(7 + 2)^2$.

Solution $(7 + 2)^2 = (9)^2$ Parentheses specify addition first.
$$= 81 \quad \checkmark$$

EXAMPLE 12   Evaluate $(16 + 2^2) \div 2 \cdot 2$.

Solution $(16 + 2^2) \div 2 \cdot 2 = (16 + 4) \div 2 \cdot 2$ $\begin{cases} \text{since exponentiation precedes} \\ \text{addition in computing the} \\ \text{value of } (16 + 2^2) \end{cases}$

$$= 20 \div 2 \cdot 2 \qquad \begin{cases} \text{after finally determining the} \\ \text{value of the quantity in} \\ \text{parentheses} \end{cases}$$

$$= 10 \cdot 2 \qquad \begin{cases} \text{since we proceed from left} \\ \text{to right with a series of} \\ \text{multiplication and division} \end{cases}$$

$$= 20 \quad \checkmark$$

EXAMPLE 13   Simplify $(16 + 2^2) \div (2 \cdot 2)$.

Solution $(16 + 2^2) \div (2 \cdot 2) = (16 + 4) \div (2 \cdot 2)$ just as in Example 12

$$= 20 \div (2 \cdot 2) \qquad \text{just as in Example 12}$$

$$= 20 \div 4 \qquad \begin{cases} \text{Parentheses overrule} \\ \text{the left-to-right order.} \end{cases}$$

$$= 5 \quad \checkmark$$

Be sure to compare Examples 12 and 13.

EXERCISES 2.1

Answers to starred exercises are given in the back of the book.

*1. Simplify each expression.

(a) $8 + 2 \cdot 6$ (b) $15 + 3 \cdot 4$ (c) $6 + 10 \div 2$

(d) $12 + 8 \div 4$ (e) $9 \div 3 + 6$ (f) $12 \div 6 \cdot 2$

(g) $20 \div 4 \cdot 5$ (h) $5 + 4 \div 2 + 1$ (i) $5 \cdot 9 + 2 \cdot 8$

(j) $5 \cdot 2^3 + 1$ (k) $2 + 3^2$ (l) $8 \cdot 10 \div 5 \cdot 2$

(m) $7 + 6 \cdot 4 - 3$ (n) $2 + 3 \cdot 7 - 4$ (o) $20 + \dfrac{4}{2} + 8$

(p) $16 + 10/1 + 1$

2. Evaluate each expression.

*(a) $7 + 3 - 4 - 2 + 1$ (b) $10 - 7 + 4 - 5 + 1 - 2$

*(c) $5 \cdot 2 + 4 + 3 \cdot 2^4 + 4$ (d) $3 + 4^2 + 5 \cdot 3^2 + 8 \cdot 2$

*(e) $6 \cdot 3 + 4 \cdot 9 \div 6 \cdot 6 + 3$ (f) $5 - 3 + 4 \cdot 6 - 2 \cdot 3/2$

*(g) $18 \cdot 2 \div 3 \cdot 6 \div 2$ (h) $20 \cdot 4 \div 2 \cdot 2 \div 2$

*3. Determine the value of each expression.

(a) $8 + 2(5 + 3)$ (b) $(4 + 1)^2$

(c) $(6 + 4) \div 2 + 8$ (d) $(6 + 4) \div (2 + 8)$

(e) $(6 + 4) \div (2 + 3)$ (f) $(4 + 2)^2/6$

(g) $(21 + 3^2) \div 5 \cdot 6$ (h) $(21 + 3^2) \div (5 \cdot 6)$

(i) $1^2 + 2^2 + 3^2 + 4^2 \div 8$ (j) $12 + 18 \div 3 + 6$

(k) $\dfrac{8 + 8}{3 + 1} + 8 - 3$ (l) $3 \cdot 4^2 + 7 - \dfrac{2 \cdot 3^2}{6}$

(m) $6 - 4 \div 2 + 5$ (n) $19 + 5^2 - 4 \cdot 2 + 5(7 + 3^4)$

CHAPTER 2 REVIEW EXERCISES

*1. Simplify each expression.

(a) $9 + 3 \cdot 5$ (b) $15 \div 3 \cdot 5$ (c) $5 + 2^2$ (d) $9 - 4 \cdot 2$

*2. Evaluate each expression.

(a) $5 \cdot 4 \div 2 \cdot 2 + 5$ (b) $5 \cdot 3^2 + 4 \cdot 3$ (c) $9 + 3(4 + 2)$

(d) $8 - 2 \div 2 + 1$ (e) $14 + \dfrac{8}{2} - 6$ (f) $9 + 3^2 - 5 \cdot 2$

*3. Determine the value of each expression.

(a) $(6 + 2) \div 4 + 4$ (b) $(6 + 2) \div (4 + 4)$

(c) $12 + 12 \div 6 + 6$ (d) $12 + 12 \div (6 + 6)$

(e) $(12 + 12) \div 6 + 6$ (f) $(12 + 12) \div (6 + 6)$

(g) $4^2 + 6^2 \div 4$ (h) $(6 - 2)^2$

(i) $5 \cdot 6 + 3$ (j) $19 - 16 \div 4 + 12$

(k) $15 - 4 + 11 - 8 + 1 - 9$ (l) $5 \cdot 2 + 6(3 + 4^2)$

3

EVALUATION OF GEOMETRIC FORMULAS

3.1 EVALUATION OF GEOMETRIC FORMULAS

Here we apply our knowledge of arithmetic to the evaluation of geometric formulas. In each case, an area, volume, or perimeter is determined by substituting numerical values for the letters given in a formula. *Such units as inches and centimeters are omitted but are discussed in detail in Chapters 8 and 9.*

The formulas use information about a geometric figure in order to determine its *perimeter* or *area* or *volume*.

---

perimeter = distance around
area = amount enclosed (2 dimensions)
volume = space enclosed (3 dimensions)

We continue, then, with six examples that demonstrate formula evaluation. A summary of the formulas used is given at the end of the section.

EXAMPLE 1 Determine the area of the rectangle whose length is 12 and width is 10.

Solution The formula for area is $A = l \cdot w$ (usually written without the dot for multiplication, as $A = lw$), where

$$A = \text{area}$$
$$l = \text{length}$$
$$w = \text{width}$$

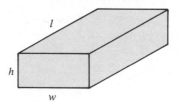

When $l = 12$ and $w = 10$, the result is

$$A = lw$$
$$= 12 \cdot 10$$
$$= 120 \quad \checkmark$$

The area is 120.

EXAMPLE 2 Determine the volume of the box whose length is 5, width is 3, and height is 2.

Solution The formula for volume is $V = lwh$ (note that lwh means $l \cdot w \cdot h$), where

$$V = \text{volume}$$
$$l = \text{length}$$
$$w = \text{width}$$
$$h = \text{height}$$

When $l = 5$, $w = 3$, and $h = 2$, the result is

$$V = lwh$$
$$= 5 \cdot 3 \cdot 2$$
$$= 30 \quad \checkmark$$

The volume is 30.

EXAMPLE 3 Determine the area of a trapezoid whose bases are 6 and 10 and whose height is 5.

Solution The formula for area is $A = \frac{1}{2}(b_1 + b_2)h$, where

$$A = \text{area}$$
$$b_1 = \text{one base}$$
$$b_2 = \text{the other base}$$
$$h = \text{height}$$

When $b_1 = 6$, $b_2 = 10$, and $h = 5$, the result is

$$A = \tfrac{1}{2}(6 + 10)5$$
$$= \tfrac{1}{2}(16)5$$
$$= 8 \cdot 5$$
$$= 40 \quad \checkmark$$

The area is 40.

EXAMPLE 4   Determine the circumference (distance around) of a circle whose radius is 92.

Solution   The formula for circumference is $C = 2\pi r$, where

$$C = \text{circumference}$$
$$\pi = \text{approximately } 3.14$$
$$r = \text{radius}$$

When $r = 92$, the result is

$$C = 2\pi r$$
$$= 2 \cdot (3.14)(92)$$
$$= 577.76 \quad \checkmark$$

The approximate circumference is 577.76.

EXAMPLE 5   Determine the perimeter of the triangle having sides of lengths 13, 17, and 21.

Solution   The perimeter (P) of a triangle with sides of lengths a, b, and c is simply $P = a + b + c$, since perimeter means distance around. Here

$$P = a + b + c$$
$$= 13 + 17 + 21$$
$$= 51 \quad \checkmark$$

EXAMPLE 6   Determine (a) the area and (b) the perimeter of the figure shown below.

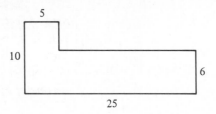

Solution   (a) One way to determine the area is to draw the dashed line shown below and consider the area to be the sum of the two rectangular areas, where $A = lw$ in each case,

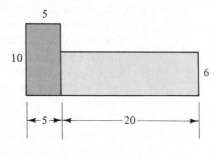

$$A = lw = 10 \cdot 5 = 50$$
$$A = lw = 20 \cdot 6 = 120$$

The total area is 50 + 120, or 170.   ✓

(b) The perimeter is the distance around. Examining the figure having vertical pieces of lengths 10 and 6, you can reason that the other vertical piece is 4 (from 10 minus 6). Also, from the reasoning in part (a), the unlabeled horizontal piece is of length 20.

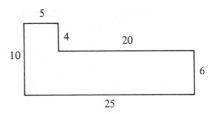

Since perimeter is the distance around,

$$P = 10 + 25 + 6 + 20 + 4 + 5$$
$$= 70 \quad ✓$$

Here is a brief summary of the formulas used in the examples of this section.

> Area of rectangle: $A = lw$ $A = \pi 2^2$
> Volume of box: $V = lwh$
> Area of trapezoid: $A = \frac{1}{2}(b_1 + b_2)h$
> Circumference of circle: $C = 2\pi r$
> Perimeter of triangle: $P = a + b + c$

EXERCISES 3.1

Answers to starred exercises are given in the back of the book.

1. Use the formulas of the examples to determine what is requested. Use 3.14 as an approximation for π.
   *(a) The area of a rectangle having length 37 and width 28.
    (b) The area of a rectangle having length 26.4 and width 15.2.
   *(c) The volume of a box having length 7, width 6, and height 10.
    (d) The volume of a box having length 12.3, width 9.1, and height 10.
   *(e) The area of a trapezoid with bases 7 and 9 and height 15.
    (f) The area of a trapezoid with bases 14 and 18 and height 13.1.
   *(g) The circumference of a circle with radius 14.
    (h) The circumference of a circle with radius 10.
   *(i) The circumference of a circle with radius 39.7.
   *(j) The perimeter of a triangle with sides 16, 20, and 23.
    (k) The perimeter of a triangle with sides 15, 19, and 27.

2. The area of a triangle is $A = \frac{1}{2}bh$, where b is the base of the triangle and h is the height.

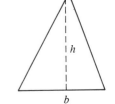

   *(a) Find the area when $b = 6$ and $h = 10$.
   *(b) Find the area when $b = 300$ and $h = 12$.
    (c) Find the area when $b = 9$ and $h = 5$.

*3. Evaluate each formula, using the values given. Use 3.14 as an approximation for π.
    (a) $P = 2l + 2w$; $l = 6, w = 5.4$. This is the perimeter (distance around) of a rectangle having length l and width w.
    (b) $A = \pi r^2$; $r = 7$. This is the area of a circle having radius r.
    (c) $S = 4\pi r^2$; $r = 9$. This is the area of the surface of a sphere having radius r.
    (d) $V = \dfrac{4\pi r^3}{3}$; $r = 5$. This is the volume of a sphere having radius r.

(e) $V = \frac{1}{3}\pi r^2 h$; $r = 8$, $h = 15$. This is the volume of a cone having radius r and height h.

*4. Determine the area and perimeter of each figure.

(a)

(b)

(c)

(d)

(e)

(f)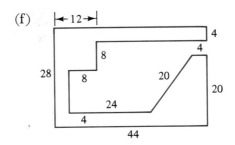

CHAPTER 3 REVIEW EXERCISES

*1. Determine each area or perimeter as requested. Use 3.14 as an approximation for π.
   (a) The area of a rectangle with length 17 and width 12.
   (b) The perimeter of a triangle with sides 23, 26, and 36.
   (c) The circumference of a circle with radius 20.
   (d) The area of a triangle with base 12 and height 30.
   (e) The area of a circle with radius 10.

4

SQUARE ROOT

4.1 FUNDAMENTALS

Up to this point we have used the arithmetic operations of addition, subtraction, multiplication, division, and exponentiation. Obtaining square roots is another operation that is important in both arithmetic and algebra.

A **square root** of a quantity is a number that when squared (that is, multiplied by itself) produces the original quantity. For example, the square root of 25 is 5, since $5 \times 5 = 25$. The square root of 81 is 9, since $9 \times 9 = 81$. Using the symbol $\sqrt{}$ for square root, we can write

$$\sqrt{25} = 5$$
$$\sqrt{81} = 9$$

Here are a few additional examples.

$$\sqrt{0} = 0 \quad \text{since} \quad 0 \times 0 = 0$$
$$\sqrt{1} = 1 \quad \text{since} \quad 1 \times 1 = 1$$
$$\sqrt{4} = 2 \quad \text{since} \quad 2 \times 2 = 4$$
$$\sqrt{9} = 3 \quad \text{since} \quad 3 \times 3 = 9$$
$$\sqrt{16} = 4 \quad \text{since} \quad 4^2 = 16$$

Numbers such as 0, 1, 4, 9, 16, and 25 are called **perfect squares.** A whole number is a perfect square if its square root is a whole number. The symbol $\sqrt{}$ is called a **radical.**

Here is a short table of square roots of perfect squares.

$\sqrt{0} = 0$	$\sqrt{49} = 7$	$\sqrt{196} = 14$
$\sqrt{1} = 1$	$\sqrt{64} = 8$	$\sqrt{225} = 15$
$\sqrt{4} = 2$	$\sqrt{81} = 9$	$\sqrt{256} = 16$
$\sqrt{9} = 3$	$\sqrt{100} = 10$	$\sqrt{289} = 17$
$\sqrt{16} = 4$	$\sqrt{121} = 11$	$\sqrt{324} = 18$
$\sqrt{25} = 5$	$\sqrt{144} = 12$	$\sqrt{361} = 19$
$\sqrt{36} = 6$	$\sqrt{169} = 13$	$\sqrt{400} = 20$

One important application of square root is in working with right triangles. A **right triangle** is one that contains a 90° angle (a *right angle,* ∟). Below are examples of right triangles, where the right angle is marked by a small square.

The side opposite the right angle is called the *hypotenuse.* The other two sides are called *legs.*

In any right triangle the following relationship is always true.

$$(\text{hypotenuse})^2 = (\text{leg 1})^2 + (\text{leg 2})^2$$

In words, the square of the length of the hypotenuse equals the sum of the squares of the lengths of the legs. This famous theorem was first proved by the Greek mathematician Pythagoras in the sixth century B.C. It is called the **Pythagorean theorem.**

We often denote the length of the hypotenuse by the letter c and the lengths of the legs by the letters a and b. Thus

Pythagorean theorem: $c^2 = a^2 + b^2$ a

EXAMPLE 1   Determine the length of the hypotenuse of a right triangle if the legs are 3 inches and 4 inches.

Solution   Since $c^2 = a^2 + b^2$
$\qquad c^2 = 4^2 + 3^2$ using 4 for a and 3 for b
$\qquad c^2 = 16 + 9$ since $4^2 = 16$ and $3^2 = 9$
$\qquad c^2 = 25$ since $16 + 9$ is 25
or$\qquad c = \sqrt{25}$
so$\qquad c = 5$ $\checkmark$

The hypotenuse is 5 inches.

Example 1 illustrates the popular 3–4–5 right triangle. Another combination is 5–12–13, as shown in the next example.

EXAMPLE 2   If the hypotenuse of a right triangle is 13 and one of the legs is 12, what is the length of the other leg?

Solution   Since $c^2 = a^2 + b^2$
$\qquad 13^2 = 12^2 + b^2$
$\qquad 169 = 144 + b^2$
or$\qquad b^2 = 25$ since $144 + \underline{25} = 169$
so$\qquad b = \sqrt{25} = 5$ $\checkmark$

The other leg is 5.

EXAMPLE 3   Determine the unknown side, x, of the triangle shown.

Solution $\qquad 10^2 = x^2 + 8^2$
$\qquad 100 = x^2 + 64$
so $x^2 = 36$ since $\underline{36} + 64 = 100$
or $x = 6$ $\checkmark$

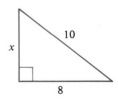

Note that this is a 6–8–10 right triangle. The relationship 6–8–10 is the same as 3–4–5. Each length of the 6–8–10 is double that of the 3–4–5.

EXAMPLE 4   Find the value of x in the triangle shown here.

Solution

$$x^2 = a^2 + b^2$$
$$x^2 = 3^2 + 2^2$$
$$x^2 = 9 + 4$$
$$x^2 = 13$$
$$x = \sqrt{13} \;\;\checkmark$$

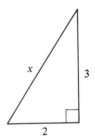

Since 13 is not a perfect square, we cannot obtain a whole number value for $\sqrt{13}$. If desired, a good approximation to the value of $\sqrt{13}$ can be obtained by using a calculator with a square root key or by looking up $\sqrt{13}$ in a table of square roots.

EXERCISES 4.1

Answers to starred exercises are given in the back of the book.

1. Study the square roots of perfect squares given at the beginning of the chapter. When you feel that you know them, evaluate each of the following without looking at that table of square roots.
   (a) $\sqrt{49}$ (b) $\sqrt{81}$ (c) $\sqrt{4}$ (d) $\sqrt{25}$ (e) $\sqrt{400}$
   (f) $\sqrt{121}$ (g) $\sqrt{36}$ (h) $\sqrt{289}$ (i) $\sqrt{0}$ (j) $\sqrt{169}$
   (k) $\sqrt{144}$ (l) $\sqrt{1}$ (m) $\sqrt{324}$ (n) $\sqrt{361}$ (o) $\sqrt{9}$
   (p) $\sqrt{225}$ (q) $\sqrt{16}$ (r) $\sqrt{100}$ (s) $\sqrt{196}$ (t) $\sqrt{256}$

2. Evaluate each square.
   (a) 9^2 (b) 11^2 (c) 19^2 (d) 2^2 (e) 0^2
   (f) 20^2 (g) 13^2 (h) 17^2 (i) 3^2 (j) 18^2
   (k) 12^2 (l) 1^2 (m) 4^2 (n) 10^2 (o) 14^2
   (p) 7^2 (q) 5^2 (r) 16^2 (s) 8^2 (t) 15^2
   (u) 6^2

3. Use the Pythagorean theorem to find the length of the side labeled x in each right triangle.
   *(a)

   *(b)

*(c)

(d)

*(e)

*(f)

(g)

(h)

*(i)

*(j)

*(k)

*(l)

4. Evaluate each formula, using the values given.

*(a) $t = \sqrt{\dfrac{2s}{a}}$, $s = 64$, $a = 32$.

(b) Redo part (a) with $s = 256$ and $a = 32$.

*(c) $x = \dfrac{7 + \sqrt{b^2 - 4ac}}{2a}$, $a = 1$, $b = 3$, $c = 2$.

(d) Redo part (c) with $a = 6$, $b = 7$, and $c = 1$.

*(e) $A = \sqrt{s(s - a)(s - b)(s - c)}$, where $s = \frac{1}{2}(a + b + c)$, $a = 3$, $b = 4$, $c = 5$.

(f) Verify that $\sqrt{576} = 24$ by squaring 24.

(g) Redo part (e) with $a = 6$, $b = 8$, and $c = 10$.

*5. A 13-foot ladder is placed against a building. If the bottom of the ladder is 5 feet from the building, how high up does the ladder touch the building?

CHAPTER 4 REVIEW EXERCISES

1. Determine the value of each square root *without* looking at the table at the beginning of the chapter. Then check your answers by using that table.

(a) $\sqrt{64}$ (b) $\sqrt{25}$ (c) $\sqrt{144}$ (d) $\sqrt{1}$ (e) $\sqrt{225}$
(f) $\sqrt{0}$ (g) $\sqrt{196}$ (h) $\sqrt{400}$ (i) $\sqrt{100}$

*2. Use the Pythagorean theorem to determine the length of the side labeled x in each right triangle.

(a)

(b)

(c)

(d)

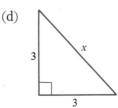

5

PROPERTIES OF ARITHMETIC

5.1 PROPERTIES OF ARITHMETIC

In this chapter we take a look at some special properties that whole numbers satisfy with respect to the arithmetic operations that we have studied. We can say that addition, subtraction, multiplication, and division are examples of *binary operations*—that is, operations in which *two* numbers are combined to produce a result.

The first property considered is the **commutative property** for addition. Here are some examples of the property.

$$2 + 3 = 3 + 2$$

This example states that $2 + 3$ and $3 + 2$ are equal; that is, both are equal to 5. Here are the rest of the examples.

$$7 + 9 = 9 + 7$$
$$x + 2 = 2 + x$$
$$p + q = q + p$$
$$5 + (6 + 7) = 5 + (7 + 6)$$

In effect, this commutative property says that if a and b are any two whole numbers, then $(a + b)$ and $(b + a)$ are the same; they are equal.

Next is the **associative property** for addition. Here are some examples.

$$2 + (3 + 1) = (2 + 3) + 1$$

This example says that $2 + (3 + 1)$ and $(2 + 3) + 1$ are equal; that is, both are equal to 6. Here are the rest of the examples.

$$12 + (6 + 5) = (12 + 6) + 5$$
$$(9 + 2) + 10 = 9 + (2 + 10)$$
$$(x + 5) + 3 = x + (5 + 3)$$
$$a + (d + x) = (a + d) + x$$

This associative property states that when three whole numbers are added, it does not matter with which number (the first or the last) you associate the middle one. Thus in

$$5 + 8 + 2$$

you can associate 8 with 5, as

$$(5 + 8) + 2$$

or associate 8 with 2, as

$$5 + (8 + 2)$$

So $(5 + 8) + 2 = 5 + (8 + 2)$. There are two ways of computing $5 + 8 + 2$.†

The associative property for addition can be written in general terms as $a + (b + c) = (a + b) + c$, where a, b, and c represent any whole numbers.

If you add two whole numbers, you obtain a whole number as the result. This fact should not be particularly surprising. The property has a name—**closure**—and it will be useful later to be able to refer to this property by name. Also included in the meaning of closure is the fact that the sum is *unique*; that is, whenever you add the same two numbers (say 4 and 5), you always get the same result (*9* in the case of 4 and 5).

Below are examples of the closure property for whole numbers using addition

$$2 + 4 = 6$$
$$3 + 5 = 8$$
$$19 = 7 + 12$$

In general, $a + b = c$; that is, if a and b are whole numbers, then the sum $(a + b)$ is a whole number; call it c.

Note that the whole numbers using division do not satisfy closure; for example, $3 \div 4 = .75$, and .75 is not a whole number.

The closure, associative, and commutative properties are true for *multiplication* as well as for addition. Here are arithmetic examples of each.

$4 \cdot (5 \cdot 7) = (4 \cdot 5) \cdot 7$	associative property for multiplication
$(8 \cdot 6) \cdot 15 = 8 \cdot (6 \cdot 15)$	associative property for multiplication
$4 \cdot 9 = 9 \cdot 4$	commutative property for multiplication
$5 \cdot (9 \cdot 2) = 5 \cdot (2 \cdot 9)$	commutative property for multiplication
$6 \cdot 3 = 18$	closure property for multiplication
$10 = 2 \cdot 5$	closure property for multiplication

---

† Recall that addition is a *binary* operation; that is, numbers must be added *two* at a time, not three at a time. Think about it and you will realize that you never add three numbers "at once"; instead, you add two of them to get an intermediate result and then immediately add the other number to get the final sum.

The last property that we will look at is called the **distributive property.** It involves both multiplication and addition. Here is an example of the distributive property.

$$5 \cdot (3 + 4) = 5 \cdot 3 + 5 \cdot 4$$

Notice what has been done.

$$\underline{5} \cdot (3 + 4) = \underline{5} \cdot 3 + \underline{5} \cdot 4$$

We *distribute* 5 to both 3 and 4. That is, 5 *times* 3 and 5 *times* 4, or $5 \cdot 3 + 5 \cdot 4$. Other examples:

$$7 \cdot (2 + 3) = 7 \cdot 2 + 7 \cdot 3$$
$$6(5 + 1) = 6 \cdot 5 + 6 \cdot 1$$
$$5(x + y) = 5 \cdot x + 5 \cdot y$$
$$4 \cdot 2 + 4 \cdot 3 = 4(2 + 3)$$
$$5 \cdot 9 + 5 \cdot 18 = 5(9 + 18)$$
$$5 \cdot a + 5 \cdot b = 5(a + b)$$

Is it obvious that this property is true? Probably not. Let's consider an example that might illustrate the point. You can make up additional examples if you are not completely convinced.

$$3(4 + 9) \overset{?}{=} 3 \cdot 4 + 3 \cdot 9$$
$$3(13) \overset{?}{=} 12 + 27$$
$$39 = 39$$

The left-hand side shows the work done without using the distributive property; the right-hand side uses the distributed form. Both results are the same, and so $3(4 + 9) = 3 \cdot 4 + 3 \cdot 9$.

Another form of the distributive property is given by

$$(3 + 5)7 = 3 \cdot 7 + 5 \cdot 7$$

Here the multiplier is on the right instead of on the left. It doesn't matter which side the multiplier is on; it is still distributed. Again,

$$(3 + 5)\underline{7} = 3 \cdot \underline{7} + 5 \cdot \underline{7}$$

We now state both distributive properties in general, for any whole numbers a, b, and c.

$$a(b + c) = a \cdot b + a \cdot c$$

and

$$(b + c)a = b \cdot a + c \cdot a$$

There is also a distributive property that involves multiplication and subtraction. See Exercises 12 and 13.

EXAMPLE 1 Name the property used in going from each step to the next one.
(a) $5(2 + 7) = 5 \cdot 2 + 5 \cdot 7$
(b) $\qquad = 2 \cdot 5 + 5 \cdot 7$
(c) $\qquad = 2 \cdot 5 + 35$

Solution (a) It is the distributive property that allows us to write $5(2 + 7)$ as $5 \cdot 2 + 5 \cdot 7$.
(b) It is the commutative property for multiplication that is used to rewrite $5 \cdot 2$ as $2 \cdot 5$.
(c) The closure property for multiplication is used to rewrite the product of the two numbers 5 and 7 as the one number 35.

EXERCISES 5.1

Answers to starred exercises are given in the back of the book.

1. Give two examples of the commutative property for addition and two examples of the commutative property for multiplication.

2. Give two examples of the associative property for addition and two examples of the associative property for multiplication.

3. Give two examples of the closure property for addition and two examples of the closure property for multiplication.

4. Give two examples of the distributive property.

*5. In each case, name the property used.
   (a) $5 + 2 = 2 + 5$ (b) $8 + (6 + 9) = (8 + 6) + 9$
   (c) $8 + (6 + 9) = 8 + (9 + 6)$ (d) $3(2 + 5) = 3 \cdot 2 + 3 \cdot 5$
   (e) $7 \cdot 2 + 7 \cdot 9 = 7(2 + 9)$ (f) $6 + 5 = 11$
   (g) $(8 + 9) + 4 = 8 + (9 + 4)$ (h) $(2 + 3) + 7 = 7 + (2 + 3)$
   (i) $29 = 14 + 15$ (j) $15 \cdot 8 = 8 \cdot 15$
   (k) $(5 + 4)7 = 5 \cdot 7 + 4 \cdot 7$ (l) $9 \cdot 3 + 14 \cdot 3 = (9 + 14)3$

6. For each, name the property used.
   (a) $4(7 + 2) = 4 \cdot 7 + 4 \cdot 2$ (b) $5 \cdot 8 \cdot 9 = 5 \cdot 9 \cdot 8$
   (c) $5(8 \cdot 9) = (5 \cdot 8)9$ (d) $2 + 1 = 3$
   (e) $3(x + y) = 3 \cdot x + 3 \cdot y$ (f) $5(3 + 1) = 5(4)$
   (g) $2 + 3 = 3 + 2$ (h) $2 \cdot 3 = 3 \cdot 2$
   (i) $2 \cdot 3 = 6$ (j) $9 + (2 + 3) = 9 + 5$

*7. Name the property used in going from each step to the next one.
   (a) $5 + (6 + 3) = (5 + 6) + 3$
   (b) $\qquad = (6 + 5) + 3$
   (c) $\qquad = 3 + (6 + 5)$
   (d) $\qquad = 3 + 11$
   (e) $\qquad = 14$

*8. Name the property used in going from each step to the next one.
   (a) $7 \cdot 2 + 7 \cdot 3 = 7(2 + 3)$
   (b) $\qquad = 7(3 + 2)$

   (c) $= 7 \cdot 3 + 7 \cdot 2$
   (d) $= 3 \cdot 7 + 7 \cdot 2$
   (e) $= 21 + 7 \cdot 2$

9. Name the property used in going from each step to the next one.
   (a) $8(9 + 4) = 8 \cdot 9 + 8 \cdot 4$
   (b) $= 9 \cdot 8 + 4 \cdot 8$
   (c) $= (9 + 4)8$
   (d) $= (13)8$
   (e) $= 104$

10. Name the property used in going from each step to the next one.
   (a) $36 = 4 \cdot 9$
   (b) $= 4(6 + 3)$
   (c) $= 4 \cdot 6 + 4 \cdot 3$
   (d) $= 24 + 12$
   (e) $= 12 + 24$

11. State the commutative, associative, and closure properties for *multiplication,* using letters to represent any numbers.

12. A distributive property for multiplication and *subtraction* can be written as $a(b - c) = a \cdot b - a \cdot c$. Verify this property in the cases below by evaluating each side and showing that they are equal.
   (a) $5(10 - 2) = 5 \cdot 10 - 5 \cdot 2$
   (b) $4(7 - 1) = 4 \cdot 7 - 4 \cdot 1$
   (c) $15(8 - 5) = 15 \cdot 8 - 15 \cdot 5$

13. Another distributive property for multiplication and subtraction can be written with the multiplier on the right side, as $(b - c)a = b \cdot a - c \cdot a$. Verify this property in the cases below by evaluating each side and showing that they are equal.
   (a) $(6 - 4)7 = 6 \cdot 7 - 4 \cdot 7$
   (b) $(12 - 2)5 = 12 \cdot 5 - 2 \cdot 5$
   (c) $(8 - 5)9 = 8 \cdot 9 - 5 \cdot 9$

CHAPTER 5 REVIEW EXERCISES

*__1.__ In each case, name the property used.
   (a) $6(7 \cdot 3) = (6 \cdot 7)3$ (b) $7 + 2 = 2 + 7$
   (c) $7(8 + 5) = 7 \cdot 8 + 7 \cdot 5$ (d) $9 \cdot 3 = 27$
   (e) $4 + 5 = 9$ (f) $(9 + 2)6 = 9 \cdot 6 + 2 \cdot 6$

*__2.__ Name the property used in going from each step to the next one.
   (a) $8(3 + 4) = 8 \cdot 3 + 8 \cdot 4$
   (b) $= 3 \cdot 8 + 8 \cdot 4$
   (c) $= 24 + 32$
   (d) $= 32 + 24$
   (e) $= 56$

6

ARITHMETIC
OF FRACTIONS

6.1 WHAT IS A FRACTION?

If we divide the rectangle below into *two* equal parts, then *one* part represents $\frac{1}{2}$ of the rectangle.

If we divide the rectangle into *three* equal parts, then *one* part represents $\frac{1}{3}$ of the whole rectangle.

If we divide the rectangle into *five* equal parts, then *two* parts represent $\frac{2}{5}$ of the whole rectangle. Thus the shaded area is $\frac{2}{5}$, the unshaded area is $\frac{3}{5}$, and the whole area (shaded plus unshaded) is $\frac{2}{5} + \frac{3}{5} = \frac{5}{5} = 1$ rectangle.

The top portion of a fraction is called the **numerator.** The bottom portion of a fraction is called the **denominator.** For example, in $\frac{2}{5}$, 2 is the numerator and 5 is the denominator.

$$\frac{2}{5} \quad \begin{matrix} \longleftarrow & \text{numerator} \\ \longleftarrow & \text{denominator} \end{matrix}$$

A fraction is called a **proper fraction** when the numerator is less than the denominator. When the numerator is greater than the denominator or equal to the denominator, the fraction is called an **improper fraction.** Here are a few examples.

$$\text{Proper fractions:} \quad \frac{1}{2}, \quad \frac{5}{6}, \quad \frac{3}{11}$$

$$\text{Improper fractions:} \quad \frac{2}{2}, \quad \frac{7}{6}, \quad \frac{17}{11}$$

Improper fractions will arise throughout the chapter and they will be given special attention in Section 6.7.

6.2 MULTIPLICATION OF FRACTIONS

Let's begin with a rule that you have surely used before.

> To multiply fractions, multiply numerator by numerator and denominator by denominator.

To help you justify this procedure, consider $\frac{1}{2} \times \frac{1}{3}$. You can visualize that half of one-third is one-sixth.

 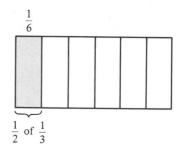

By the rule for multiplication of fractions,

$$\frac{1}{2} \times \frac{1}{3} = \frac{1 \times 1}{2 \times 3} = \frac{1}{6}$$

EXAMPLE 1   Multiply $\frac{2}{3} \times \frac{1}{7}$.

Solution

$$\frac{2}{3} \times \frac{1}{7} = \frac{2 \times 1}{3 \times 7} = \frac{2}{21} \quad \checkmark$$

EXAMPLE 2   Multiply $\frac{5}{6} \times \frac{7}{8}$.

Solution

$$\frac{5}{6} \times \frac{7}{8} = \frac{5 \times 7}{6 \times 8} = \frac{35}{48} \quad \checkmark$$

EXAMPLE 3   Multiply $3 \times \frac{2}{7}$.

Solution

$$3 \times \frac{2}{7} = \frac{3}{1} \times \frac{2}{7} = \frac{3 \times 2}{1 \times 7} = \frac{6}{7} \quad \checkmark$$

Writing the whole number 3 as the fraction $\frac{3}{1}$ is merely a convenient notation that may make it easier to perform the multiplication.

EXAMPLE 4   Multiply $0 \times \frac{3}{8}$.

Solution   Since 0 times any number is 0, it follows that

$$0 \times \frac{3}{8} = 0 \quad \checkmark$$

EXERCISES 6.2

Answers to starred exercises are given in the back of the book.

1. Multiply.

*(a) $\frac{3}{5} \times \frac{1}{2}$ *(b) $\frac{4}{9} \times \frac{2}{3}$ *(c) $\frac{6}{7} \times \frac{1}{5}$ (d) $\frac{1}{6} \times \frac{5}{8}$

*(e) $\frac{3}{4} \times \frac{5}{7}$ (f) $\frac{3}{11} \times \frac{5}{7}$ *(g) $\frac{1}{9} \times \frac{1}{9}$ *(h) $\frac{4}{5} \times \frac{3}{5}$

(i) $\frac{9}{14} \times \frac{3}{4}$ *(j) $\frac{10}{13} \times \frac{6}{7}$ *(k) $\frac{15}{23} \times \frac{12}{13}$ (l) $\frac{11}{15} \times \frac{11}{17}$

*(m) $7 \times \frac{1}{8}$ (n) $6 \times \frac{1}{7}$ *(o) $4 \times \frac{2}{9}$ *(p) $\frac{3}{8} \times 1$

(q) $\frac{9}{100} \times 7$ *(r) $\frac{10}{99} \times 5$ *(s) $0 \times \frac{3}{5}$ (t) $\frac{7}{8} \times 0$

2. Use the formula $A = l \cdot w$ to determine the area of a rectangle whose length (l) and width (w) are given below.

*(a) $l = \dfrac{2}{3}, w = \dfrac{5}{7}$ (b) $l = \dfrac{1}{5}, w = \dfrac{1}{5}$

*(c) $l = \dfrac{4}{7}, w = \dfrac{2}{5}$ (d) $l = \dfrac{5}{9}, w = \dfrac{2}{3}$

6.3 REDUCTION OF FRACTIONS

The multiplication process can be reversed in order to reduce fractions. Also used in the reducing process is the fact that any nonzero number divided by itself is equal to 1. For instance,

$$\frac{5}{5} = 1, \qquad \frac{4}{4} = 1, \qquad \frac{17}{17} = 1$$

To see how fractions are reduced, observe the following example in which the fraction $\frac{20}{36}$ is reduced.

$$\frac{20}{36} = \frac{5 \times 4}{9 \times 4} \qquad \text{writing 20 as } 5 \times 4 \text{ and 36 as } 9 \times 4$$

$$= \frac{5}{9} \times \frac{4}{4} \qquad \left\{ \begin{array}{l} \text{separating } \dfrac{4}{4} \text{ by reversing the steps of} \\ \text{multiplication of fractions} \end{array} \right.$$

$$= \frac{5}{9} \times 1 \qquad \text{since } \frac{4}{4} \text{ is equal to 1}$$

$$= \frac{5}{9} \qquad \text{since } \frac{5}{9} \times 1 = \frac{5}{9}$$

The reason that $\frac{20}{36}$ can be reduced to $\frac{5}{9}$ is that both numerator 20 and denominator 36 have a common factor—4. And so we chose to write 20 as 5×4 and 36 as 9×4. If a *factor* appears in both the numerator and denominator, then both can be removed because they form a fraction equal to one and multiplication by one has no effect. This point can be seen once again in the next examples.

EXAMPLE 5   Reduce $\dfrac{12}{15}$.

Solution

$$\frac{12}{15} = \frac{3 \times 4}{3 \times 5}$$

$$= \frac{3}{3} \times \frac{4}{5}$$

$$= 1 \times \frac{4}{5}$$

$$= \frac{4}{5} \quad \checkmark$$

In Example 5 we had

$$\frac{3}{3} = 1$$

To show that the 3's divide out, we can use the "slash" symbol

$$\frac{\not{3}}{\not{3}}$$

This notation is demonstrated in the next example.

EXAMPLE 6   Reduce $\dfrac{18}{72}$.

Solution

$$\frac{18}{72} = \frac{9 \times 2}{9 \times 8} = \frac{\not{9}}{\not{9}} \times \frac{2}{8} = \frac{2}{8}$$

But $\frac{2}{8}$ can be reduced!

$$\frac{2}{8} = \frac{2 \times 1}{2 \times 4} = \frac{2}{2} \times \frac{1}{4} = \frac{1}{4} \quad \checkmark$$

Actually, this could have been done in one step, as

$$\frac{18}{72} = \frac{18 \times 1}{18 \times 4} = \frac{18}{18} \times \frac{1}{4} = \frac{1}{4} \quad \checkmark$$

Either approach is correct; it pays to look for shortcuts.

EXAMPLE 7   Multiply $\dfrac{4}{5} \times \dfrac{3}{4}$.

Solution

$$\frac{4}{5} \times \frac{3}{4} = \frac{4 \times 3}{5 \times 4} = \frac{12}{20}$$

And $\frac{12}{20}$ can be reduced.

$$\frac{12}{20} = \frac{4 \times 3}{4 \times 5} = \frac{4}{4} \times \frac{3}{5} = \frac{3}{5} \quad \checkmark$$

Actually, reduction earlier in the process was possible.

$$\frac{4}{5} \times \frac{3}{4} = \frac{4 \times 3}{5 \times 4} = \frac{3 \times 4}{5 \times 4} = \frac{3}{5} \times \frac{4}{4} = \frac{3}{5} \quad \checkmark$$

For convenience, we often write

$$\frac{4}{5} \times \frac{3}{4} = \frac{\overset{1}{\not{4}}}{5} \times \frac{3}{\underset{1}{\not{4}}} = \frac{3}{5}$$

Note the common factor 4 in both numerator and denominator.

where it is *observed* that we could manipulate the numbers to get $\frac{4}{4}$, which is 1, but we don't actually write all the steps.

EXAMPLE 8   Multiply $\dfrac{15}{14} \times \dfrac{7}{11}$.

Solution
$$\frac{15}{14} \times \frac{7}{11} = \frac{15}{\overset{1}{\cancel{7}} \times 2} \times \frac{\overset{1}{\cancel{7}}}{11} = \frac{15 \times 1}{1 \times 2 \times 11} = \frac{15}{22} \quad \checkmark$$

EXERCISES 6.3

1. Reduce each fraction.

*(a) $\dfrac{4}{6}$ *(b) $\dfrac{15}{15}$ *(c) $\dfrac{6}{8}$ *(d) $\dfrac{9}{12}$ (e) $\dfrac{10}{15}$

(f) $\dfrac{8}{10}$ *(g) $\dfrac{20}{30}$ *(h) $\dfrac{14}{28}$ (i) $\dfrac{14}{21}$ (j) $\dfrac{24}{30}$

*(k) $\dfrac{24}{32}$ *(l) $\dfrac{20}{48}$ *(m) $\dfrac{18}{54}$ (n) $\dfrac{14}{105}$

2. Multiply; reduce if possible.

*(a) $\dfrac{4}{9} \times \dfrac{1}{4}$ *(b) $\dfrac{7}{16} \times \dfrac{4}{5}$ *(c) $\dfrac{1}{9} \times \dfrac{3}{4}$ *(d) $\dfrac{3}{5} \times \dfrac{6}{7}$

*(e) $\dfrac{12}{15} \times \dfrac{5}{6}$ *(f) $2 \times \dfrac{1}{2}$ (g) $\dfrac{2}{3} \times \dfrac{3}{8}$ *(h) $\dfrac{2}{5} \times \dfrac{10}{11}$

(i) $\dfrac{2}{5} \times \dfrac{10}{44}$ *(j) $\dfrac{3}{4} \times 4$ *(k) $\dfrac{8}{5} \times \dfrac{5}{16}$ (l) $18 \times \dfrac{2}{9}$

*(m) $\dfrac{18}{24} \times \dfrac{16}{81}$ (n) $\dfrac{24}{32} \times 40$ *(o) $\dfrac{13}{35} \times \dfrac{21}{26}$ (p) $\dfrac{33}{48} \times \dfrac{32}{68}$

6.4 DIVISIBILITY TESTS

In this section we present special tests that will make it easier for you to reduce fractions. The tests will also be useful in later applications as well.

We begin with the concept of *divisibility*. Using your present terminology, you would say that 3 is a **factor** of 15, since $15 = 3 \cdot 5$; that is, 15 can be written as 3 times some whole number. That 3 is a factor of 15 can also be viewed as 3 is a **divisor** of 15; that is, when 15 is divided by 3, the result is a whole number and no remainder. We say that 15 is **divisible** by 3. Observe a few examples.

- 15 is divisible by 3,
  since $15 \div 3 = 5$ (no remainder in the division)

- 14 is divisible by 7,
  since $14 \div 7 = 2$ (no remainder in the division)

- 14 is divisible by 2,
  since $14 \div 2 = 7$ (no remainder in the division)

But

- 15 is *not* divisible by 6, since $15 \div 6 \neq$ whole number. In fact, $15 \div 6 = 2$ with remainder 3.

- 14 is not divisible by 5, since $14 \div 5 \neq$ whole number. In fact, $14 \div 5 = 2$ with remainder 4.

It should be clear that factors and divisors are the same thing. The factors of 12 are 1, 2, 3, 4, 6, and 12. The divisors of 12 are also 1, 2, 3, 4, 6, and 12. We have merely changed our terminology from factors to divisors because the tests that we will be using are traditionally called **divisibility tests.** And here they are.

2   *A number is divisible by* 2 *if it is an even number—that is, if its rightmost digit is* 0, 2, 4, 6, *or* 8.
<u>Example.</u>   84, 32, 5376, 100, and 88 are each divisible by 2; their rightmost digits are 4, 2, 6, 0, and 8, respectively.

3   *A number is divisible by* 3 *if the sum of its digits is divisible by* 3.
<u>Example.</u>   825 is divisible by 3 because $8 + 2 + 5 = 15$, and 15 is divisible by 3. Similarly, 414 is divisible by 3, since $4 + 1 + 4 = 9$, and 9 is divisible by 3.

4   *A number is divisible by* 4 *if the number formed by its rightmost two digits is divisible by* 4.
<u>Example.</u>   740 is divisible by 4 because 40 is divisible by 4. Also, 512 is divisible by 4 because 12 is divisible by 4.

5   *A number is divisible by* 5 *if its rightmost digit is* 0 *or* 5.
<u>Example.</u>   20, 685, 700, 810, and 415 are each divisible by 5.

6   *A number is divisible by* 6 *if it is divisible by both* 2 *and* 3.
<u>Example.</u>   714 is divisible by 6 because it is divisible by 2 (it ends in 4, which is even) and is also divisible by 3 ($7 + 1 + 4 = 12$, which is divisible by 3).

9   *A number is divisible by* 9 *if the sum of its digits is divisible by* 9.
<u>Example.</u>   864 is divisible by 9 because $8 + 6 + 4 = 18$, and 18 is divisible by 9.

10   *A number is divisible by* 10 *if its rightmost digit is* 0.
<u>Example.</u>   50, 90, 300, and 1040 are each divisible by 10.

The test for divisibility suggests the following. *If a number is divisible by two numbers having no common factors, then it is divisible by the product of those two numbers.* For instance, if a number is divisible by 2 and divisible by 3, then it is divisible by $2 \cdot 3$, or 6, because 2 and 3 have no common factors. On the other hand, if a number is divisible by 2 and by 4, it is not necessarily divisible by $2 \cdot 4$, or 8, because 2 and 4 have a common factor—2. Consider that 12 is divisible by 2 and by 4, but 12 is not divisible by 8. Let's apply the divisibility tests to the reduction of fractions.

EXAMPLE 9   Reduce $\dfrac{27}{126}$.

Solution   To reduce a fraction, you must find a factor (or divisor) common to both numerator and denominator. In this instance, both the numerator and denominator are divisible by 9 (check this). So divide both 27 and 126 by 9 to see what the factored form will be.

$$\overset{3}{9)\overline{27}}, \qquad \overset{14}{9)\overline{126}}$$

This means that $27 = 3 \cdot 9$ and $126 = 14 \cdot 9$. Thus

$$\frac{27}{126} = \frac{3 \times 9}{14 \times 9}$$

$$= \frac{3}{14} \times \frac{9}{9}$$

$$= \frac{3}{14} \times 1$$

$$= \frac{3}{14} \quad \checkmark$$

EXAMPLE 10   Reduce $\dfrac{14}{21}$.

Solution   Unfortunately, our divisibility tests don't help us directly on this one, although they do lead us to $14 = 2 \cdot 7$ and $21 = 3 \cdot 7$ as we try various tests. And this, of course, demonstrates that both numerator and denominator are divisible by 7.

$$\frac{14}{21} = \frac{2 \times 7}{3 \times 7}$$

$$= \frac{2}{3} \times \frac{7}{7}$$

$$= \frac{2}{3} \times 1$$

$$= \frac{2}{3} \quad \checkmark$$

EXAMPLE 11   Reduce $\dfrac{84}{132}$.

Solution   Both 84 and 132 are divisible by 4. (Check this.) Also, both 84 and 132 are divisible by 3. (Check this.) The numbers 4 and 3 have no common factors; so 84 and 132 are each divisible by $4 \cdot 3$, or 12. Moreover,

$$12\overline{)84}^{\,7} \quad \text{and} \quad 12\overline{)132}^{\,11}$$

So $84 = 7 \cdot 12$ and $132 = 11 \cdot 12$. Thus

$$\frac{84}{132} = \frac{7 \cdot 12}{11 \cdot 12}$$

$$= \frac{7}{11} \times \frac{12}{12}$$

$$= \frac{7}{11} \times 1$$

$$= \frac{7}{11} \quad \checkmark$$

EXERCISES 6.4

1. Apply all the divisibility tests to determine by which of the numbers 2, 3, 4, 5, 6, 9, and 10 the given number is divisible.
*(a) 20   *(b) 30   *(c)   45   (d)   48   (e) 28
*(f) 51   *(g) 19   *(h)   22   (i)   65   *(j) 42
 (k) 72   *(l) 75   *(m)   23   (n) 1000   *(o) 111
*(p) 96   *(q) 252   (r)   414   *(s) 188   (t) 315

2. Reduce each fraction, if possible.
*(a) $\dfrac{4}{10}$ *(b) $\dfrac{21}{33}$ *(c) $\dfrac{2}{6}$ *(d) $\dfrac{28}{30}$ *(e) $\dfrac{12}{15}$

(f) $\dfrac{3}{21}$ *(g) $\dfrac{35}{40}$ (h) $\dfrac{20}{65}$ *(i) $\dfrac{18}{72}$ *(j) $\dfrac{10}{240}$

(k) $\dfrac{14}{33}$ (l) $\dfrac{78}{90}$ *(m) $\dfrac{15}{81}$ *(n) $\dfrac{58}{135}$ (o) $\dfrac{50}{80}$

*(p) $\dfrac{42}{48}$ (q) $\dfrac{85}{90}$ *(r) $\dfrac{117}{153}$

3. Reduce each fraction.
*(a) $\dfrac{60}{105}$ *(b) $\dfrac{12}{300}$ *(c) $\dfrac{30}{135}$ *(d) $\dfrac{105}{120}$ (e) $\dfrac{90}{126}$

*(f) $\dfrac{84}{132}$ (g) $\dfrac{42}{210}$ *(h) $\dfrac{78}{144}$ *(i) $\dfrac{135}{720}$ (j) $\dfrac{210}{600}$

*(k) $\dfrac{160}{620}$ (l) $\dfrac{82}{205}$

4. Multiply; reduce when possible.
*(a) $\dfrac{30}{31} \times \dfrac{7}{30}$ *(b) $\dfrac{30}{35} \times \dfrac{1}{7}$ *(c) $\dfrac{30}{35} \times \dfrac{5}{7}$ (d) $\dfrac{30}{35} \times \dfrac{1}{2}$

*(e) $\dfrac{21}{5} \times \dfrac{10}{42}$ *(f) $\dfrac{1}{3} \times \dfrac{114}{19}$ (g) $\dfrac{4}{105} \times \dfrac{15}{22}$ *(h) $\dfrac{17}{100} \times \dfrac{5}{34}$

$$\text{(i)}\ \frac{3}{7}\times\frac{14}{48}\qquad \text{*(j)}\ \frac{30}{72}\times\frac{9}{144}\qquad \text{*(k)}\ 15\times\frac{32}{1200}\qquad \text{(l)}\ \frac{27}{35}\times\frac{21}{81}$$

6.5 DIVISION OF FRACTIONS

To divide two fractions, invert (that is, flip over) the second fraction and then multiply.

EXAMPLE 12   Divide $\dfrac{3}{5}\div\dfrac{4}{7}$.

Solution

$$\frac{3}{5}\div\frac{4}{7}=\frac{3}{5}\times\frac{7}{4}\qquad \begin{cases}\text{invert:}\ \dfrac{4}{7}\ \longrightarrow\ \dfrac{7}{4}\\[4pt]\text{and change}\div\text{to}\times\end{cases}$$

$$=\frac{3\times 7}{5\times 4}$$

$$=\frac{21}{20}\quad\checkmark$$

Note that you must invert the *second* fraction, not the first one. To see why, consider another approach to the example.

$$\frac{3}{5}\div\frac{4}{7}=\frac{\frac{3}{5}}{\frac{4}{7}}\qquad\text{another way of writing it}$$

$$=\frac{\frac{3}{5}\cdot\frac{7}{4}}{\frac{4}{7}\cdot\frac{7}{4}}\qquad\begin{cases}\text{multiplying both numerator}\left(\dfrac{3}{5}\right)\text{and denominator}\\[6pt]\left(\dfrac{4}{7}\right)\text{by the same number}\left(\dfrac{7}{4}\right)\end{cases}$$

$$=\frac{\frac{3}{5}\cdot\frac{7}{4}}{\frac{4}{7}\cdot\frac{7}{4}}$$

$$=\frac{\frac{3}{5}\cdot\frac{7}{4}}{1}\qquad\text{since}\ \frac{4}{7}\cdot\frac{7}{4}=1$$

$$=\frac{3}{5}\cdot\frac{7}{4}\qquad\begin{cases}\text{This completes the steps showing why we invert}\\\text{and change to multiplication.}\end{cases}$$

$$=\frac{21}{20}\quad\checkmark$$

EXAMPLE 13   Divide $\dfrac{4}{7}\div\dfrac{4}{3}$.

Solution

$$\frac{4}{7}\div\frac{4}{3}=\frac{4}{7}\times\frac{3}{4}$$

$$=\frac{\overset{1}{\cancel{4}}}{7}\times\frac{3}{\underset{1}{\cancel{4}}}$$

$$= \frac{1 \times 3}{7 \times 1}$$

$$= \frac{3}{7} \quad \checkmark$$

EXAMPLE 14 Divide $\frac{7}{13} \div \frac{6}{8}$.

Solution

$$\frac{7}{13} \div \frac{6}{8} = \frac{7}{13} \div \frac{3}{4} \qquad \text{The } \frac{6}{8} \text{ can be reduced to } \frac{3}{4}.$$

$$= \frac{7}{13} \times \frac{4}{3} \qquad \begin{cases} \text{invert and change to} \\ \text{multiplication} \end{cases}$$

$$= \frac{7 \times 4}{13 \times 3}$$

$$= \frac{28}{39} \quad \checkmark$$

EXAMPLE 15 Divide $\frac{5}{28} \div \frac{1}{5}$.

Solution

$$\frac{5}{28} \div \frac{1}{5} = \frac{5}{28} \times \frac{5}{1}$$

$$= \frac{5 \times 5}{28 \times 1}$$

$$= \frac{25}{28} \quad \checkmark$$

Note that these fives cannot be removed. You cannot reduce across the ÷ symbol.

EXERCISES 6.5

1. Divide each and reduce, if possible.

*(a) $\frac{5}{7} \div \frac{3}{4}$ *(b) $\frac{1}{3} \div \frac{1}{2}$ (c) $\frac{1}{5} \div \frac{1}{4}$ *(d) $\frac{5}{9} \div \frac{2}{3}$

(e) $\frac{11}{16} \div \frac{4}{5}$ *(f) $\frac{7}{8} \div \frac{3}{5}$ *(g) $\frac{5}{6} \div \frac{7}{12}$ *(h) $\frac{2}{7} \div \frac{7}{12}$

(i) $\frac{2}{7} \div \frac{12}{7}$ *(j) $\frac{5}{9} \div \frac{5}{9}$ (k) $\frac{4}{9} \div \frac{1}{2}$ *(l) $\frac{24}{30} \div \frac{5}{4}$

(m) $\frac{8}{25} \div \frac{2}{5}$ *(n) $\frac{2}{7} \div 3$ (o) $\frac{6}{11} \div 6$ *(p) $\frac{17}{12} \div 2$

*(q) $12 \div \frac{4}{5}$ (r) $4 \div \frac{2}{3}$

6.6 ADDITION AND SUBTRACTION OF FRACTIONS

You probably know that $\frac{2}{8} + \frac{3}{8} = \frac{5}{8}$. But do you know why we can merely add the numerators 2 and 3 to get 5 and why the denominator remains an 8? The following illustration may provide some insight.

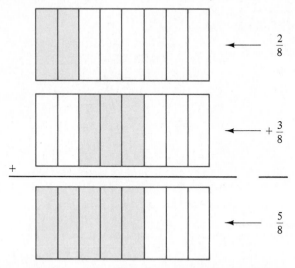

Mathematically, we can apply the distributive property (recall from Chapter 5) to see how the addition can be done.

$$\frac{2}{8} + \frac{3}{8} = 2 \times \frac{1}{8} + 3 \times \frac{1}{8} \qquad \begin{cases} \dfrac{2}{8} \text{ means } 2 \times \dfrac{1}{8} \quad \text{(two-eighths)} \\[2mm] \dfrac{3}{8} \text{ means } 3 \times \dfrac{1}{8} \quad \text{(three-eighths)} \end{cases}$$

$$= (2 + 3) \times \frac{1}{8}$$

using the distributive property

$$= (5) \times \frac{1}{8}$$

$$= \frac{5}{8} \quad \checkmark$$

As another example, consider

$$\frac{2}{13} + \frac{5}{13} + \frac{4}{13} = 2 \times \frac{1}{13} + 5 \times \frac{1}{13} + 4 \times \frac{1}{13}$$

$$= (2 + 5 + 4) \times \frac{1}{13}$$

$$= (11) \times \frac{1}{13}$$

$$= \frac{11}{13} \quad \checkmark$$

As you can see, to add fractions: add the numerators to obtain the new numerator, and keep the same denominator. This addition can be expressed symbolically. If a, b, and c are used to represent whole numbers, then

$$\frac{a}{c} + \frac{b}{c} = \frac{a+b}{c}$$

Since addition of fractions uses the distributive property, the denominators must be the same in order to complete the process. We cannot add $\frac{1}{3} + \frac{1}{5}$ unless we can change their denominators so that both are the same; then we can apply the distributive property. The "same denominator" we seek for fractions being added is called the **common denominator.**

One sure way to always make the denominators the same is to change each to a denominator equal to the product of the two denominators. With $\frac{1}{3}$ and $\frac{1}{5}$, the desired denominator is 3×5 or 15.

As you know, if a number is multiplied by 1 its value is unchanged. Let us take the fractions $\frac{1}{3}$ and $\frac{1}{5}$ that we wish to add and multiply each by an appropriate form of 1.

$$\frac{1}{3} \times 1 = \frac{1}{3} \times \frac{5}{5} = \frac{1 \times 5}{3 \times 5} = \frac{5}{15}$$

$$\frac{1}{5} \times 1 = \frac{1}{5} \times \frac{3}{3} = \frac{1 \times 3}{5 \times 3} = \frac{3}{15}$$

Note that 1 can be written as $\frac{5}{5}$ and as $\frac{3}{3}$, since any nonzero number divided by itself is equal to 1.

The forms of 1 that were chosen, $\frac{5}{5}$ and $\frac{3}{3}$, were selected so that the end result in each case would be a fraction with a denominator of 15, 5 times 3. The fractions can now be added.

$$\frac{1}{3} + \frac{1}{5} = \frac{1}{3} \cdot \frac{5}{5} + \frac{1}{5} \cdot \frac{3}{3}$$

$$= \frac{5}{15} + \frac{3}{15}$$

$$= \frac{5+3}{15}$$

$$= \frac{8}{15} \quad \checkmark$$

EXAMPLE 16 Add $\dfrac{1}{7} + \dfrac{1}{4}$.

Solution Make each denominator 28—that is, 7×4.

$$\frac{1}{7} + \frac{1}{4} = \frac{1}{7} \cdot \frac{4}{4} + \frac{1}{4} \cdot \frac{7}{7}$$

$$= \frac{4}{28} + \frac{7}{28}$$

$$= \frac{11}{28} \quad \checkmark$$

EXAMPLE 17   Add $\frac{1}{4} + \frac{3}{8}$.

Solution   Note that the denominator 4 can be changed to 8 very easily. And so 8 need not be changed. Although 4×8, or 32, will work as a denominator, there is no need to use the larger number, since 8 is sufficient. Keep in mind that you seek as common denominator the smallest number into which both 4 and 8 will divide exactly, and 8 is that number.

$$\frac{1}{4} + \frac{3}{8} = \frac{1}{4} \cdot \frac{2}{2} + \frac{3}{8}$$

$$= \frac{2}{8} + \frac{3}{8}$$

$$= \frac{5}{8} \quad \checkmark$$

EXAMPLE 18   Add $\frac{1}{4} + \frac{1}{3} + \frac{2}{5}$.

Solution   To add three fractions, all denominators must be the same; that is, the fractions must have a common denominator. The common denominator here is $4 \times 3 \times 5 = 60$, the product of the three denominators.

$$\frac{1}{4} + \frac{1}{3} + \frac{2}{5} = \frac{1}{4} \cdot \frac{15}{15} + \frac{1}{3} \cdot \frac{20}{20} + \frac{2}{5} \cdot \frac{12}{12}$$

$$= \frac{15}{60} + \frac{20}{60} + \frac{24}{60}$$

$$= \frac{59}{60} \quad \checkmark$$

Subtraction of fractions resembles addition, as shown in examples below.

EXAMPLE 19   Combine $\frac{5}{8} - \frac{4}{8}$.

Solution

$$\frac{5}{8} - \frac{4}{8} = 5 \times \frac{1}{8} - 4 \times \frac{1}{8}$$

$$= (5 - 4) \times \frac{1}{8}$$

$$= 1 \times \frac{1}{8}$$

$$= \frac{1}{8} \quad \checkmark$$

EXAMPLE 20   Combine $\frac{5}{9} - \frac{1}{4}$.

Solution

$$\frac{5}{9} - \frac{1}{4} = \frac{5}{9} \cdot \frac{4}{4} - \frac{1}{4} \cdot \frac{9}{9}$$

$$= \frac{20}{36} - \frac{9}{36}$$

$$= \frac{11}{36} \quad \checkmark$$

In the next section we develop techniques that will help us to add and subtract fractions in cases where the common denominator is more difficult to determine.

EXERCISES 6.6

*1. Add the fractions.

(a) $\frac{4}{13} + \frac{2}{13}$ (b) $\frac{3}{7} + \frac{1}{7}$ (c) $\frac{1}{9} + \frac{4}{9}$ (d) $\frac{2}{5} + \frac{2}{5}$

*2. Subtract.

(a) $\frac{5}{7} - \frac{2}{7}$ (b) $\frac{10}{11} - \frac{8}{11}$ (c) $\frac{2}{3} - \frac{1}{3}$ (d) $\frac{8}{9} - \frac{7}{9}$

3. Add the fractions; reduce the answer, if possible.

*(a) $\frac{2}{5} + \frac{3}{5}$ (b) $\frac{4}{7} + \frac{3}{7}$ *(c) $\frac{1}{4} + \frac{1}{4}$

(d) $\frac{1}{6} + \frac{1}{6}$ *(e) $\frac{1}{3} + \frac{1}{6}$ (f) $\frac{5}{12} + \frac{1}{3}$

*(g) $\frac{2}{5} + \frac{3}{4}$ (h) $\frac{3}{10} + \frac{1}{2}$ *(i) $\frac{4}{13} + \frac{3}{7}$

(j) $\frac{1}{9} + \frac{3}{4}$ *(k) $\frac{3}{11} + \frac{5}{12}$ (l) $\frac{2}{9} + \frac{2}{5}$

*(m) $\frac{3}{8} + \frac{5}{16} + \frac{3}{4}$ (n) $\frac{1}{3} + \frac{5}{12} + \frac{1}{4}$ *(o) $\frac{4}{7} + \frac{2}{3} + \frac{1}{2}$

(p) $\frac{2}{9} + \frac{1}{3} + \frac{2}{5}$

4. Subtract; reduce the answer, if possible.

*(a) $\frac{4}{9} - \frac{1}{9}$ (b) $\frac{7}{10} - \frac{3}{10}$ *(c) $\frac{6}{7} - \frac{2}{5}$ (d) $\frac{4}{5} - \frac{2}{3}$

*(e) $\dfrac{3}{4} - \dfrac{2}{3}$ (f) $\dfrac{7}{9} - \dfrac{3}{7}$ *(g) $\dfrac{11}{12} - \dfrac{2}{3}$ (h) $\dfrac{11}{18} - \dfrac{4}{9}$

*(i) $\dfrac{11}{12} - \dfrac{3}{4}$ (j) $\dfrac{13}{16} - \dfrac{3}{8}$

*5. Find the perimeter of each geometric figure.

(a)   (b)

(c) A square, each side of which is $\frac{2}{11}$.

6.7 IMPROPER FRACTIONS AND MIXED NUMBERS

The number $1\frac{7}{30}$ is an example of a **mixed number.** It is a whole number plus a fraction combined. $1\frac{7}{30}$ is an abbreviated form of $1 + \frac{7}{30}$.

Mixed Number	Its Meaning
$1\dfrac{7}{30}$	$1 + \dfrac{7}{30}$
$2\dfrac{1}{2}$	$2 + \dfrac{1}{2}$
$3\dfrac{7}{8}$	$3 + \dfrac{7}{8}$

As mentioned in Section 6.1, a fraction is called an *improper fraction* if the numerator is greater than the denominator or equal to the denominator. If the numerator is less than the denominator, the fraction is a *proper fraction*.

Some proper fractions: $\dfrac{1}{2}, \dfrac{5}{6}, \dfrac{3}{11}$

Some improper fractions: $\dfrac{2}{2}, \dfrac{7}{6}, \dfrac{13}{11}$

Often it is desirable to change an improper fraction to the form of a mixed number. If you divide the denominator into the numerator, the quotient will be the whole-number part of the mixed number and the remainder will be the numerator of the fractional part.

EXAMPLE 21   Change $\dfrac{14}{3}$ to a mixed number.

Solution

$$\begin{array}{r} 4 \\ 3\overline{)14} \\ \underline{12} \\ 2 \end{array}\ \begin{array}{l}\text{quotient}\\ \\ \\ \text{remainder}\end{array} \quad\longrightarrow\quad 4\dfrac{2}{3} \ \ \checkmark$$

Since the quotient is 4, the whole-number part is 4. The remainder 2 is the numerator of the fraction. The denominator, of course, is 3, since the original fraction $\frac{14}{3}$ has a denominator of 3.

Consider how we might change a mixed number to an improper fraction.

$$7\frac{2}{5} = 7 + \frac{2}{5} = 7 \cdot \frac{5}{5} + \frac{2}{5} = \frac{35}{5} + \frac{2}{5} = \frac{35 + 2}{5} = \frac{37}{5}$$

A shorter way to change a mixed number to an improper fraction, based on the procedure just shown, is as follows: Multiply the denominator of the fraction by the whole number and then add the numerator of the fraction. Place the total over the denominator.

EXAMPLE 22   Change $7\dfrac{2}{5}$ to an improper fraction.

Solution

$$7\frac{2}{5} \quad\longrightarrow\quad 5 \times 7 + 2 = 37$$

So

$$7\frac{2}{5} = \frac{37}{5} \ \ \checkmark$$

EXAMPLE 23   Multiply $2\dfrac{7}{8} \times 1\dfrac{2}{3}$.

Solution   To multiply (or divide) mixed numbers, begin by changing the mixed numbers to improper fractions.

$$2\frac{7}{8} \times 1\frac{2}{3} = \frac{23}{8} \times \frac{5}{3}$$
$$= \frac{115}{24} \ \ \checkmark$$

If you prefer a mixed-number answer, change the improper fraction to a mixed number.

$$\frac{115}{24} = 4\frac{19}{24} \ \ \checkmark$$

EXAMPLE 24   Add $5\frac{3}{4} + 3\frac{1}{2}$.

Solution   When addition (or subtraction) of mixed numbers is required, change them to improper fractions and then add (or subtract) the improper fractions.

$$5\frac{3}{4} + 3\frac{1}{2} = \frac{23}{4} + \frac{7}{2}$$

$$= \frac{23}{4} + \frac{14}{4}$$

$$= \frac{37}{4} \; \checkmark \quad \text{or} \quad 9\frac{1}{4} \; \checkmark$$

EXERCISES 6.7

1. Change each improper fraction to a mixed number.

   *(a) $\frac{7}{6}$ (b) $\frac{5}{3}$ *(c) $\frac{9}{2}$ (d) $\frac{15}{7}$ *(e) $\frac{29}{4}$

   (f) $\frac{22}{4}$ *(g) $\frac{19}{5}$ (h) $\frac{14}{3}$ *(i) $\frac{51}{2}$ (j) $\frac{25}{7}$

2. Change each mixed number to an improper fraction.

   *(a) $1\frac{2}{3}$ (b) $1\frac{4}{5}$ *(c) $2\frac{6}{7}$ *(d) $3\frac{1}{9}$ *(e) $5\frac{2}{5}$

   *(f) $8\frac{1}{2}$ *(g) $4\frac{5}{12}$ *(h) $7\frac{2}{9}$ *(i) 12 *(j) $9\frac{3}{7}$

3. Multiply or divide as indicated.

   *(a) $3\frac{1}{2} \times 2\frac{5}{8}$ (b) $3\frac{2}{3} \times 2\frac{5}{8}$ *(c) $1\frac{2}{3} \times 1\frac{2}{5}$

   (d) $4\frac{1}{8} \times 5\frac{1}{3}$ *(e) $3\frac{3}{7} \div 1\frac{1}{3}$ (f) $7\frac{1}{4} \div 2\frac{3}{8}$

   *(g) $5\frac{3}{4} \div 5\frac{1}{9}$ (h) $3\frac{6}{7} \div 2\frac{2}{3}$

4. Add or subtract as indicated; reduce if possible.

   *(a) $2\frac{1}{2} + 1\frac{3}{4}$ (b) $3\frac{5}{8} + 2\frac{1}{4}$ *(c) $4\frac{5}{12} + 7\frac{1}{6}$

   (d) $5\frac{2}{3} + 1\frac{1}{5}$ *(e) $2\frac{2}{5} - 2\frac{1}{7}$ (f) $6\frac{1}{2} - 4\frac{2}{3}$

   *(g) $5\frac{2}{3} - 3\frac{3}{4}$ (h) $3\frac{7}{10} - 1\frac{1}{5}$

*5. In each exercise you are given the price of a share of stock and the change that occurs in that price during a day of trading on the stock market. Determine the stock price after the change.

(a) Pan Am began at $5\frac{3}{4}$ and gained $\frac{3}{8}$.
(b) Chrysler began at $25\frac{5}{8}$ and gained $\frac{1}{2}$.
(c) IBM began at $112\frac{1}{8}$ and lost $2\frac{3}{4}$.
(d) Marriott began at $28\frac{7}{8}$ and lost $1\frac{1}{4}$.

6.8 PRIME FACTORIZATION AND LCM

The terminology and techniques developed in this section will be applied to the addition of fractions.

Such numbers as 2, 3, 5, 7, 11, 13, 17, 19, and 23 are **prime numbers**; none can be written as the product of two whole numbers other than 1 and the number itself. On the other hand, 4, 6, 8, 9, 10, 12, 14, 15, 16, 18, 20, 21, and 22 are *not prime*. For instance, $4 = 2 \cdot 2, 6 = 2 \cdot 3, 15 = 3 \cdot 5$, and so on. Numbers that can be written in this factored form, as the product of two whole numbers other than 1 and the number itself, are called **composite numbers.** The number 1 is not considered to be prime.

EXAMPLE 25   Write 30 as the product of prime numbers.

Solution   One approach to this problem is to examine 30 in search of a divisor (factor), preferably one that is a prime number. The divisibility tests should prove helpful in such problems. Since 30 is an even number, 2 is a divisor. So divide 30 by 2 and write the factored form.

$$30 = 2 \cdot 15$$

But 15 is not a prime number; so we must next write 15 as a product of primes. Since 15 ends in 5, it is divisible by 5. In fact, $15 = 3 \cdot 5$. Thus the factored form of 30 becomes

$$30 = 2 \cdot 3 \cdot 5 \quad \checkmark$$

Because all of the factors are prime numbers, this then is the desired **prime factorization.** The number (30) is now written as the product of prime numbers.

EXAMPLE 26   Write 56 as the product of prime numbers.

Solution   Here are the steps in one approach to finding the prime factorization. Although you may prefer to go in a different direction, we will both end up with the same prime factorization. Such factorization is unique for any given number.

$$\begin{aligned} 56 &= 2 \cdot 28 \\ &= 2 \cdot 2 \cdot 14 \\ &= 2 \cdot 2 \cdot 2 \cdot 7 \end{aligned}$$

We can use exponent notation to compact the $2 \cdot 2 \cdot 2$ as 2^3. Thus the desired prime factorization is

$$56 = 2^3 \cdot 7 \quad \checkmark$$

Now let's look at two other prime factorizations.

$$12 = 2^2 \cdot 3$$
$$15 = 3 \cdot 5$$

Suppose that we select and multiply all the different prime factors present in the above factorizations (2, 3, 5) and include the largest exponents that appear on any factors with exponents (2^2). The result is

$$2^2 \cdot 3 \cdot 5, \text{ or } 60$$

This number 60 is a multiple of 12 (it is $5 \cdot 12$), and it is a multiple of 15 (it is $4 \cdot 15$). It is also the *smallest* number that is a multiple of both 12 and 15. As a result, it is called the **least common multiple,** or **LCM** for short. That the least common multiple of 12 and 15 is 60 can be written in the following notation:

$$\text{LCM } (12, 15) = 60$$

In general, the LCM of two or more numbers is the product of all the different prime factors of those numbers, including the highest power of each that appears.

To determine the LCM:

1. Obtain the prime factorization of each number.
2. Select all the different prime factors that appear, including the highest power of each, and multiply them together.

EXAMPLE 27   Determine the LCM of 35 and 10.

Solution   The prime factorizations:

$$35 = 5 \cdot 7$$
$$10 = 2 \cdot 5$$

So

$$\text{LCM } (35, 10) = 2 \cdot 5 \cdot 7 = 70 \quad \checkmark$$

EXAMPLE 28   Determine the LCM of 30 and 56.

Solution   From prime factorizations determined earlier in the section we know that

$$30 = 2 \cdot 3 \cdot 5$$
$$56 = 2^3 \cdot 7$$

So

$$\text{LCM } (30, 56) = 2^3 \cdot 3 \cdot 5 \cdot 7 = 840 \quad \checkmark$$

EXAMPLE 29 Determine LCM (24, 30, 45).

Solution The prime factorizations:

$$24 = 2^3 \cdot 3$$
$$30 = 2 \cdot 3 \cdot 5$$
$$45 = 3^2 \cdot 5$$

So

$$\text{LCM } (24, 30, 45) = 2^3 \cdot 3^2 \cdot 5 = 360 \quad \checkmark$$

EXERCISES 6.8

1. Write each number as a product of prime numbers.
   *(a) 21   *(b) 10   (c) 15   *(d) 33   *(e) 42
   *(f) 105   (g) 20   *(h) 40   (i) 18   *(j) 50
   (k) 100   *(l) 36   *(m) 64   (n) 81   (o) 98
   *(p) 90   (q) 120   (r) 150   *(s) 144   *(t) 216
   (u) 210

2. Determine the least common multiple (LCM) of each set of numbers.
   *(a) 6,10   *(b) 6, 35   *(c) 20, 25
   *(d) 30, 32   *(e) 21, 35   (f) 96, 108
   *(g) 4, 20, 30   *(h) 2, 3, 18   *(i) 10, 24, 30
   (j) 15, 20, 25   (k) 18, 40, 50   *(l) 20, 36, 60
   *(m) 50, 75, 120   (n) 4, 18, 20, 35   *(o) 12, 15, 20, 36
   (p) 2, 4, 6, 8, 10

6.9 THE LCM AND ADDITION OF FRACTIONS

Examine the following *addition* exercises to determine what should be used as the common denominator in each case. The lowest (smallest) common denominator for each exercise is then given after exercise (f).

$$\text{(a) } \frac{1}{6} + \frac{3}{6}$$

$$\text{(b) } \frac{1}{2} + \frac{3}{8}$$

$$\text{(c) } \frac{2}{3} + \frac{1}{4}$$

$$\text{(d) } \frac{1}{2} + \frac{2}{3} + \frac{5}{7}$$

$$\text{(e) } \frac{5}{12} + \frac{2}{15} + \frac{3}{8}$$

$$\text{(f) } \frac{5}{8} + \frac{3}{10} + \frac{5}{6}$$

The common denominators are: (a) 6 (b) 8 (c) 12 (d) 42 (e) 120 (f) 120. And you will probably agree that the last two were the most difficult to determine. Were you able to determine them? Your knowledge of LCM should have helped you. *After all, what you seek as a common denominator is the smallest number that is a multiple of each of the denominators.* In other words, the least common multiple (LCM) of the denominators is the smallest common denominator.

EXAMPLE 30 Determine the common denominator to be used in adding:

$$\frac{5}{8} + \frac{3}{10} + \frac{5}{6}$$

Solution The prime factorizations of the denominators:

$$8 = 2^3$$
$$10 = 2 \cdot 5$$
$$6 = 2 \cdot 3$$

So LCM $(8, 10, 6) = 2^3 \cdot 5 \cdot 3 = 120$. The common denominator is 120. ✓

EXAMPLE 31 Add $\dfrac{4}{15} + \dfrac{5}{6} + \dfrac{1}{4}$.

Solution The prime factorization of the denominators:

$$15 = 3 \cdot 5$$
$$6 = 2 \cdot 3$$
$$4 = 2^2$$

The LCM $(15, 6, 4)$ is $2^2 \cdot 3 \cdot 5$, or 60. So the common denominator is 60. We now proceed to change all denominators to 60 and then add the fractions.

$$\frac{4}{15} + \frac{5}{6} + \frac{1}{4} = \frac{4}{15} \cdot \frac{4}{4} + \frac{5}{6} \cdot \frac{10}{10} + \frac{1}{4} \cdot \frac{15}{15}$$

$$= \frac{16}{60} + \frac{50}{60} + \frac{15}{60}$$

$$= \frac{81}{60} \quad \text{which can be reduced}$$

$$= \frac{3 \cdot 27}{3 \cdot 20}$$

$$= \frac{27}{20}, \text{ or } 1\frac{7}{20} \quad \checkmark$$

EXAMPLE 32 Add $\dfrac{7}{12} + 5 + \dfrac{11}{14}$.

Solution The whole number 5 is what makes this example different. One approach is to rewrite the 5 as the fraction $\frac{5}{1}$ and then determine the least common multiple of the denominators 12, 1, and 14. And LCM (12, 1, 14) = $2^2 \cdot 3 \cdot 7 = 84$. So we proceed as

$$\frac{7}{12} + \frac{5}{1} + \frac{11}{14} = \frac{7}{12} \cdot \frac{7}{7} + \frac{5}{1} \cdot \frac{84}{84} + \frac{11}{14} \cdot \frac{6}{6}$$

$$= \frac{49}{84} + \frac{420}{84} + \frac{66}{84}$$

$$= \frac{535}{84}, \text{ or } 6\frac{31}{84} \quad \checkmark$$

Another approach is to combine the whole number 5 with one of the two fractions to make a mixed number, change the mixed number to an improper fraction, and proceed to add. This process is shown next.

$$\frac{7}{12} + 5 + \frac{11}{14} = 5\frac{7}{12} + \frac{11}{14}$$

$$= \frac{67}{12} + \frac{11}{14}$$

$$= \frac{67}{12} \cdot \frac{7}{7} + \frac{11}{14} \cdot \frac{6}{6}$$

$$= \frac{469}{84} + \frac{66}{84}$$

$$= \frac{535}{84}, \text{ or } 6\frac{31}{84} \quad \checkmark$$

EXERCISES 6.9

1. Add the fractions; reduce all results to simplest form.

*(a) $\dfrac{3}{18} + \dfrac{1}{9}$ *(b) $\dfrac{17}{30} + \dfrac{3}{40}$ *(c) $\dfrac{5}{18} + \dfrac{3}{24}$

(d) $\dfrac{4}{9} + \dfrac{11}{30}$ (e) $\dfrac{5}{12} + \dfrac{4}{15}$ *(f) $\dfrac{7}{30} + \dfrac{11}{24}$

(g) $\dfrac{10}{21} + \dfrac{5}{14}$ *(h) $\dfrac{2}{15} + \dfrac{5}{28}$ (i) $\dfrac{13}{48} + \dfrac{13}{35}$

*(j) $\dfrac{5}{36} + \dfrac{7}{30}$ (k) $\dfrac{11}{40} + \dfrac{5}{16}$ *(l) $\dfrac{11}{27} + \dfrac{5}{18}$

2. Add the fractions; reduce all results to simplest form.

*(a) $\dfrac{3}{10} + \dfrac{3}{20} + \dfrac{7}{25}$ (b) $\dfrac{1}{5} + \dfrac{7}{20} + \dfrac{13}{30}$ *(c) $\dfrac{5}{16} + \dfrac{17}{40} + \dfrac{9}{10}$

(d) $\dfrac{3}{4} + \dfrac{7}{18} + \dfrac{5}{36}$ *(e) $\dfrac{1}{4} + \dfrac{5}{9} + \dfrac{19}{30}$ (f) $\dfrac{11}{24} + \dfrac{1}{32} + \dfrac{1}{4}$

*(g) $\dfrac{1}{12} + \dfrac{7}{50} + \dfrac{4}{15}$ (h) $\dfrac{5}{16} + \dfrac{1}{24} + \dfrac{1}{6}$ *(i) $\dfrac{13}{45} + 6 + \dfrac{7}{18}$

(j) $\dfrac{5}{16} + \dfrac{7}{12} + 3$ *(k) $2 + \dfrac{1}{15} + \dfrac{3}{20}$ (l) $1 + \dfrac{7}{18} + \dfrac{7}{30}$

3. Subtract the fractions; reduce all results to simplest form.

*(a) $\dfrac{11}{12} - \dfrac{4}{15}$ (b) $\dfrac{13}{24} - \dfrac{5}{18}$ *(c) $\dfrac{7}{24} - \dfrac{1}{16}$

(d) $\dfrac{13}{20} - \dfrac{7}{30}$ *(e) $\dfrac{7}{30} - \dfrac{1}{40}$ (f) $\dfrac{15}{32} - \dfrac{5}{18}$

*4. Determine the perimeter of a triangle ($P = a + b + c$) whose sides a, b, and c are $\frac{2}{9}$, $\frac{5}{6}$, and $\frac{7}{15}$.

5. Determine the perimeter of a triangle whose sides are $\frac{3}{4}$, $\frac{2}{3}$, and $\frac{1}{6}$.

*6. Determine the perimeter of a rectangle ($P = 2l + 2w$) whose length is $\frac{4}{11}$ and whose width is $\frac{1}{7}$.

7. Determine the perimeter of a rectangle whose length is $\frac{11}{20}$ and whose width is $\frac{5}{24}$.

6.10 A NOTE ON DECIMALS IN MULTIPLICATION AND DIVISION

In Chapter 1 a rule was given for multiplication of numbers containing decimal points. The product will have after its decimal point the total number of digits found after the decimal points of the numbers being multiplied. Here is an example.

$$14.37 \quad \longleftarrow \quad \text{two digits after the decimal}$$
$$\underline{21.2} \quad \longleftarrow \quad \text{one digit after the decimal}$$
$$304.644 \quad \longleftarrow \quad \text{three (two + one) digits after decimal}$$

To see *why* this rule works, begin by writing 14.37 and 21.2 as fractions and multiplying them that way.

$$14.37 = 14\,\dfrac{37}{100} = \dfrac{1437}{100}$$

$$21.2 = 21\,\dfrac{2}{10} = \dfrac{212}{10}$$

and $\dfrac{1437}{100} \times \dfrac{212}{10} = \dfrac{1437 \times 212}{100 \times 10} = \dfrac{304644}{1000} = 304.644$

Notice why the number of digits after the decimal point is determined to be the total number of digits after the decimal points of the two numbers being multiplied. When 14.37 is written as a fraction with denominator 100, you can see by the 100 that there are two decimal places. Similarly, when 21.2 is written as a fraction with denominator 10, you can see by the 10 that there is one decimal place. So when the two fractions are multiplied, you can see the three decimal places evolving as two from the first fraction and one from the second.

In Chapter 1 a rule was given for shifting the decimal points in order to make the divisor a whole number. Here is an example of such a shift. Consider the division

$$7.42\overline{)269.3516}$$

The mechanical process is to shift the decimal two places to the right in both numbers. This step will make the divisor a whole number, 742.

$$7.42\overline{)269.3516}$$

or $$742\overline{)26935.16}$$

To see why this new form is equivalent to the original, observe the following use of fractions. First,

$$7.42\overline{)269.3516}$$

is the same as

$$\frac{269.3516}{7.42}$$

And the value of this fraction will not change if it is multiplied by 1. Choose 1 in the form

$$1 = \frac{100}{100}$$

because this will eliminate the decimal point in the denominator. So let's proceed to multiply the fraction by $\frac{100}{100}$.

$$\frac{269.3516}{7.42} \times \frac{100}{100} = \frac{26935.16}{742}$$

And this resulting fraction is the same as the division

$$742\overline{)26935.16}$$

CHAPTER 6 REVIEW EXERCISES

1. Reduce each fraction, if possible.

*(a) $\dfrac{15}{25}$ *(b) $\dfrac{22}{80}$ *(c) $\dfrac{24}{40}$ (d) $\dfrac{48}{64}$ *(e) $\dfrac{48}{120}$

(f) $\dfrac{34}{57}$ *(g) $\dfrac{21}{98}$ (h) $\dfrac{64}{81}$ *(i) $\dfrac{112}{150}$

2. Multiply; reduce if possible.

*(a) $\dfrac{2}{3} \times \dfrac{3}{2}$ (b) $\dfrac{3}{5} \times \dfrac{7}{9}$ *(c) $\dfrac{12}{15} \times \dfrac{5}{8}$

(d) $\dfrac{15}{24} \times \dfrac{6}{5}$ *(e) $\dfrac{32}{50} \times \dfrac{25}{14}$ (f) $\dfrac{22}{9} \times \dfrac{27}{33}$

$*$(g) $\dfrac{17}{13} \times \dfrac{23}{29}$ (h) $\dfrac{26}{39} \times \dfrac{15}{11}$ $*$(i) $\dfrac{14}{98} \times \dfrac{14}{35}$

(j) $\dfrac{64}{81} \times \dfrac{27}{12}$ $*$(k) $5 \times \dfrac{13}{44}$ (l) $\dfrac{7}{20} \times 5$

$*$(m) $2\dfrac{3}{4} \times 3\dfrac{2}{5}$ $*$(n) $1\dfrac{2}{7} \times 2\dfrac{4}{9}$

3. Divide; reduce if possible.

$*$(a) $\dfrac{2}{3} \div \dfrac{2}{3}$ (b) $\dfrac{3}{5} \div \dfrac{5}{3}$ $*$(c) $\dfrac{15}{17} \div \dfrac{25}{34}$

(d) $\dfrac{100}{120} \div \dfrac{20}{15}$ $*$(e) $\dfrac{65}{85} \div \dfrac{39}{34}$ (f) $\dfrac{31}{37} \div \dfrac{11}{19}$

$*$(g) $\dfrac{38}{39} \div \dfrac{11}{26}$ (h) $\dfrac{45}{78} \div \dfrac{20}{19}$ $*$(i) $\dfrac{46}{54} \div \dfrac{69}{48}$

(j) $\dfrac{124}{105} \div \dfrac{88}{21}$ $*$(k) $\dfrac{15}{31} \div 4$ (l) $7 \div \dfrac{14}{19}$

$*$(m) $5\dfrac{3}{4} \div 2\dfrac{1}{2}$ $*$(n) $3\dfrac{2}{3} \div 2\dfrac{3}{7}$

4. Write each number as a product of prime numbers.
   $*$(a) 135   (b) 72 $*$(c) 88   (d) 84
   $*$(e) 96   (f) 54 $*$(g) 110   (h) 125
   $*$(i) 243

5. Determine the least common multiple (LCM) of each set of numbers.
   $*$(a) 2, 3, 5   (b) 15, 24 $*$(c) 30, 40
   (d) 26, 65 $*$(e) 8, 12, 20   (f) 18, 30, 36
   $*$(g) 35, 45, 63   (h) 36, 72, 81

6. Add or subtract as indicated; reduce all results to simplest form.

$*$(a) $\dfrac{11}{20} + \dfrac{1}{10}$ (b) $\dfrac{13}{22} + \dfrac{10}{33}$ $*$(c) $\dfrac{13}{24} - \dfrac{9}{32}$

(d) $\dfrac{12}{42} + \dfrac{11}{12}$ $*$(e) $\dfrac{15}{56} + \dfrac{16}{35}$ (f) $\dfrac{17}{50} - \dfrac{13}{55}$

$*$(g) $\dfrac{15}{84} + \dfrac{17}{63}$ (h) $\dfrac{15}{34} + \dfrac{1}{6} + \dfrac{6}{17}$ $*$(i) $\dfrac{5}{18} + \dfrac{4}{21} + \dfrac{7}{15}$

(j) $\dfrac{17}{28} + \dfrac{3}{14} + \dfrac{9}{16}$ $*$(k) $7 + \dfrac{3}{20} + \dfrac{5}{36}$ (l) $\dfrac{11}{49} + 3 + \dfrac{4}{21}$

$*$(m) $5\dfrac{3}{4} + 4\dfrac{1}{8}$ $*$(n) $3\dfrac{2}{5} - 1\dfrac{2}{3}$

FRACTIONS, DECIMALS, AND PERCENT

In this chapter we explore the relationships between fractions and decimals, decimals and percent, and fractions and percent. We begin with a look at fractions and decimals.

7.1 FORMS OF FRACTIONS AND DECIMALS

The first place to the right of the decimal point is the *tenths* place. The second place is *hundredths*. The third is *thousandths*. The fourth is *ten thousandths*. The fifth is *hundred thousandths*. The sixth is *millionths*. Thus

.7	is read	7 tenths
.23	is read	23 hundredths
.785	is read	785 thousandths
.6409	is read	6409 ten thousandths

All this is consistent with the way you would read or write the decimal in fraction form.

$$.7 = \frac{7}{10} \qquad \text{one zero in the denominator}$$

$$.23 = \frac{23}{100} \qquad \text{two zeros in the denominator}$$

$$.785 = \frac{785}{1000} \qquad \text{three zeros in the denominator}$$

$$.6409 = \frac{6409}{10,000} \qquad \text{four zeros in the denominator}$$

It should be clear how to write a decimal form as a fraction. The numerator of the fraction will be the original number without the decimal point, and there are as many zeros after the one in the denominator as there are digits after the decimal of the original.

EXAMPLE 1   Write .49 and .9521 as fractions.

Solution
$$.49 = \frac{49}{100} \checkmark$$

$$.9521 = \frac{9521}{10,000} \checkmark$$

Often the fractional form can be simplified; we can *reduce* it. For instance,

$$.5 = \frac{5}{10} = \frac{1}{2}$$

$$.75 = \frac{75}{100} = \frac{3}{4}$$

$$.24 = \frac{24}{100} = \frac{6}{25}$$

$$.720 = \frac{720}{1000} = \frac{18}{25}$$

Consider the reverse process, that of changing a fraction to its decimal form.

EXAMPLE 2   Change $\frac{3}{8}$ to its decimal equivalent.

Solution   The fraction $\frac{3}{8}$ can be considered as $3 \div 8$. And division of 3 by 8 gives a decimal result.

$$\frac{3}{8} = 3 \div 8 \longrightarrow \begin{array}{r} .375 \\ 8\overline{)3.000} \\ \underline{24} \\ 60 \\ \underline{56} \\ 40 \\ \underline{40} \end{array}$$

So $\frac{3}{8} = .375$, exactly.

EXAMPLE 3   Change $\frac{5}{19}$ to a decimal.

Solution

$$\frac{5}{19} = 5 \div 19 \longrightarrow$$

$$
\begin{array}{r}
.26315 \\
19\overline{)5.00000} \\
\underline{38} \\
120 \\
\underline{114} \\
60 \\
\underline{57} \\
30 \\
\underline{19} \\
110 \\
\underline{95}
\end{array}
$$

We have decided to quit with five digits after the decimal point. When the result does not come out exact after a few decimal places, you must decide when to quit. This decision usually depends on the application. Here the decision was arbitrary.

EXERCISES 7.1

Answers to starred exercises are given in the back of the book.

1. Change each decimal to a fraction and reduce if possible.

  *(a) .17   *(b) .31   *(c) .50   *(d) .34
  *(e) .75   (f) .44   *(g) .2   *(h) .315
  *(i) .704   (j) .9   *(k) .95   (l) .85
  *(m) .48   *(n) .3000   *(o) .5655   (p) .454
  *(q) .825   (r) .285   *(s) .950   *(t) .999999

2. Change each fraction to a decimal form. Use four decimal places.

  *(a) $\frac{1}{8}$ *(b) $\frac{3}{5}$ *(c) $\frac{7}{9}$ (d) $\frac{5}{12}$ *(e) $\frac{4}{7}$

  *(f) $\frac{2}{3}$ *(g) $\frac{11}{13}$ (h) $\frac{5}{6}$ *(i) $\frac{5}{8}$ (j) $\frac{3}{14}$

  *(k) $\frac{2}{11}$ (l) $\frac{3}{7}$ *(m) $\frac{1}{9}$ *(n) $\frac{4}{15}$

7.2 PERCENT

Percents are fractions in disguise. As you probably know, 50% means the same as $\frac{1}{2}$; 75% and $\frac{3}{4}$ are the same. Actually, **percent** *means hundredths.*

$$50\% = \frac{50}{100} \quad \text{which reduces to } \frac{1}{2}$$

$$75\% = \frac{75}{100} \quad \text{which reduces to } \frac{3}{4}$$

$$40\% = \frac{40}{100} \quad \text{which reduces to } \frac{2}{5}$$

$$35\% = \frac{35}{100} \quad \text{which reduces to } \frac{7}{20}$$

$$29\% = \frac{29}{100} \quad \text{which cannot be reduced}$$

EXAMPLE 4 Change 64% to a fraction.

Solution
$$64\% = \frac{64}{100} \quad \text{which can be reduced to } \frac{16}{25} \quad \checkmark$$

Percents are easily represented as decimals, since the fraction form of the percent has 100 in the denominator.

$$30\% = \frac{30}{100} = .30$$

$$89\% = \frac{89}{100} = .89$$

$$50\% = \frac{50}{100} = .50$$

$$25\% = \frac{25}{100} = .25$$

Dividing by 100 merely moves the decimal point two places to the left. This is shown in more detail below.

$$89\% = \frac{89}{100} = 89 \div 100 \longrightarrow \begin{array}{r} .89 \\ 100\overline{)89.00} \\ \underline{80\ 0} \\ 9\ 00 \\ \underline{9\ 00} \end{array}$$

Thus

> To change a percent to a decimal, move the decimal point two places to the left and remove the % sign.

Again, a percent is a fraction with denominator 100.

EXAMPLE 5 Change 15% to a decimal.

Solution
$$15\% = \frac{15}{100} = .15 \quad \checkmark$$

EXAMPLE 6   Change 2% to a decimal.

Solution
$$2\% = \frac{2}{100} = .02 \quad \checkmark$$

EXAMPLE 7   Change 17% to a decimal.

Solution
$$17\% = \frac{17}{100} = .17 \quad \checkmark$$

You can change a decimal to a percent by reversing this process.

> To change a decimal to a percent, multiply by 100 (or move the decimal point two places to the right) and attach a % sign.

Here are some examples.

EXAMPLE 8   Change .29 to a percent.

Solution $.29 = .29 = 29\% \quad \checkmark$ As a check, note that $29\% = \dfrac{29}{100} = .29.$

EXAMPLE 9   Change .84 to a percent.

Solution
$$.84 = .84 = 84\% \quad \checkmark$$

EXAMPLE 10   Change .816 to a percent.

Solution
$$.816 = .816 = 81.6\% \quad \checkmark$$

Note that the percent form can include a decimal within it.

How can we change a fraction to a percent? If the fraction has 100 as its denominator (or can be made to have 100 as its denominator), then the corresponding numerator is the percent.

EXAMPLE 11   Change $\dfrac{53}{100}$ to a percent.

Solution $\dfrac{53}{100} = 53\% \quad \checkmark$ Recall, percent means hundredths.

EXAMPLE 12   Change $\dfrac{9}{20}$ to a percent.

Solution   The denominator 20 is easily changed to 100.

$$\frac{9}{20} = \frac{9}{20} \times \frac{5}{5} = \frac{45}{100} = 45\% \quad \checkmark$$

EXAMPLE 13   Change $\frac{3}{10}$ to a percent.

Solution   The denominator 10 is easily changed to 100.

$$\frac{3}{10} = \frac{3}{10} \times \frac{10}{10} = \frac{30}{100} = 30\% \quad \checkmark$$

But what if the denominator is 17 or 83 or 42? In such cases, change the fraction to a decimal and then change the decimal to a percent. Often the result will be approximate, because the division does not come out exact, yet we choose to quit after three or four decimal places. We will be using the symbol $\approx$ to mean "approximately equal to."

EXAMPLE 14   Change $\frac{5}{12}$ to a percent.

Solution

$$\frac{5}{12} = 5 \div 12 \longrightarrow$$

$$\begin{array}{r} .4166 \\ 12\overline{)5.0000} \\ \underline{4\ 8} \\ 20 \\ \underline{12} \\ 80 \\ \underline{72} \\ 80 \\ \underline{72} \end{array} \approx .4166 \approx 41.66\% \quad \checkmark$$

EXAMPLE 15   Change $\frac{1}{3}$ to a percent.

Solution

$$\frac{1}{3} = 1 \div 3 \longrightarrow \quad 3\overline{)1.000}^{\;.333} \approx .333 \approx 33.3\% \quad \checkmark$$

EXAMPLE 16   Change $\frac{7}{80}$ to a percent.

Solution

$$\frac{7}{80} = 7 \div 80 \longrightarrow$$

$$\begin{array}{r} .0875 \\ 80\overline{)7.0000} \\ \underline{6\ 40} \\ 600 \\ \underline{560} \\ 400 \\ \underline{400} \end{array} = .0875 = 8.75\% \quad \checkmark$$

EXERCISES 7.2

1. Change each percent to a fraction and reduce it if possible.
   *(a) 35%   (b) 40%   *(c) 28%   (d) 47%   *(e) 20%
   (f) 65%   *(g) 36%   *(h) 5%   *(i) 3%   *(j) 64%
   *(k) 42%   (l) 75%   *(m) 85%   (n) 12%   *(o) 8%
   (p) 4%

2. Change each percent to a decimal.
   *(a) 40%   (b) 60%   *(c) 25%   (d) 18%   *(e) 15%
   *(f) 6%   *(g) 19%   (h) 24%   *(i) 54%   (j) 88%

3. Change each decimal to a percent.
   *(a) .73   (b) .49   *(c) .65   (d) .58   *(e) .07
   (f) .03   *(g) .08   *(h) .80   *(i) .8   (j) .3
   *(k) .723   (l) .518   *(m) .062   (n) .091   *(o) .901
   (p) .003   *(q) .007   (r) .807

4. Change each fraction to a percent. If you use division and decimals arise, continue the division to three decimal places only.

   *(a) $\dfrac{4}{5}$ (b) $\dfrac{3}{10}$ *(c) $\dfrac{1}{3}$ (d) $\dfrac{2}{9}$ *(e) $\dfrac{4}{15}$

   (f) $\dfrac{5}{12}$ *(g) $\dfrac{6}{7}$ *(h) $\dfrac{4}{11}$ *(i) $\dfrac{2}{5}$ (j) $\dfrac{5}{6}$

   *(k) $\dfrac{7}{12}$ (l) $\dfrac{3}{4}$ *(m) $\dfrac{9}{13}$ (n) $\dfrac{9}{10}$ *(o) $\dfrac{3}{14}$

   *(p) $\dfrac{3}{8}$

7.3 MORE PERCENTS

A few important cases were omitted in the preceding section. Let's consider them now by using examples.

The next two examples show that even if a percent involves a number having a decimal point, you can move the decimal point two places to the left in order to change it from a percent to a decimal.

EXAMPLE 17   Change 58.4% to a decimal.

Solution
$$58.4\% = 58.4 = .584 \quad \checkmark$$

EXAMPLE 18   Change 2.3% to a decimal.

Solution
$$2.3\% = 02.3 = .023 \quad \checkmark$$

EXAMPLE 19   Change $2\frac{1}{4}$% to a decimal.

Solution
$$2\tfrac{1}{4}\% = 2.25\% = 02.25 = .0225 \quad \checkmark$$

To change such percents to fractions, proceed as in Examples 17 and 18. Then, once the decimal form is obtained, write the decimal as a fraction. This is demonstrated in the next examples.

EXAMPLE 20   Change 58.4% to a fraction.

Solution
$$58.4\% = 58.4 = .584 = \frac{584}{1000} = \frac{73}{125}$$

EXAMPLE 21   Change 2.3% to a fraction.

Solution
$$2.3\% = 02.3 = .023 = \frac{23}{1000} \quad \checkmark$$

Sometimes you will encounter percents that are greater than 100%. What do such percents mean? For instance, what is meant by 200%, 180%, or 354%? The following example will show you.

EXAMPLE 22   Change 250% to a fraction and to a decimal.

Solution
$$250\% = \frac{250}{100} = \frac{5}{2} \quad \checkmark$$

The fraction 5/2 is, of course, the same as the mixed number $2\frac{1}{2}$.

$$250\% = \frac{250}{100} = 2.5 \quad \checkmark$$

This result may seem strange. If so, note that percents between 0% and 100% represent numbers between 0 and 1; specifically, if we change the number 1 to a percent, the result is 100%.

$$1 = \frac{1}{1} = \frac{100}{100} = 100\%$$

So you should expect *any percent larger than* 100% *to be equivalent to a fraction larger than 1.*

Also, *any fraction greater than 1 is equivalent to a percent greater than* 100%.

EXAMPLE 23   Change $3\frac{3}{4}$ to a percent.

Solution
$$3\frac{3}{4} = 3.75 = 3.75 = 375\% \quad \checkmark$$

EXAMPLE 24   Change $\frac{9}{5}$ to a percent.

Solution
$$\frac{9}{5} = 1.80 = 1.80 = 180\% \quad \checkmark$$

EXERCISES 7.3

1. Change each percent to a fraction. Then reduce the fraction, if possible.
   *(a) 5.9%   *(b) 200%   *(c) 15.3%   (d) 32.5%
   (e) 175%   *(f) 0%   (g) 50.8%   *(h) 1.6%
   *(i) 400%   (j) 365%   *(k) .45%   *(l) .03%
   *(m) 240%   (n) 10.9%   *(o) 134.5%   (p) 152.7%
   *(q) 313%   (r) 100%   *(s) 1000%   (t) 1300%

2. Change each percent to a decimal.

   *(a) $3\frac{1}{2}\%$ (b) $5\frac{1}{2}\%$ *(c) $7\frac{1}{4}\%$ (d) $8\frac{1}{4}\%$

   *(e) $2\frac{3}{4}\%$ (f) $7\frac{3}{4}\%$ *(g) $5\frac{3}{8}\%$ (h) $4\frac{1}{8}\%$

3. Change each percent to a decimal.
   *(a) 300%   (b) 700%   *(c) 150%   (d) 160%
   *(e) 173%   (f) 942%   *(g) 1350%   (h) 1150%
   *(i) 1425%   (j) 1392%

4. Change each fraction or decimal to a percent.

   *(a) $2\frac{3}{10}$ (b) $4\frac{1}{2}$ *(c) $\frac{3}{2}$ (d) 1.6

   *(e) 3.5   (f) $6\frac{1}{4}$ *(g) $\frac{1}{100}$ (h) $\frac{1}{1000}$

   *(i) .2   (j) .02   *(k) .002   (l) 2
   *(m) .071   (n) .086   *(o) 9.99   (p) 1.01

   *(q) $7\frac{2}{5}$ (r) $5\frac{3}{5}$ *(s) $2\frac{1}{8}$ (t) $3\frac{5}{8}$

7.4 APPLICATIONS OF PERCENT

In this section you will see how to solve a few basic types of percent problems. Others will be solved in Chapter 13, after you have studied some algebra.

EXAMPLE 25   What number is 16% of 62?

Solution   Here you are asked to compute 16% of 62. As you know, in such an arithmetic setting, "of" means "times." Thus

$$16\% \text{ of } 62 = 16\% \times 62$$
$$= .16 \times 62$$
$$= 9.92 \quad \checkmark$$

Thus 16% of 62 is 9.92. Notice that 16% was changed to the decimal form .16 before being used in the calculation. It would also be correct to use the fractional form $\frac{16}{100}$ in the calculation, but the form 16% cannot be used in calculations.

EXAMPLE 26   16% of what number is 62?

Solution   This problem is best solved by using algebra, and that will be done in Chapter 13. For now, let us see how the problem can be solved by using arithmetic.

When compared with the preceding example, we see that here we must determine which number should be multiplied by .16 in order to produce 62.

$$.16 \times (?) = 62$$

The number that we seek can be determined by dividing 62 by .16.

$$(?) = \frac{62}{.16}$$

or

$$(?) = 387.5 \quad \checkmark$$

To convince you that this division does, in fact, produce the correct number, let's check it. Is 16% of 387.5 equal to 62? Indeed it is.

$$.16 \times 387.5 = 62$$

Once you have studied the algebra of Chapters 12 and 13, the reason for dividing will be more apparent.

EXAMPLE 27   What percent of 8 is 3?

Solution   Percentages are related to fractions. So begin by changing the question temporarily to "What fraction of 8 is 3?" Of course, the answer is

$$\frac{3}{8}$$

But we want a percentage, not a fraction. So change $\frac{3}{8}$ to a percent.

$$\frac{3}{8} \longrightarrow 8\overline{)3.000}^{.375}$$

So

$$\frac{3}{8} = .375 = 37.5\% \quad \checkmark$$

Thus 3 is 37.5% of 8.

The following word problem involves one of the three basic types of percent problems explained in this section. The exercises include word problems involving all three types.

EXAMPLE 28 Bill bought 40 light bulbs and 3 were defective. What percent of the 40 bulbs were defective?

Solution Since 3 of the 40 were defective, it follows that the fraction of the 40 that were defective is

$$\frac{3}{40}$$

So to find the percent of the 40 bulbs that were defective, merely change this fraction to a percent. The result is

$$7.5\% \quad \checkmark$$

EXERCISES 7.4

1. Solve each problem after first deciding which of the three types it is. When decimals are involved, three decimal places is sufficient for all answers.
    *(a) What number is 43% of 75?
    *(b) 58% of 80 is ____.
    *(c) 25% of what number is 41?
    *(d) 18% of ____ is 90.
    *(e) What percent of 72 is 9?
    *(f) 15 is ____% of 105.
    *(g) 19% of what number is 50?
     (h) What number is 24% of 91?
     (i) What percent of 42 is 5?
     (j) 29 is ____% of 74.
     (k) 52% of ____ is 100.
     (l) 79% of 18 is ____.
    (m) 15 is ____% of 5.

2. Solve each word problem.
    *(a) A team has won 14 of the 20 games it has played. What percent of its games has it won?
     (b) A team has lost 5 of the 8 games it has played. What percent of its games has it lost?
    *(c) Carol's house has 15 windows and she has cleaned 40% of them. How many windows has she cleaned?
     (d) Joan invested 24% of her $15,000 in U.S. Treasury notes. How much money does she have invested in Treasury notes?
    *(e) Bill has invested 15% of his life savings in a municipal bond fund. The dollar amount of his investment is $6000. How much is his life savings?
     (f) A father decides to give his son 18% of the land he owns. If this means that his son will get 7.2 acres, how much land does the father own?

*(g) A quarterback has completed 20 out of 32 passes he has thrown. What percent of his passes were completed?

*(h) What percent of the passes thrown by the quarterback [see part (g)] were incomplete?

(i) A field goal kicker made 28 of 40 attempts this season. What percent did he make?

*(j) If a student gets 17 out of 20 points on a quiz, what is her percent score?

(k) If you get 16 out of 25 on a quiz, what is your grade in terms of percent?

*(l) How many points would you need to obtain on a 40-point quiz in order to have a score of 85%?

*(m) Suppose that you own shares of a stock. If you bought the stock for $5.00 a share and it is now worth $5.50 a share, what percent has the value of the stock increased?

CHAPTER 7 REVIEW EXERCISES

1. Change each number to a decimal (four decimal places if needed) and then to a percent.

   *(a) $\frac{4}{5}$ *(b) $\frac{1}{6}$ (c) $\frac{5}{7}$

   (d) $\frac{3}{11}$ *(e) 6   (f) $7\frac{1}{5}$

   *(g) $\frac{5}{3}$ (h) .68   *(i) 9.514

2. Change each percent to a decimal and to a fraction. Reduce the fraction if possible.
   *(a) 55%   (b) 32%   *(c) 72%
   (d) 88%   *(e) 725%   *(f) 16.3%
   (g) .51%   *(h) .09%   (i) 214.8%

3. Solve each problem. When decimals are involved, three decimal places is sufficient for all answers.
   *(a) What percent of 81 is 9?
   *(b) What number is 35% of 116?
   (c) 16 is ____% of 120.
   (d) 30% of what number is 17?

*4. Solve each word problem.
   (a) Susan has sold 12 of the 40 paintings she has made. What percent of the paintings has she sold?
   (b) A town has 40,000 men and 70% of them are married. How many men are married?
   (c) Donna has invested 35% of her life savings in a certificate of deposit. The amount of this investment is $3850. How much is her life savings?

8

UNITS OF MEASURE

8.1 BASIC CONVERSIONS

If you were given a measurement in inches and asked to change it to feet, you would simply divide by 12 to get the result. For example,

$$39 \text{ inches} = 39 \div 12 = 3.25 \text{ feet}$$

To change from feet to inches, multiply by 12. For example,

$$7 \text{ feet} = 7 \cdot 12 = 84 \text{ inches}$$

To change from hours to minutes, multiply by 60. And there are other situations in which you can easily change from one unit to another. However, there are also situations in which you would have considerable difficulty. For instance, how would you change miles per hour to feet per second? When using inches in computations, when is the result inches, when is it square inches, and when is it cubic inches? Our purpose here is to answer these questions and others like them.

Metric units will not be used until Chapter 9, where the metric system will be explained. Our approach is demonstrated in the two examples that follow.

EXAMPLE 1   Change 39 inches to feet.

Solution   We know that 1 ft = 12 in. or that there is 1 foot in every 12 inches. That is,

$$\frac{1 \text{ ft}}{12 \text{ in.}} = 1$$

So if we write 39 inches as

$$\frac{39 \text{ in.}}{1}$$

and then multiply it by 1 in the form $\frac{1 \text{ ft}}{12 \text{ in.}}$, we get

$$39 \text{ in.} = \frac{39 \text{ in.}}{1} \times \frac{1 \text{ ft}}{12 \text{ in.}}$$

$$= \frac{39 \cancel{\text{ in.}}}{1} \times \frac{1 \text{ ft}}{12 \cancel{\text{ in.}}} \qquad \left\{ \begin{array}{l} \text{The inches} \\ \text{can be divided out.} \end{array} \right.$$

$$= \frac{39 \times 1 \text{ ft}}{1 \times 12}$$

$$= \frac{39 \text{ ft}}{12}$$

$$= 3.25 \text{ ft} \quad \checkmark$$

EXAMPLE 2   Change 7 feet to inches.

Solution
$$7 \text{ ft} = \frac{7 \text{ ft}}{1}$$

$$= \frac{7 \text{ ft}}{1} \times \frac{12 \text{ in.}}{1 \text{ ft}} \qquad 12 \text{ inches per foot}$$

$$= \frac{7 \cancel{\text{ ft}}}{1} \times \frac{12 \text{ in.}}{1 \cancel{\text{ ft}}} \qquad \text{The feet divide out.}$$

$$= 84 \text{ in.} \quad \checkmark$$

In summary, the method to be used to change units is presented in the box below.

> When changing units, get the units to line up
> so that only the desired ones are left; that is,
> all others divide out.

Here are a few more examples.

EXAMPLE 3   Change 5 hours to seconds.

Solution   1 hr = 60 min,   1 min = 60 sec.

$$5 \text{ hr} = \frac{5 \text{ hr}}{1}$$

$$= \frac{5 \text{ hr}}{1} \times \frac{60 \text{ min}}{1 \text{ hr}} \times \frac{60 \text{ sec}}{1 \text{ min}} \qquad \begin{cases} 60 \text{ minutes per hour} \\ 60 \text{ seconds per minute} \end{cases}$$

$$= \frac{5 \cancel{\text{ hr}}}{1} \times \frac{60 \cancel{\text{ min}}}{1 \cancel{\text{ hr}}} \times \frac{60 \text{ sec}}{1 \cancel{\text{ min}}} \qquad \begin{cases} \text{The hours and minutes} \\ \text{divide out.} \end{cases}$$

$$= \frac{5 \times 60 \times 60 \text{ sec}}{1}$$

$$= 18{,}000 \text{ sec} \quad \checkmark$$

EXAMPLE 4   Change 17 miles to feet.

Solution   1 mi = 5280 ft.

$$17 \text{ mi} = \frac{17 \text{ mi}}{1}$$

$$= \frac{17 \text{ mi}}{1} \times \frac{5280 \text{ ft}}{1 \text{ mi}} \qquad 5280 \text{ ft per mile}$$

$$= \frac{17 \cancel{\text{ mi}}}{1} \times \frac{5280 \text{ ft}}{1 \cancel{\text{ mi}}} \qquad \text{The miles divide out.}$$

$$= \frac{17 \times 5280 \text{ ft}}{1}$$

$$= 89{,}760 \text{ ft} \quad \checkmark$$

EXAMPLE 5   Change 30 miles per hour (mph) to feet per second.

Solution   Here we combine the concepts developed in Example 3 and 4. In Example 3 we changed hours to seconds. In Example 4 we changed miles to feet.

$$30 \text{ mph} = 30 \text{ mph}$$

$$= \frac{30 \text{ mi}}{1 \text{ hr}}$$

changes hours to seconds

$$= \frac{30 \text{ mi}}{1 \text{ hr}} \times \frac{5280 \text{ ft}}{1 \text{ mi}} \times \frac{1 \text{ hr}}{60 \text{ min}} \times \frac{1 \text{ min}}{60 \text{ sec}}$$

changes miles to feet

$$= \frac{30 \cancel{\text{ mi}}}{1 \cancel{\text{ hr}}} \times \frac{5280 \text{ ft}}{1 \cancel{\text{ mi}}} \times \frac{1 \cancel{\text{ hr}}}{60 \cancel{\text{ min}}} \times \frac{1 \cancel{\text{ min}}}{60 \text{ sec}}$$

The miles, hours, and minutes divide out.

$$= \frac{30 \times 5280 \text{ ft}}{60 \times 60 \text{ sec}}$$

$$= 44 \text{ ft/sec} \quad \checkmark$$

The fractions were chosen to eliminate unwanted units and introduce the desired units.

When you *add* inches and inches, you get a sum of *inches*. Yet when you *multiply* inches by inches, you get *square inches*. Let's see why this happens.

EXAMPLE 6   (a) Add 5 inches + 6 inches.
(b) Multiply 5 inches × 6 inches.

Solution   (a) 5 in. + 6 in. = (5 + 6)in.   using the "distributive property"
= 11 in. $\checkmark$

(b) 5 in. × 6 in. = 5 × 6 × in. × in.
= 30 in.², or 30 sq in. $\checkmark$

Linear measure, such as lengths and distances, are expressed in units such as inches, feet, centimeters, and meters. Since **perimeters** are merely sums of lengths or sums of distances, from Example 6(a) you can see that perimeters are expressed in linear units such as inches, feet, centimeters, and meters.

EXAMPLE 7   Determine the perimeter of a square with sides 21 centimeters each.

Solution   The square:

21 cm

21 cm

The perimeter can be determined as

$$P = 4s = 4(21 \text{ cm}) = 84 \text{ cm} \quad \checkmark$$

or as

$$P = s + s + s + s$$
$$= 21 \text{ cm} + 21 \text{ cm} + 21 \text{ cm} + 21 \text{ cm}$$
$$= 84 \text{ cm} \quad \checkmark$$

The perimeter is 84 centimeters.

Areas are expressed in "square" units, such as square inches, square feet, square centimeters, and square meters. The next example shows how a typical area computation results in square units, just as square units were produced in Example 6(b).

EXAMPLE 8 The area A of a triangle is computed as $A = \frac{1}{2}bh$, where b is the base and h is the height. Determine the area of the triangle that has a base of 24 feet and a height of 10 feet.

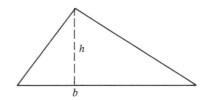

Solution

$$A = \frac{1}{2}bh \qquad \text{(Note: } \tfrac{1}{2}bh \text{ means } \tfrac{1}{2} \cdot b \cdot h)$$

$$= \frac{1}{2} \times 24 \text{ ft} \times 10 \text{ ft}$$

$$= \frac{1}{2} \times 24 \times 10 \times \text{ft} \times \text{ft}$$

$$= 120 \text{ ft}^2 \quad \checkmark$$

The area of the triangle is 120 square feet.

Volumes are expressed in "cubic" units such as cubic inches and cubic centimeters, as demonstrated by the next example.

EXAMPLE 9 The volume of a box is computed by multiplying the length times the width times the height ($V = lwh$). Find the volume of a box that has length 10 inches, width 5 inches, and height 4 inches.

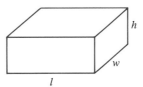

Solution

$$V = lwh \qquad \text{(Note: } lwh = l \cdot w \cdot h)$$
$$= 10 \text{ in.} \times 5 \text{ in.} \times 4 \text{ in.}$$
$$= 10 \times 5 \times 4 \times \text{in.} \times \text{in.} \times \text{in.}$$
$$= 200 \text{ in.}^3 \quad \checkmark$$

The volume of the box is 200 cubic inches.

Suppose the length and width of the box above were again given as 10 inches and 5 inches, but the height was given as 2 *feet*. To use the formula $V = l \cdot w \cdot h$, we must have the same units for all dimensions l, w, and h. The easiest way to accomplish this would be to change the height of 2 feet to 24 inches. Then

$$
\begin{aligned}
V &= l \cdot w \cdot h \\
&= 10 \text{ in.} \times 5 \text{ in.} \times 24 \text{ in.} \\
&= 10 \times 5 \times 24 \times \text{in.} \times \text{in.} \times \text{in.} \\
&= 1200 \text{ in.}^3 \quad \checkmark
\end{aligned}
$$

EXAMPLE 10   Change 7 cubic feet to cubic inches.

Solution

$$
\begin{aligned}
7 \text{ cubic feet} &= 7 \text{ ft}^3 \\
&= 7 \times \text{ft} \times \text{ft} \times \text{ft} \\
&= 7 \times \text{ft} \times \text{ft} \times \text{ft} \times \frac{12 \text{ in.}}{1 \text{ ft}} \times \frac{12 \text{ in.}}{1 \text{ ft}} \times \frac{12 \text{ in.}}{1 \text{ ft}} \\
&= 7 \times 12 \times 12 \times 12 \times \text{in.} \times \text{in.} \times \text{in.} \\
&= 12{,}096 \text{ in.}^3 \quad (12{,}096 \text{ cubic inches}) \quad \checkmark
\end{aligned}
$$

Formulas

$$
\begin{aligned}
C &= \text{circumference} \\
A &= \text{area} \\
P &= \text{perimeter} \\
V &= \text{volume}
\end{aligned}
$$

1. Circle:

$$
\begin{aligned}
C &= 2\pi r \\
A &= \pi r^2 \\
(\pi &\approx 3.14)
\end{aligned}
$$

2. Rectangle:

$$
\begin{aligned}
A &= lw \\
P &= 2l + 2w
\end{aligned}
$$

3. Triangle:

$$
A = \tfrac{1}{2}bh
$$

4. Rectangular solid:

$$V = lwh$$

5. Cone:

$$V = \tfrac{1}{3}\pi r^2 h$$
$$(\pi \approx 3.14)$$

6. Cylinder:

$$V = \pi r^2 h$$
$$(\pi \approx 3.14)$$

7. Sphere:

$$V = \tfrac{4}{3}\pi r^3$$
$$(\pi \approx 3.14)$$

Here is a brief table of units of measure that will be useful in doing the exercises that follow.

12 inches	= 1 foot
3 feet	= 1 yard
5280 feet	= 1 mile
60 seconds	= 1 minute
60 minutes	= 1 hour
24 hours	= 1 day
7 days	= 1 week
16 ounces	= 1 pound
2000 pounds	= 1 ton
16 fluid ounces	= 1 pint
2 pints	= 1 quart
4 quarts	= 1 gallon

EXERCISES 8.1

Answers to starred exercises are given in the back of the book.

1. Change the units as requested.
   *(a) Change 1476 inches to feet.
   *(b) Change 23 feet to inches.
   *(c) Change 12 hours to minutes.
   *(d) Change 5640 minutes to hours.
   (e) Change 73,920 feet to miles.
   (f) Change 54,000 pounds to tons.
   *(g) Change 20 gallons to quarts.
   *(h) Change 18 yards to feet.
   (i) Change 496 ounces to pounds.

2. Change the units as requested.
   *(a) Change 3 hours to seconds.
   *(b) Change 5 miles to inches.
   *(c) Change 10 weeks to hours.
   *(d) Change 8 tons to ounces.
   *(e) Change 25,200 seconds to hours.
   (f) Change 150,000 ounces to tons.
   (g) Change 300 minutes to seconds.
   *(h) Change 65,536 fluid ounces to gallons.
   (i) Change 1,108,800 inches to miles.

*3. Change the units as requested.
   (a) Change 180 miles per hour to miles per minute.
   (b) Change 180 miles per hour to feet per minute.
   (c) Change 180 miles per hour to feet per second.
   (d) Change 66 feet per second to miles per hour.
   (e) Change 34 inches per second to feet per minute.
   (f) Change 85 feet per minute to inches per second.

4. Determine each perimeter or circumference. Be sure to include the *units* in your answer. (Use $\pi \approx 3.14$.)
   *(a) The perimeter of a square with side 9 inches.
   *(b) The perimeter of a triangle with sides 8 centimeters, 10 centimeters, and 7 centimeters.
   *(c) The perimeter of a rectangle with length 10 meters and width 8 meters.
   (d) The perimeter of a rectangle with length 13 feet and width 9 feet.
   *(e) The circumference of a circle with radius 4 inches.
   (f) The circumference of a circle with radius 7 centimeters.

5. Determine each area. Be sure to include the *units* in your answer. (Use $\pi \approx 3.14$.)
   *(a) The area of a rectangle with length 14 feet and width 6 feet.
   (b) The area of a rectangle with length 12.3 centimeters and width 2.7 centimeters.

*(c) The area of a triangle with base 18 inches and height 11 inches.

(d) The area of a triangle with base 20 inches and height 6.2 inches.

*(e) The area of a square with side 1.4 meters.

*(f) The area of a circle with radius 2.3 centimeters.

(g) The area of a circle with radius 15 meters.

6. Determine each volume. Be sure to include the *units* in your answer. (Use $\pi \approx 3.14$.)

*(a) The volume of a rectangular box having length 10 centimeters, width 6.8 centimeters, and height 4 centimeters.

(b) The volume of a rectangular box having length 5 inches, width 3 inches, and height 2 inches.

*(c) The volume of a cylinder with radius 10 inches and height 5 inches.

(d) The volume of a cylinder with radius 3 feet and height 9.3 feet.

*(e) The volume of a sphere having radius 4 inches.

(f) The volume of a sphere having radius 7 centimeters.

*(g) The volume of a cone having a radius of 6 centimeters and a height of 10 centimeters.

(h) The volume of a cone having a radius of 12 inches and a height of 4 inches.

7. Determine the requested perimeter, area, or volume. Include units.

*(a) The perimeter of a fence that encloses a garden which has a length of 40 feet and a width of 24 feet.

*(b) The area of the garden mentioned in part (a).

*(c) The volume of a cement wall that is 60 feet long, 6 feet high, and 2 feet thick.

*(d) The volume of soup in a can having a radius of 1.5 inches and a height of 5 inches.

(e) The volume held by a cone-shaped drinking cup having a radius of 2 inches and a height of 6 inches.

(f) The volume held by a cylindrical drinking cup having a radius of 2 inches and a height of 6 inches.

8. Change the units as requested.

*(a) Change 5 square feet to square inches.

*(b) Change 2304 square inches to square feet.

(c) Change 135 square yards to square feet.

(d) Change 135 square feet to square yards.

*(e) Change 17 cubic feet to cubic inches.

(f) Change 90 cubic feet to cubic yards.

CHAPTER 8 REVIEW EXERCISES

*1. Change the units as requested.

(a) Change 522 feet to yards.

(b) Change 1140 seconds to minutes.

(c) Change 9 weeks to seconds.

(d) Change 846,720 seconds to weeks.

(e) Change 90 cubic yards to cubic feet.

*2. Determine the requested perimeter, area, or volume. Be sure to include units.

(a) The area of a triangle with base 34 centimeters and height 14 centimeters.

(b) The volume of a sphere having a radius of 5.2 inches.

(c) The perimeter of a square with side 3.7 meters.

(d) The area of a square with side 1 inch.

(e) The volume of a rectangular box with length 13 centimeters, width 6 centimeters, and height 5 centimeters.

9

METRIC SYSTEM

9.1 BASIC CONCEPTS

The metric system of units is widely used throughout the world and is used to a limited extent here in the United States. You are probably more familiar with the English system of units. In the English system length is measured in inches, feet, yards, and miles. Which unit is used depends on the length or distance measured. In the metric system one basic unit is used, and all lengths are measured in terms of that unit. The basic unit of length in the metric system is the *meter* (abbreviated m). A meter is about 39.37 inches, a little more than a yard. The symbol ≈ is used to mean *approximately equal to.* Thus

$$1 \text{ meter} \approx 39.37 \text{ inches}$$

This section continues with examples that demonstrate how to change units when metric units are involved. You will see that the methods presented in Chapter 8 are applied here.

EXAMPLE 1   Change 27 inches to meters.

Solution   We'll use the techniques of Chapter 8 and the fact that 1 meter is approximately 39.37 inches.

$$27 \text{ in.} \approx \frac{27 \text{ in.}}{1} \times \frac{1 \text{ m}}{39.37 \text{ in.}}$$

$$\approx \frac{27}{39.37} \text{ m}$$

$$\approx .69 \text{ m} \quad \checkmark$$

EXAMPLE 2   Change 3.4 meters to inches.

Solution $$3.4 \text{ m} \approx \frac{3.4 \text{ m}}{1} \times \frac{39.37 \text{ in.}}{1 \text{ m}}$$

$$\approx 3.4 \times 39.37 \text{ in.}$$

$$\approx 133.86 \text{ in.} \quad \checkmark$$

EXAMPLE 3   Change 35 feet to meters.

Solution $$35 \text{ ft} \approx \frac{35 \text{ ft}}{1} \times \frac{12 \text{ in.}}{1 \text{ ft}} \times \frac{1 \text{ m}}{39.37 \text{ in.}}$$

$$\approx \frac{35 \times 12}{39.37} \text{ m}$$

$$\approx 10.67 \text{ m} \quad \checkmark$$

Prefixes are used in the metric system to specify lengths that are much larger or much smaller than a meter. The most common prefixes used in the metric system are given here, together with their meanings. Three others (hecto, deca, and deci) are introduced in Exercise 8.

kilo	1000	thousand
centi	$\frac{1}{100}$	hundredth
milli	$\frac{1}{1000}$	thousandth

A kilometer (km) is a thousand meters, which is about six-tenths of a mile. The kilometer is the metric unit used on road signs and maps.

$$\boxed{1 \text{ kilometer} \approx .6 \text{ mile}}$$

EXAMPLE 4   Change 50 kilometers to miles.

Solution

$$50 \text{ km} \approx \frac{50 \text{ km}}{1} \times \frac{.6 \text{ mi}}{1 \text{ km}}$$
$$\approx 50 \times .6 \text{ mi}$$
$$\approx 30 \text{ mi} \quad \checkmark$$

EXAMPLE 5   Change 48 miles per hour to kilometers per hour.

Solution

$$48 \frac{\text{mi}}{\text{hr}} \approx \frac{48 \text{ mi}}{1 \text{ hr}} \times \frac{1 \text{ km}}{.6 \text{ mi}}$$
$$\approx \frac{48}{.6} \times \frac{\text{km}}{\text{hr}}$$
$$\approx 80 \text{ km/hr} \quad \checkmark$$

A centimeter (cm) is one hundredth of a meter or about the diameter of an aspirin tablet. A millimeter (mm) is a tenth of a centimeter or a thousandth of a meter.

$$\boxed{1 \text{ inch} \approx 2.54 \text{ centimeters}}$$

EXAMPLE 6   Change a measure of 19 centimeters to inches.

Solution

$$19 \text{ cm} \approx \frac{19 \text{ cm}}{1} \times \frac{1 \text{ in.}}{2.54 \text{ cm}}$$
$$\approx \frac{19}{2.54} \text{ in.}$$
$$\approx 7.5 \text{ in.} \quad \checkmark$$

The next two examples point out the simplicity of conversions done within the metric system. This is a real advantage of using the metric system.

EXAMPLE 7   Change 3 meters to millimeters.

Solution   Since 1 meter equals 1000 millimeters,

$$3 \text{ m} = \frac{3 \text{ m}}{1} \times \frac{1000 \text{ mm}}{1 \text{ m}}$$
$$= 3000 \text{ mm} \quad \checkmark \quad \text{exactly}$$

Notice the simplicity of this conversion within the metric system. If you become familiar enough with the metric system so that you need not rely on the English system for reference, conversions between units will be simple.

EXAMPLE 8 Change 5 centimeters to meters.

Solution Since 1 meter equals 100 centimeters,

$$5 \text{ cm} = \frac{5 \text{ cm}}{1} \times \frac{1 \text{ m}}{100 \text{ cm}}$$

$$= \frac{5}{100} \text{ m, or .05 m} \quad \checkmark \quad \text{exactly}$$

So far we have been concerned with units of length. In the metric system the *liter* is the basic unit of volume. A liter (1) is approximately 1.06 quarts (qt) in volume. In the English system fluid ounces (oz), quarts, and gallons (gal) are used. In the metric it's liters, milliliters (ml), and kiloliters (kl). *One fluid ounce is about 30 milliliters.*

$$\boxed{1 \text{ liter} \approx 1.06 \text{ quarts}}$$

$$\boxed{1 \text{ ounce} \approx 30 \text{ milliliters}}$$

EXAMPLE 9 Change 15 quarts to liters.

Solution

$$15 \text{ qt} \approx \frac{15 \text{ qt}}{1} \times \frac{1 \text{ liter}}{1.06 \text{ qt}}$$

$$\approx \frac{15}{1.06} \text{ liters}$$

$$\approx 14.15 \text{ liters} \quad \checkmark$$

EXAMPLE 10 Change 85 milliliters to fluid ounces.

Solution

$$85 \text{ ml} \approx \frac{85 \text{ ml}}{1} \times \frac{1 \text{ oz}}{30 \text{ ml}}$$

$$\approx \frac{85}{30} \text{ oz}$$

$$\approx 2.83 \text{ oz} \quad \checkmark$$

The next two examples show conversions within metric. Again, note the simplicity of such conversions.

EXAMPLE 11 Change 450 milliliters to liters.

Solution A milliliter is one-thousandth of a liter. Thus

$$450 \text{ ml} = \frac{450 \cancel{\text{ml}}}{1} \times \frac{1 \text{ liter}}{1000 \cancel{\text{ml}}}$$

$$= \frac{450}{1000} \text{ liter}$$

$$= .450 \text{ liter} \quad \checkmark \quad \text{exactly}$$

EXAMPLE 12   Change 12 liters to milliliters.

Solution
$$12 \text{ liters} = \frac{12 \cancel{\text{liters}}}{1} \cdot \frac{1000 \text{ ml}}{1 \cancel{\text{liter}}}$$

$$= 12,000 \text{ ml} \quad \checkmark \quad \text{exactly}$$

In the English system weights are measured in ounces (oz), pounds (lb), and tons (ton). In metric it's *grams* (g) [and milligrams (mg) and kilograms (kg)]. An ounce is equal to about 28.35 grams; a pound is about 454 grams; a kilogram is about 2.2 pounds.

$$\boxed{\begin{array}{l} 1 \text{ ounce} \approx 28.35 \text{ grams} \\ 1 \text{ pound} \approx 454 \text{ grams} \end{array}}$$

$$\boxed{1 \text{ kilogram} \approx 2.2 \text{ pounds}}$$

EXAMPLE 13   Change 15 kilograms to pounds.

Solution   Use the fact that 1 kg ≈ 2.2 lb.

$$15 \text{ kg} \approx \frac{15 \cancel{\text{kg}}}{1} \times \frac{2.2 \text{ lb}}{1 \cancel{\text{kg}}}$$

$$\approx 15 \times 2.2 \text{ lb}$$

$$\approx 33 \text{ lb} \quad \checkmark$$

The next three examples show conversions within metric. Such conversions are simple—involving only a shift of the decimal point.

EXAMPLE 14   Change 5 kilograms to grams.

Solution
$$5 \text{ kg} = \frac{5 \cancel{\text{kg}}}{1} \times \frac{1000 \text{ g}}{1 \cancel{\text{kg}}}$$

$$= 5000 \text{ g} \quad \checkmark \quad \text{exactly}$$

EXAMPLE 15   Change 7 kilograms to milligrams.

Solution $$7 \text{ kg} = \frac{7 \cancel{\text{kg}}}{1} \times \frac{1000 \cancel{\text{g}}}{1 \cancel{\text{kg}}} \times \frac{1000 \text{ mg}}{1 \cancel{\text{g}}}$$

$$= 7{,}000{,}000 \text{ mg} \quad \checkmark \quad \text{exactly}$$

EXAMPLE 16   Change 38 milligrams to grams.

Solution $$38 \text{ mg} = \frac{38 \cancel{\text{mg}}}{1} \times \frac{1 \text{ g}}{1000 \cancel{\text{mg}}}$$

$$= \frac{38}{1000} \text{ g, or } .038 \text{ g} \quad \checkmark \quad \text{exactly}$$

Users of the English system usually measure temperature in degrees *Fahrenheit* (*F*). Water freezes at 32°F and boils at 212°F. Users of the metric system usually measure temperature in degrees *Celsius* (*C*). Water freezes at 0°C and boils at 100°C. The relationship between Fahrenheit and Celsius is given next.

$$\boxed{\begin{aligned} C &= \frac{5}{9}(F - 32) \\ F &= \frac{9}{5}C + 32 \end{aligned}}$$

EXAMPLE 17   Change 68°F to Celsius.

Solution $$C = \frac{5}{9}(F - 32°)$$

$$= \frac{5}{9}(68° - 32°)$$

$$= \frac{5}{9}(36°)$$

$$= \frac{5 \cdot 36°}{9}$$

$$= 20° \quad \checkmark$$

EXAMPLE 18   Change 15°C to Fahrenheit.

Solution $$F = \frac{9}{5}C + 32°$$

$$= \frac{9}{5} \cdot 15° + 32°$$

$$= 27° + 32°$$

$$= 59° \quad \checkmark$$

EXERCISES 9.1

Answers to starred exercises are given in the back of the book.

In Exercises 1 through 6, two decimal places is sufficient for answers that require digits after the decimal point. You will probably need to refer to the conversions given throughout this chapter and in the table on page 89.

1. Change each length from English to metric.
   - *(a) 53 inches to meters.
   - (b) 93 inches to meters.
   - *(c) 29 feet to meters.
   - (d) 153 feet to meters.
   - *(e) 90 miles to kilometers.
   - (f) 123 miles to kilometers.
   - *(g) 19 inches to centimeters.
   - (h) 34 inches to centimeters.
   - *(i) 3 inches to millimeters.
   - (j) 7 inches to millimeters.

2. Change each length from metric to English.
   - *(a) 7 meters to inches.
   - (b) 9.7 meters to inches.
   - *(c) 43 meters to feet.
   - (d) 80 meters to feet.
   - *(e) 24 centimeters to inches.
   - (f) 76.3 centimeters to inches.
   - *(g) 140 kilometers to miles.
   - (h) 103 kilometers to miles.
   - *(i) 235 millimeters to inches.
   - (j) 254 millimeters to inches.

3. Change each volume from English to metric.
   - *(a) 23 quarts to liters.
   - (b) 50 quarts to liters.
   - *(c) 12 fluid ounces to milliliters.
   - (d) 34 fluid ounces to milliliters.
   - *(e) 428 fluid ounces to liters.
   - (f) 168 fluid ounces to liters.
   - *(g) 18 gallons to liters.
   - (h) 26 gallons to liters.
   - *(i) 300 gallons to kiloliters.
   - (j) 245 gallons to kiloliters.

4. Change each volume from metric to English.
   - *(a) 37 liters to quarts.
   - (b) 56 liters to quarts.
   - *(c) 90 milliliters to fluid ounces.
   - (d) 214 milliliters to fluid ounces.
   - *(e) 15 liters to fluid ounces.
   - (f) 23 liters to fluid ounces.
   - *(g) 23 liters to gallons.
   - (h) 19 liters to gallons.
   - *(i) 9 kiloliters to gallons.
   - (j) 17 kiloliters to gallons.

5. Change each weight from English to metric.
   - *(a) 23 pounds to kilograms.
   - (b) 65 pounds to kilograms.
   - *(c) 7 ounces to grams.
   - (d) 12 ounces to grams.
   - *(e) 3 ounces to milligrams.
   - (f) 10 ounces to milligrams.

6. Change each weight from metric to English.
   - *(a) 28 kilograms to pounds.
   - (b) 34 kilograms to pounds.
   - *(c) 97 grams to ounces.
   - (d) 115 grams to ounces.
   - *(e) 753 milligrams to ounces.
   - (f) 1987 milligrams to ounces.

7. Change each measure within metric.
   - *(a) 7 meters to millimeters.
   - (b) 16 meters to centimeters.
   - *(c) 158 millimeters to meters.
   - (d) 387 centimeters to meters.
   - *(e) 45 liters to milliliters.
   - (f) 852 liters to kiloliters.
   - *(g) 700 milliliters to liters.
   - *(h) 19 kilograms to grams.

*(i) 70 grams to milligrams.   (j) 43 kilograms to milligrams.
(k) 60 milligrams to grams.   *(l) 3 kiloliters to milliliters.
*(m) 7 kilometers to centi-   *(n) 7 kilometers to millimeters.
meters.
*(o) 400 centimeters to meters.   (p) 564 millimeters to meters.
*(q) 200 kilograms to grams.   (r) 100 liters to milliliters.
*(s) 12 kiloliters to milliliters.   (t) 120 grams to milligrams.
*(u) 35 kilometers to millime-
ters.   (v) 2000 milligrams to grams.
*(w) 500 liters to kiloliters.   (x) 50 meters to millimeters.
*(y) 10 meters to centimeters.   (z) 40 milliliters to liters.

*8. Here are three other prefixes used in the metric system.

hecto	100	hundred
deca	10	ten
deci	$\frac{1}{10}$	tenth

This means that a hectogram is 100 grams, a decagram is 10 grams, and a decigram is one-tenth of a gram. Change each measure within metric.

(a) 3 hectograms to grams.   (b) 19 hectograms to grams.
(c) 5 decagrams to grams.   (d) 17 decagrams to grams.
(e) 250 decigrams to grams.   (f) 33 decigrams to grams.
(g) 74 decaliters to liters.   (h) 150 liters to decaliters.
(i) 200 decimeters to meters.   (j) 300 meters to decimeters.
(k) 400 hectoliters to decaliters.   (l) 3 decaliters to deciliters.

9. Change each Fahrenheit temperature to Celsius and each Celsius temperature to Fahrenheit. If decimals arise, round answers to the nearest degree.

*(a) 95°F   (b) 59°F   *(c) 41°F
(d) 122°F   *(e) 20°C   (f) 80°C
*(g) 45°C   (h) 30°C   *(i) 75°F
(j) 35°F   *(k) 59°C   (l) 99°C

CHAPTER 9 REVIEW EXERCISES

*1. Change from English to metric.
(a) 5 inches to centimeters.   (b) 19 ounces to grams.
(c) 13 quarts to liters.   (d) 300 miles to kilometers.
*2. Change from metric to English.
(a) 40 kilograms to pounds.   (b) 5 liters to fluid ounces.
(c) 320 millimeters to inches.   (d) 17 meters to feet.

*3. Change within metric.
  (a) 852 centimeters to millimeters.   (b) 852 millimeters to centi-meters.
  (c) 52,000 millimeters to kilometers.   (d) 5000 milligrams to kilo-grams.

*4. Change from Fahrenheit to Celsius or from Celsius to Fahrenheit, as appropriate. If decimals arise, round answers to the nearest degree.
  (a) 52°F   (b) 71°C

REVIEW EXERCISES FOR CHAPTERS 1–9

1. Perform each operation.
  *(a) Add   294   (b) Add   152.9   *(c) Subtract   8937
           865   761.1   7959
           473   605.4

  (d) Subtract   42.85   *(e) Multiply   951   (f) Multiply   19.25
             38.99   827   6.3

  *(g) $62\overline{)74196}$ (h) $3.8\overline{)29.46}$ *(i) 7^3 (j) 4^5

2. Evaluate each expression.
  *(a) $9 + 4 \cdot 6 + 2 + 3^2$
  *(b) $15 - 16 \div 2 + 2$
  (c) $(9 + 2)^2$

3. Evaluate each formula, using the values given.
  *(a) $A = \frac{1}{2}bh$; $b = 10$, $h = 4$
  *(b) $S = 4\pi r^2$; $r = 6$, $\pi = 3.14$
  (c) $A = \frac{1}{2}(b_1 + b_2)h$; $b_1 = 4$, $b_2 = 22$, $h = 11$

4. Use the Pythagorean theorem to determine the length of the hypote-nuse c when the lengths of the two legs a and b are those given.
  *(a) $a = 5, b = 12$
  (b) $a = 9, b = 12$
  *(c) $a = 1, b = 1$

5. Use the Pythagorean theorem to determine the length of the missing side (a or b) of the triangle, given the length of the hypotenuse c and one of the legs (a or b).
  *(a) $c = 13, b = 5$
  (b) $c = 15, a = 12$
  *(c) $c = 10, a = 7$

6. Name the property used in going from each step to the next one.
  *(a) $(5 + 3)6 = (3 + 5)6$
  (b) $= 3 \cdot 6 + 5 \cdot 6$
  *(c) $= 18 + 5 \cdot 6$
  *(d) $= 18 + 6 \cdot 5$
  (e) $= 6 \cdot 5 + 18$

7. Carry out each operation; reduce when possible.

*(a) $\dfrac{8}{9} \times \dfrac{18}{17}$ (b) $\dfrac{4}{15} \times \dfrac{25}{10}$

*(c) $\dfrac{18}{32} \div \dfrac{5}{16}$ (d) $\dfrac{35}{40} \div \dfrac{11}{12}$

*(e) $\dfrac{77}{10} + \dfrac{4}{25}$ (f) $\dfrac{1}{6} + \dfrac{3}{8} + \dfrac{5}{12}$

*(g) $\dfrac{9}{14} - \dfrac{11}{42}$ (h) $\dfrac{15}{24} - \dfrac{7}{32}$

8. Determine the least common multiple (LCM) of each set of numbers.
   *(a) 3, 5, 7   (b) 6, 8
   *(c) 24, 90   (d) 32, 40

9. Change each fraction to a percent.

   *(a) $\dfrac{2}{5}$ *(b) $\dfrac{7}{8}$ (c) $\dfrac{17}{25}$

10. Change each percent to a fraction.
    *(a) 45%   *(b) 83%   (c) 1.3%

*11. Change each percent to a decimal.
    (a) 34%   (b) 321%   (c) 8.5%

12. Change units as requested.
    *(a) Change 5 hours to seconds.
    (b) Change 2304 inches to feet.
    *(c) Change 240 miles per hour to miles per minute.
    (d) Change 270 cubic feet to cubic yards.

13. Change units as requested.
    *(a) Change 15 gallons to liters.
    *(b) Change 500 grams to pounds.
    *(c) Change 5 meters to millimeters.
    (d) Change 120 kilometers to miles.
    (e) Change 135 centimeters to meters.

10

SIGNED NUMBERS

10.1 INTRODUCTION

On the thermometer shown on the next page there are temperatures both above zero degrees and below zero degrees. The temperatures below zero are preceded by minus (or negative) signs to differentiate them from temperatures above zero. Thus 5 degrees below zero is $-5°$, whereas 5 degrees above zero is just $5°$ or perhaps $+5°$ for emphasis.

Signed numbers are used in other settings, too. Positive numbers are used to indicate gain, increase, profit, and similar items. Negative numbers are used to indicate loss, decrease, deficit, and so forth.

The use of signed numbers in algebra is so common that you must understand them if you are to progress in the study of algebra.

Before beginning the study of signed numbers, it is worthwhile to distinguish between different kinds of numbers—natural numbers (counting numbers), whole numbers, and integers.

Natural numbers:	1, 2, 3, 4, 5, 6, etc.
Counting numbers:	1, 2, 3, 4, 5, 6, etc.
Whole numbers:	0, 1, 2, 3, 4, 5, 6, etc.
Integers:	0, 1, -1, 2, -2, 3, -3, 4, -4, etc.
Positive integers:	1, 2, 3, 4, 5, 6, etc.
Negative integers:	-1, -2, -3, -4, -5, -6, etc.

The thermometer represents one form of the **number line.** We shall use the number line to introduce basic signed-number concepts. The number line can be pictured as

Positive numbers above the "0," negative numbers below.

or as

Positive numbers to the right of "0," negative numbers to the left.

We shall use the second line, the horizontal one.

EXERCISES 10.1

Answers to starred exercises are given in the back of the book.

1. Use a signed number to represent each change.
   *(a) The Bengals gained 12 yards on the pass play.
   *(b) The Dow Jones Industrial Average lost 79 points today.
   *(c) Mount Everest is 29,028 feet above sea level.
   *(d) The greatest known depth of the Pacific Ocean is 10,918 meters.
   (e) Bill got a raise of $1.50 an hour.
   (f) The temperature is now 3 degrees below zero.
   *(g) My favorite stock gained $2 yesterday.
   (h) The Standard and Poor average went up 8 points yesterday.
   *(i) It was a great vacation, but I gained 5 pounds.
   *(j) What a fantastic diet! I have lost 14 pounds already.
   *(k) The government reported a trade deficit of 8 billion dollars.
   (l) The defense blitzed and sacked the quarterback. As a result, the Raiders lost 11 yards on the play.

10.2 ADDITION OF SIGNED NUMBERS

Signed numbers are also called **directed numbers.** The sign ($+$ or $-$) indicates direction on the number line; $+$ is right and $-$ is left. Plus and minus are opposite signs; they indicate opposite directions. The unsigned portion of a number is called the **magnitude** of the number. For example, the magnitude of $+17$ is 17. The magnitude of -17 is also 17.

We can add two signed numbers by placing one of them on a number line and moving (from that point) the number of units in the direction indicated by the second number. Here are some examples.

EXAMPLE 1   Add $+3$ and $+4$.

Solution   First, place $+3$ on the line.

Next, move $+4$ units from the point $+3$. This means move 4 units to the right, since $+$ indicates the right direction.

Moving $+4$ takes us to $+7$. So $+7$ is the sum of $+3$ and $+4$.

$$\begin{array}{r} +3 \\ +4 \\ \hline +7 \end{array} \ \checkmark$$

EXAMPLE 2   Add -3 and -4.

Solution   First, place -3 on the line.

Next, move -4 units from the point -3. That is, move 4 units to the left, since $-$ indicates the left direction.

Moving −4 takes us to −7. So −7 is the sum of −3 and −4.

$$\begin{array}{r} -3 \\ \underline{-4} \\ -7 \end{array} \checkmark$$

EXAMPLE 3   Add −3 and +4.

Solution   First, place −3 on the line.

Next, move +4 units from the point −3. In other words, move 4 units to the right.

Moving +4 takes us to +1. So the sum of −3 and +4 is +1.

$$\begin{array}{r} -3 \\ \underline{+4} \\ +1 \end{array} \checkmark$$

EXAMPLE 4   Add +3 and −4.

Solution   First, place +3 on the line.

Next, move −4 units from the point +3. That is, move 4 units to the left.

Moving −4 takes us to −1. Thus the sum of +3 and −4 is −1.

$$\begin{array}{r} +3 \\ \underline{-4} \\ -1 \end{array} \checkmark$$

The number line is useful in gaining an understanding of addition of signed numbers, but it is slow and not always practical. For example, to add $+1945$ and -2813 on a number line would be difficult. And how about adding -243.79 and $+175.85$? So we need a mechanical method that does not use a number line.

After studying the preceding four examples, we can make some important observations about the addition of signed numbers.

Addition of Signed Numbers

1. When both numbers have the same sign (both $+$ or both $-$), the result has that same sign. The magnitude of the result is simply the sum of the magnitudes.

Two examples:

$$\begin{array}{r} +3 \\ +4 \\ \hline +7 \end{array} \qquad \begin{array}{r} -3 \\ -4 \\ \hline -7 \end{array}$$

Addition of Signed Numbers

2. When two numbers have opposite signs (one $+$, one $-$), the sign of the sum is the sign of the number that is larger in magnitude. The magnitude of the sum is the magnitude of the difference of the two numbers.

Two examples:

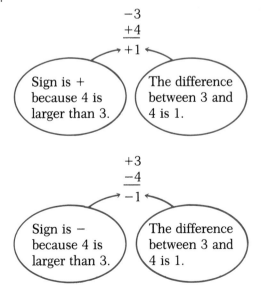

EXAMPLE 5   Add each.

$$
\begin{array}{cccc}
+17 & -29 & +39 & -\ 6 \\
+\ 5 & +15 & -12 & -19
\end{array}
$$

Solution

$$
\begin{array}{cccc}
+17 & -29 & +39 & -\ 6 \\
+\ 5 & +15 & -12 & -19 \\
\hline
+22 & -14 & +27 & -25
\end{array}
$$

Note that we added in each case. The negative (minus) signs indicate negative numbers, not subtraction.

Often two numbers to be added are written on the same line. The addition of -6 and $+17$ would be so written as

$$(-6) + (+17)$$

Similarly, the addition of $+12$ and -10 is specified as

$$(+12) + (-10)$$

Verify the following additions.

$$
\begin{aligned}
(-6) + (+17) &= +11 \\
(+12) + (-10) &= \ +2 \\
(-5) + \ (+1) &= \ -4 \\
(-8) + \ (-3) &= -11
\end{aligned}
$$

If there are more than two numbers to add, add them two at a time.

EXAMPLE 6   Combine $(+6) + (-3) + (+5) + (-6) + (+3)$.

Solution   Here we'll add the numbers two at a time, proceeding from left to right.

$$
\begin{aligned}
(+6) + (-3) + (+5) + (-6) + (+3) &= (+3) + (+5) + (-6) + (+3) \\
&= (+8) + (-6) + (+3) \\
&= (+2) + (+3) \\
&= +5 \quad \checkmark
\end{aligned}
$$

You may prefer to look for shortcuts if they exist. One shortcut is based on using opposites. The **opposite** of a number is a number that has the same magnitude but an opposite sign. The word "opposite" is short for "opposite-signed number." An important property is given in the box below.

$$\boxed{\text{number} + \text{opposite} = 0}$$

Consider these examples.

Number	Opposite
+5	−5
−5	+5
−7	+7
+3.2	−3.2
+8¼	−8¼
−19	+19
x	$-x$
$-x$	x

Observe that −5 is the opposite of +5; they add to zero.

$$\begin{array}{r} +5 \\ \underline{-5} \\ 0 \end{array}$$

EXAMPLE 7 Combine $(+6) + (-3) + (+5) + (-6) + (+3)$. Look for shortcuts.

Solution Let's look for opposites and use them to simplify the computations. If we examine the expression $(+6) + (-3) + (+5) + (-6) + (+3)$, we find +6 and −6. They are opposites; so they add up to zero. Similarly, −3 and +3 are present, and they add up to zero. All that is left, then, is +5. Therefore the sum is +5.

NOTE
Recall that adding 0 to any number does not change the number. In symbols, for any number x,

$$0 + x = x \qquad and \qquad x + 0 = x$$

Another approach to adding signed numbers is to add all the positive numbers together and add all the negative numbers together. Then add the two sums. This method is demonstrated in the next example. The advantage of this method is that you only add numbers of unlike sign once.

EXAMPLE 8 Combine $(+7) + (-4) + (-5) + (+12) + (-1)$.

Solution Separate $(+7) + (-4) + (-5) + (+12) + (-1)$ into positive and negative numbers and compute each sum separately. Then add the two sums.

Positive: $(+7) + (+12)$ $= +19$
Negative: $(-4) + (-5) + (-1) = \underline{-10}$
$+9$ is the sum ✓

EXERCISES 10.2

*1. Add.

    (a) $+19$ (b) $+15$ (c) $+13$ (d) $+16$
        $+14$ $+24$ $- 8$ -25

    (e) -19 (f) -35 (g) -15 (h) -12
        $+23$ $+28$ -12 -15

    (i) $+14$ (j) -18 (k) -49 (l) -33
        -19 $+23$ -28 $+15$

2. Add.

    *(a) $(+8) + (+3)$ *(b) $(-7) + (-4)$
    *(c) $(-6) + (+2)$ *(d) $(-3) + (+9)$
    *(e) $(+7) + (-7)$ (f) $(-12) + (-19)$
    *(g) $(-6) + (+17)$ (h) $(+13) + (-5)$
    *(i) $(-16) + (-7)$ (j) $(-19) + (-11)$
    *(k) $(+16) + (-9)$ (l) $(-12) + (-8)$
    *(m) $(+17) + (+28)$ (n) $(-19) + (+3)$
    *(o) $(-23) + (+14)$ (p) $(+30) + (-14)$
    *(q) $(-16) + (-15)$ (r) $(+13) + (+18)$
    *(s) $(-6) + (-25)$ (t) $(-12) + (-12)$

*3. Add.

    (a) $\left(-\dfrac{2}{3}\right) + \left(+\dfrac{3}{4}\right)$ (b) $\left(-\dfrac{1}{2}\right) + \left(+\dfrac{2}{5}\right)$

    (c) $\left(-\dfrac{3}{5}\right) + \left(-\dfrac{2}{3}\right)$ (d) $\left(+\dfrac{6}{7}\right) + \left(-\dfrac{1}{4}\right)$

    (e) $\left(+\dfrac{3}{4}\right) + \left(-\dfrac{5}{8}\right)$ (f) $\left(-\dfrac{7}{12}\right) + \left(-\dfrac{3}{4}\right)$

    (g) $\left(+\dfrac{1}{6}\right) + \left(+\dfrac{1}{4}\right)$ (h) $\left(-\dfrac{3}{8}\right) + \left(+\dfrac{5}{6}\right)$

4. Determine the sum; that is, combine the numbers.

    *(a) $(+16) + (+2) + (-5) + (+3)$
    *(b) $(-7) + (+1) + (+3) + (-4) + (-3) + (+7)$
    *(c) $(+9) + (+2) + (+4) + (+6) + (-7) + (-3) + (-5)$
    *(d) $(+6) + (-6) + (+9) + (-9) + (-4) + (+4)$
    *(e) $(+3) + (+4) + (-6) + (-7) + (0)$
    (f) $(+3) + (-2) + (+1) + (-4) + (+5) + (-2) + (+1) + (-1)$
    *(g) $(+15) + (-12) + (+60) + (-5) + (-30) + (-17)$
    (h) $(+52) + (+13) + (-42) + (+12) + (-1) + (+49)$
    *(i) $(0) + (-5) + (+1) + (-2) + (+3) + (-1) + (-2) + (+1)$
    (j) $(+16) + (-19) + (-6) + (+9) + (-28) + (-34) + (-12)$
    *(k) $(-18) + (-3) + (+12) + (-6) + (-1) + (+9) + (-20)$
    (l) $(-5) + (0) + (+17) + (-10) + (-9) + (+5) + (-1)$

*5. Combine the numbers.
 (a) $(+3.6) + (+2.1) + (-2.8) + (+1.3) + (-7.3)$
 (b) $(-1.05) + (3.46) + (-2.91) + (8.2)$
 (c) $(-4.56) + (+2.95) + (+5.06) + (-1.08)$
 (d) $(+6.7) + (-7.8) + (-4.1) + (+2.3)$
 (e) $(-1.02) + (-2.33) + (-9.87) + (+15.58)$
 (f) $(+7.23) + (-9.45) + (+3.45) + (-9.53)$

10.3 SUBTRACTION OF SIGNED NUMBERS

If you add 3 to 8, the result is 11. But if you subtract 3 from 8, the result is 5. On the number line you would *add* + 3 to +8 by placing the +8 on the line and moving 3 units to the right. The result is, of course, +11.

In order to obtain a result of +5 when *subtracting* +3 from +8 on the number line, place +8 on the line and move in the opposite direction as you would if adding. In other words, move 3 units to the left rather than to the right.

In this sense, subtraction can be considered the opposite of addition. When subtracting a positive number, move in the negative (left) direction. When subtracting a negative number, move in the positive (right) direction.

EXAMPLE 9 Use a number line to show the result of subtracting -5 from -6.

Solution Place -6 on the line and subtract -5 from it.

The result is -1. Notice that in the subtraction of -5 we moved in the opposite direction we would have if we were adding -5. We moved right for -5 instead of left, as we would have done in addition.

Moving in the opposite direction on the number line is the same as adding the opposite signed number. In other words,

---

To subtract a signed number (from another), add its opposite.

or

To subtract a signed number (from another), change its sign and add it.

---

subtract	*add*	
+8	+8	
+3	−3	change +3 to −3 and add
	+5 $\checkmark$	

subtract	*add*	
−6	−6	
−5	+5	change −5 to +5 and add
	−1 $\checkmark$	

EXAMPLE 10   Subtract −35 from +180.

Solution

subtract	*add*
+180	+180
− 35	+ 35
	+215 $\checkmark$

The difference between a and b, $a - b$, can be *defined* as the sum of a and the opposite of b. That is, $a - b = a +$ opposite of b, or

$$(a) - (b) = (a) + (-b)$$

where $-b$ means the opposite of b.

In subtraction of signed numbers, you change two things:

1. Change the operation to addition.
2. Change the sign of the number you were subtracting.

EXAMPLE 11   Subtract +5 from −2.

Solution   Since we are subtracting +5 from −2, this should be written, for calculation purposes, as

$$(-2) - (+5)$$

Now, instead of subtracting $+5$, let us add -5.

$$(-2) + (-5)$$

The result is

$$-7 \quad \checkmark$$

EXAMPLE 12 Compute $(+5) - (-3)$.

Solution This is, of course, a subtraction. So proceed accordingly.

$$(+5) - (-3)$$
$$= (+5) + (+3)$$
$$= +8 \quad \checkmark$$

EXAMPLE 13 Compute $(-8) - (-5)$.

Solution
$$(-8) - (-5)$$
$$= (-8) + (+5)$$
$$= -3 \quad \checkmark$$

EXAMPLE 14 Compute $(+5) - (+1)$.

Solution
$$(+5) - (+1)$$
$$= (+5) + (-1)$$
$$= +4 \quad \checkmark$$

Addition can be used as a check for subtraction, and it can also be used indirectly to perform the subtraction. Consider the subtraction $(5) - (2)$. The result is 3. And 3 checks. That is, 3 is the number that when added to 2 produces 5. $(5) - (2) = 3$, since $2 + 3 = 5$. Similarly, $(+7) - (-6) = +13$, since $(-6) + (+13) = +7$. Consider another example. $(-8) - (+6) = ?$ The result is the number that when added to $+6$ produces -8. In equation form, $(+6) + (\quad) = -8$. The number is -14. Thus $(-8) - (+6) = -14$.

In these last examples we have performed the subtraction indirectly by determining what number would check. If this method is applied to the earlier examples of the section, the results will be the same as those obtained by "adding the opposite" or "changing the sign of the number being subtracted." Yet this procedure is less mechanical and may perhaps be more convincing. It is shown here for understanding, but the mechanical approaches are generally more efficient.

EXERCISES 10.3

*1. Subtract.

(a) $+30$
$+ 5$

(b) -30
$+ 5$

(c) -43
$+17$

(d)	-15	(e)	$+32$	(f)	$+54$
	$\underline{-17}$		$\underline{-19}$		$\underline{+25}$
(g)	-57	(h)	-15	(i)	-43
	$\underline{+41}$		$\underline{-94}$		$\underline{+72}$
(j)	$+17$	(k)	-34	(l)	$+387$
	$\underline{+29}$		$\underline{+42}$		$\underline{-524}$
(m)	-341	(n)	-196	(o)	$+385$
	$\underline{-102}$		$\underline{+145}$		$\underline{+290}$

*2. Subtract.

(a)	$+68.52$	(b)	$+359.16$	(c)	-157.23
	$\underline{-49.23}$		$\underline{+267.06}$		$\underline{-251.95}$

3. Perform the indicated subtractions.

    *(a) $(+7) - (+4)$ *(b) $(+9) - (+18)$
    *(c) $(+6) - (-3)$ *(d) $(-9) - (+6)$
    *(e) $(+5) - (-12)$ *(f) $(-7) - (-4)$
    (g) $(-6) - (-10)$ *(h) $(+7) - (-8)$
    (i) $(+3) - (-1)$ (j) $(-1) - (-3)$
    *(k) $(-16) - (+19)$ (l) $(+176) - (+761)$

*4. Perform the indicated subtractions.

    (a) $(-1.67) - (+3.45)$ (b) $(+78.45) - (-34.98)$

    (c) $\left(+\frac{3}{7}\right) - \left(-\frac{2}{7}\right)$ (d) $\left(-\frac{1}{2}\right) - \left(+\frac{1}{4}\right)$

    (e) $\left(+\frac{2}{5}\right) - \left(+\frac{2}{3}\right)$ (f) $\left(-\frac{2}{3}\right) - \left(-\frac{3}{4}\right)$

5. Compute.

    *(a) $(-5) - (-9)$ (b) $(-4) - (+7)$
    *(c) $(+7) - (-2)$ (d) $(+5) - (+9)$
    *(e) $(-9) - (+4)$ (f) $(+3) - (-11)$
    *(g) $(+7) - (+3)$ (h) $(-15) - (+3)$
    *(i) $(-25) - (-17)$ (j) $(+2) - (-56)$
    *(k) $(-16) - (+24)$ (l) $(+16) - (+24)$
    *(m) $(+2) - (-2)$ (n) $(+2) - (+2)$

10.4 MULTIPLICATION OF SIGNED NUMBERS

Four cases are possible in the multiplication of two signed numbers. Using the notation "$(+)$" for *positive-signed number* and "$(-)$" for *negative-signed number,* we list the four cases.

$$(+)(+) = ?$$
$$(+)(-) = ?$$
$$(-)(+) = ?$$
$$(-)(-) = ?$$

Let us consider each of these cases.

(+)(+)

Since positive numbers are the same as unsigned numbers, the product

$$(+6)(+7)$$

is the same as the product

$$(6)(7)$$

And $(6)(7) = 42$; so $(+6)(+7) = +42$.

In general,

$$\boxed{(+)(+) = (+)}$$

(+)(−)

The product

$$(+4)(-3)$$

is the same as

$$(4)(-3)$$

since $+4$ is the same as 4. This second form can be read as four -3's, which can be written and computed as

$$
\left.
\begin{array}{r}
-3 \\
-3 \\
-3 \\
-3 \\
\end{array}
\right\}
\quad \text{the sum of four } -3\text{'s}
$$

$$-12$$

Thus

$$(+4)(-3) = -12$$

In general,

$$\boxed{(+)(-) = (-)}$$

(−)(+)

The product

$$(-3)(+4)$$

is the same as

$$(+4)(-3)$$

by the commutative property $(a \cdot b = b \cdot a)$. So

$$(-3)(+4) = -12$$

as determined in the previous case. Thus

$$(-)(+) = (-)$$

The next case to be presented will make use of the fact that if a number is multiplied by zero the result is zero. This can be stated as

$$x \cdot 0 = 0$$
$$0 \cdot x = 0$$

for any number x. For example, $(5)(0) = 0$ and $(0)(-8.1) = 0$.

$(-)(-)$

This case must be shown less directly than the others. Consider some negative number, say -3, times 0. Since $x \cdot 0 = 0$ for any number x,

$$(-3)(0) = 0$$

and rewrite 0 as the sum of a number and its opposite, say $(+2) + (-2)$, since a number and its opposite add to zero.

$$(-3)[(+2) + (-2)] = 0$$

Applying the distributive property (given on page 42), which says $a(b + c) = a \cdot b + a \cdot c$,

$$(-3)[(+2) + (-2)] = 0$$

becomes

$$(-3)(+2) + (-3)(-2) = 0$$

Since $(-)(+) = (-)$, we know that $(-3)(+2) = -6$. So we can replace $(-3)(+2)$ by -6 to obtain

$$(-6) + (-3)(-2) = 0$$

which tells us that $(-3)(-2)$ must multiply to give us the opposite of -6, since the sum of (-6) and $(-3)(-2)$ must add to 0. This means that $(-3)(-2) = +6$.

In general,

$$(-)(-) = (+)$$

In summary,

$$(+)(+) = (+)$$
$$(+)(-) = (-)$$
$$(-)(+) = (-)$$
$$(-)(-) = (+)$$

EXAMPLE 15   Observe the multiplication of signed numbers.

Solution

$$(+7)(+4) = +28$$
$$(+6)(-2) = -12$$
$$(-5)(-4) = +20$$
$$(+5)(-3) = -15$$
$$(-7)(+2) = -14$$
$$(-3)(-4) = +12$$
$$(+2)(+5) = +10$$
$$(-5)(-3) = +15$$

In the event you have more than two numbers to multiply, multiply them two at a time, as shown in the next example.

EXAMPLE 16   Multiply $(+6)(-2)(-4)(+3)$.

Solution

$$(+6)(-2)(-4)(+3)$$
$$(-12)(-4)(+3)$$
$$(+48)(+3)$$
$$+144 \quad \checkmark$$

You can determine the *sign* of the product of several signed numbers by counting the number of negatives involved in the multiplication. Keeping in mind that 0, 2, 4, 6, 8, 10, etc. are *even numbers* and that 1, 3, 5, 7, 9, etc. are *odd numbers,* we have

> *Multiplication*
>
> even number of negatives → result is positive
> odd number of negatives → result is negative

To see that this is true, note that $(-)(-) = (+)$; that is, each pair of negatives (negative numbers) has no effect on the sign. So if there are four negatives, then we get $(-)(-)$ times $(-)(-)$, or $(+)$ times $(+)$.

$$(-)(-)(-)(-)$$
$$(+) \quad (+)$$
$$(+)$$

Similarly, six negatives, eight negatives—any even number of negatives—produce a positive result.

EXAMPLE 17   Multiply $(-5)(+2)(+3)(-1)(-2)(-4)$.

Solution   There are *four* negatives, an even number; so the sign of the result is $+$. The magnitude is $5 \cdot 2 \cdot 3 \cdot 1 \cdot 2 \cdot 4 = 240$, so the product is $+240$.

Determining the sign in advance minimizes the chance for error; it's simpler.

EXAMPLE 18   Multiply $(-1)(+6)(-2)(+1)(-5)(+3)(-2)(+1)(-3)$.

Solution   There are *five* negatives, an odd number; so the sign of the result is $-$. The magnitude is $1 \cdot 6 \cdot 2 \cdot 1 \cdot 5 \cdot 3 \cdot 2 \cdot 1 \cdot 3 = 1080$, and therefore the product is -1080.

What happens if we raise a negative number to a power? Let's use -2 as an example.[†]

$$(-2)^1 = (-2) = -2$$
$$(-2)^2 = (-2)(-2) = +4$$
$$(-2)^3 = (-2)(-2)(-2) = -8$$
$$(-2)^4 = (-2)(-2)(-2)(-2) = +16$$
$$(-2)^5 = (-2)(-2)(-2)(-2)(-2) = -32$$
$$(-2)^6 = (-2)(-2)(-2)(-2)(-2)(-2) = +64$$

The conclusion can be stated symbolically as

$$\boxed{\begin{aligned} (-)^{\text{odd}} &= (-) \\ (-)^{\text{even}} &= (+) \end{aligned}}$$

In words, if a negative number is raised to an odd power, the result is negative; if it is raised to an even power, the result is positive. This is to be expected, since the exponent indicates the number of signs that we have in the product, and in this case they are all negatives. So here the exponent tells us how many negatives are in the product.

On the other hand, if a *positive* number is raised to *any* power, the result is positive, since there won't be any negative numbers in the product.

$$\boxed{(+)^{\text{any}} = (+)}$$

EXAMPLE 19   Find the value of $(-3)^3 + (-4)^4 + (+2)^3$.

Solution
$$(-3)^3 = -27$$
$$(-4)^4 = +256$$
$$(+2)^3 = +8$$

so

$$(-3)^3 + (-4)^4 + (+2)^3 = (-27) + (+256) + (+8) = +237 \quad \checkmark$$

---

† The meaning of exponent and power was explained in Chapter 1. If these examples of powers of -2 are not clear, you might want to refer to page 20 for an in-depth explanation of exponents.

As a final note on multiplication, consider that if any number is multiplied by 1, the result is the same number. This can be stated as

$$1 \cdot x = x$$
$$x \cdot 1 = x$$

for any number x. For example, $(1)(-6) = -6$ and $(8.43)(1) = 8.43$.

EXERCISES 10.4

*1. Multiply.
  (a) $(+3)(-2)$ (b) $(-5)(+2)$ (c) $(+7)(+5)$
  (d) $(-3)(+6)$ (e) $(-8)(-3)$ (f) $(+9)(+7)$
  (g) $(-1)(-9)$ (h) $(+8)(-5)$ (i) $(+2)(+10)$
  (j) $(-4)(-9)$ (k) $(+4)(-1)$ (l) $(-6)(+8)$
  (m) $(-3)(-3)$ (n) $(+7)(-7)$ (o) $(-9)(+10)$
  (p) $(-5)(+16)$ (q) $(-12)(-7)$ (r) $(+19)(+4)$
  (s) $(+15)(-8)$ (t) $(-16)(-4)$ (u) $(-19)(+3)$
  (v) $(-11)(-9)$ (w) $(+16)(+15)$ (x) $(+15)(-7)$

*2. Multiply.
  (a) $(-1.6)(+3.2)$ (b) $(+6.4)(+5.2)$ (c) $(+9.65)(-7.7)$

  (d) $(-87.3)(-7.6)$ (e) $\left(+\frac{1}{3}\right)\left(-\frac{1}{7}\right)$ (f) $\left(+\frac{2}{5}\right)\left(+\frac{3}{11}\right)$

  (g) $\left(-\frac{2}{3}\right)\left(-\frac{3}{4}\right)$ (h) $\left(-\frac{5}{6}\right)\left(+\frac{7}{8}\right)$

3. Multiply.
  *(a) $(+4)(+3)(-2)$ *(b) $(+1)(-2)(+3)$
  *(c) $(-8)(+9)(-3)$ *(d) $(-10)(-3)(-2)$
  *(e) $(-4)(+8)(+3)$ (f) $(+5)(+7)(+6)$
  *(g) $(-2)(-2)(+2)(+2)$ *(h) $(-1)(+1)(-1)(+1)(-1)$
  *(i) $(+3)(-5)(+6)(-4)$ (j) $(-2)(+4)(-3)(+5)$
  *(k) $(-4)(-3)(-2)(+1)$ (l) $(+3)(-5)(+2)(+1)$
  *(m) $(+3)(+4)(-4)(-3)$ (n) $(-7)(-2)(+1)(-5)$
  *(o) $(-2)(-1)(-3)(-4)$ (p) $(+9)(+1)(-2)(+1)$

*4. Determine the value of the expression.
  (a) $(-1)^5$ (b) $(-2)^4$ (c) $(-3)^3$ (d) $(+2)^5$
  (e) $(-3)^4$ (f) $(-3)^5$ (g) $(-2)^7$ (h) $(-1)^{50}$
  (i) $(-1)^{101}$ (j) $(-4)^3$ (k) $(+1)^5$ (l) $(+3)^3$

*5. Determine the value of each expression; that is, simplify it.
  (a) $(-1)^3 + (-2)^4$ (b) $(-2)^3 + (-1)^6$
  (c) $(-4)^3 + (+2)^3 + (-2)^3$ (d) $(-5)^3 + (-4)^2 + (+3)^3$
  (e) $(-1)^5 + (-1)^3 + (-1)^4$ (f) $(-9)^2 + (-4)^3 + (-10)^3$
  (g) $(-2)^3 + (-1)^2 + (-2)^2$ (h) $(-1)^4 + (-5)^2 + (-2)^3$
  (i) $(-4)^2 + (+3)^2 + (-1)^5$ (j) $(-2)^5 + (-1)^2 + (-1)^7$

10.5 NOTATION

Frequently it is too cumbersome to write

$$(+6) + (+2)$$

to specify the addition of +6 and +2. Instead it is written less formally as

$$+6 + 2$$

with the understanding that this is the *addition* of the two numbers +6 and +2. Similarly,

$$(+6) + (-2)$$

can be written

$$+6 - 2$$

with the understanding that this is the *addition* of the two numbers +6 and −2.

> +6 − 2 is the addition of +6 and −2.
> +6 + 2 is the addition of +6 and +2.

As an extension of this notation to the addition of several numbers,

$$-8 + 5 - 4 + 19$$

specifies the *sum* of the four numbers −8, +5, −4, and +19.

EXAMPLE 20   Combine −8 + 5 − 4 + 19.

Solution   There are four numbers to be added. As done previously, we will separate the sum into positive and negative numbers and compute each sum separately. Then we'll add the two sums.

$$\begin{aligned}\text{Positive:} \quad &+5 + 19 = +24\\ \text{Negative:} \quad &-8 - 4 = \underline{-12}\\ & +12 \ \checkmark\end{aligned}$$

The sum is +12.

As you know, to carry out subtraction such as

$$(+8) - (+3)$$

and

$$(+9) - (-7)$$

we change to addition. With this in mind and in consideration of the new notation, we have

$$(+8) - (+3) = (+8) + (-3)$$
$$= +8 - 3$$
$$= +5$$

and

$$(+9) - (-7) = (+9) + (+7)$$
$$= +9 + 7$$
$$= +16$$

An easy way to handle such simplifications is to realize that a *minus* in front of parentheses results in the addition of the opposite of the number inside parentheses. A *plus* in front of parentheses results in the addition of the number inside parentheses. In general, if a represents any number, then

$$-(+a) = -a$$
$$-(-a) = +a$$
$$+(+a) = +a$$
$$+(-a) = -a$$

Here are some examples.

$$(+9) - (+8) = +9 - 8 = +1$$
$$(+9) - (-8) = +9 + 8 = +17$$
$$(+9) + (+8) = +9 + 8 = +17$$
$$(+9) + (-8) = +9 - 8 = +1$$

Next we consider other examples in which this notation arises.

EXAMPLE 21 Compute the value of $+7 - (-2)^3$.

Solution

$$+7 - (-2)^3 = +7 - (-8) \qquad \text{since } (-2)^3 = -8$$
$$= +7 + 8 \qquad \text{since } -(-8) = +8$$
$$= +15 \quad \checkmark$$

EXAMPLE 22 Simplify $+6 + (-1)(+2)$.

Solution

$$+6 + (-1)(+2) = +6 + (-2) \qquad \text{since } (-1)(+2) = -2$$
$$= +6 - 2 \qquad \text{since } +(-2) = -2$$
$$= +4 \quad \checkmark$$

Keep in mind the order in which different arithmetic operations are to be carried out. *Unless parentheses specify otherwise*, the order of operations is exponents first, then multiplication and division, then addition and subtraction.

> *Order of Operations*
> 1. Exponents
> 2. Multiplication and division
> 3. Addition and subtraction

EXAMPLE 23 Simplify $(-8) + (-3)(+4)$.

Solution

$$(-8) + (-3)(+4) = -8 + (-12) \qquad \left\{ \begin{array}{l} \text{since multiplication is} \\ \text{done before addition} \end{array} \right.$$

$$= -8 - 12 \qquad \text{simplification } +(-12) = -12$$

$$= -20 \;\checkmark \qquad \text{the addition of } -8 \text{ and } -12$$

EXAMPLE 24 Simplify $(-3)(-2)^3$.

Solution

$$(-3)(-2)^3 = (-3)(-8) \qquad \left\{ \begin{array}{l} \text{since exponentiation is} \\ \text{done before multiplication} \end{array} \right.$$

$$= 24 \;\checkmark$$

EXAMPLE 25 Compute the value of $(-4)(-3) - (+6)(-2)$.

Solution

$$(-4)(-3) - (+6)(-2) = +12 - (-12) \qquad \text{Multiply before subtracting.}$$

$$= 12 + 12 \qquad \text{using } -(-12) = +12$$

$$= 24 \;\checkmark$$

EXAMPLE 26 Simplify $(-3)(+2) + (-8) - 7$.

Solution The multiplication $(-3)(+2)$ should be done first. Also, note that $(-8) - 7$ is *not a multiplication*. It would have to be $(-8)(-7)$ or $-7(-8)$ in order to be a multiplication. Here then is the simplification, step by step.

$$(-3)(+2) + (-8) - 7 = -6 + (-8) - 7$$

$$= -6 - 8 - 7$$

$$= -21 \;\checkmark$$

EXAMPLE 27 Determine the value of the expression $-2^4 + (-2)^4$.

Solution $-2^4 = -16$, since the 2 is raised to the fourth power before the sign is attached. On the other hand, $(-2)^4$ is 16 because the parentheses specify that it is -2 that is raised to the fourth power. Without parentheses, only the 2 is raised to the fourth power. Thus

$$-2^4 + (-2)^4 = -16 + 16$$

$$= 0 \;\checkmark$$

Another way to simplify -2^4 is to realize that -2^4 means $- \, 2 \cdot 2 \cdot 2 \cdot 2$. Thus

$$-2^4 = -2 \cdot 2 \cdot 2 \cdot 2$$
$$= -16$$

Still more examples of order of operation can be found in Chapter 2.

EXERCISES 10.5

*1. Simplify.
    (a) $+8 - 7 + 3 - 9$ (b) $-8 - 5 - 3 - 2$
    (c) $+4 + 3 + 8 + 7$ (d) $+9 - 7 - 5 - 2$
    (e) $7 - 1 - 9 + 8 + 5$ (f) $16 + 3 - 5 + 2 + 0$
    (g) $-2 + 4 - 5 + 19 - 2$ (h) $+15 - 12 + 45 - 16 + 2$

2. Simplify.
    *(a) $-9 + 7 - 1 + 6 - 15 + 8 - 2$
    (b) $+10 - 6 + 4 + 9 + 3 - 5 - 2 + 7$
    *(c) $-11 + 6 + 3 - 5 + 11 - 3 + 7 - 6$
    (d) $-5 + 17 - 3 + 10 + 5 + 3 - 17 + 6$

*3. Simplify.
    (a) $-7 + (-6)$ (b) $+8 - (-7)$
    (c) $-3 + (+13)$ (d) $-9 - (+6)$
    (e) $+8 + (+9)$ (f) $+5 + (-2)$
    (g) $-4 - (-10)$ (h) $+1 - (+3)$
    (i) $19 + (-7)$ (j) $12 - (+17)$
    (k) $6 + (-9)$ (l) $15 - (+7)$
    (m) $-8 - (-10)$ (n) $-15 - (-12)$
    (o) $+7 - (+2)$ (p) $+6 - (+6)$

4. Determine the value of each expression.
    *(a) $(-5) + (-2)(+6)$
    *(b) $+4 - (-3)(-2)$
    *(c) $7 - (-3)(+5)$
    (d) $18 + (-5)(-7)$
    *(e) $(-3)(-2) + (-9)(-8)$
    (f) $(-3)(+2) - (+3)(-7)$
    *(g) $7(-3) + (+5)(-2)$
    (h) $(-4)(-1)(-3) - (-9)$
    *(i) $-8 - (+7) - (-8)(+1)$
    *(j) $29 + 6(-5) + (-3)$
    *(k) $(-5)(-3) - (-7) + (-6)$
    (l) $-4 + (+4)(-2) - (+10)$
    *(m) $-3(+6) + (0)(+5) - (-5)$
    (n) $(-3)(-1)(-4) - (-5)(+2)(-1)$
    *(o) $-2(-5) + (-5)(-8) + (-9)$
    (p) $-4(+5) + (-4)(0) - (-3)$
    *(q) $(-6)(-1)(+2) - (-2)(-1)(-3)$
    (r) $10 - (+3)(+2) - (-6)$

5. Compute the value of each expression.

*(a) $(-4)(-2)^3$
*(c) $(+5)^2 - (-5)^2$
*(e) $(-3)(-1)^2(-2)$
*(g) $(-2)^2(-3)^3 + (-1)^9(-5)$
*(i) $-5^2 + (-5)^2$
*(k) $(-5)(-4)(-2) - (-3)^3$
*(m) $-3^2 - 3^2 + (-4)^2$
*(o) $(-4^2)(-4)(+1) - 0(-7)$
*(q) $+3^3 - 5(-2)^3 + (-3) - 1$

*(b) $(-3)^2 - 3^2$
(d) $(-7)(+9) - 6^2 + (-8)$
(f) $(8)(-2)(+3) - (-4)(+16)$
*(h) $(-2)(3)(0) + (-4)(+5)(-1)$
*(j) $(-4)^2 + 7 - 4^2 - 9$
(l) $(-5)^2 - 5^2 + 5(-2)$
(n) $-2(-3)^2 - (-4)(-3)(-1)$
(p) $(-3)(-2)^3 + (-4)(-7) - 3$
(r) $-2^3 + 2(-1)^3 - (-5) - 7$

10.6 DIVISION OF SIGNED NUMBERS

Since we already know the rules for multiplication of signed numbers

$$(+)(+) = (+)$$
$$(+)(-) = (-)$$
$$(-)(+) = (-)$$
$$(-)(-) = (+)$$

we can easily determine the rules for division of signed numbers.

Consider the case $\dfrac{(+)}{(+)}$, which can also be written $(+) \div (+)$ or

$$(+)\overline{)(+)}$$

Let's use $(+5)\overline{)(+10)}$ as an example. Since the result must check in order to be the correct quotient, we must have

$$(+5)\overline{)(+10)}$$
$$\underline{(+10)}$$

so that subtracting $+10$ from $+10$ produces a zero remainder. This means the quotient must be positive $(+)$ to produce it.

$$(+5)\overline{)\overset{(+\ 2)}{(+10)}} \quad \text{or} \quad \frac{(+)}{(+)} = (+)$$
$$\underline{(+10)}$$

Similarly,

$$\frac{(+)}{(-)} = (-) \quad \text{since} \quad (-5)\overline{)\overset{(-2)}{(+10)}}$$
$$\underline{(+10)}$$

$$\frac{(-)}{(+)} = (-) \quad \text{since} \quad (+5)\overline{)\overset{(-2)}{(-10)}}$$
$$\underline{(-10)}$$

$$\frac{(-)}{(-)} = (+) \quad \text{since} \quad (-5)\overline{)(-10)}^{\,(+2)}_{\,(-10)}$$

Thus the rules for division of signed numbers are the same as those for multiplication.

$$\frac{(+)}{(+)} = (+) \qquad \qquad \frac{(-)}{(-)} = (+)$$

$$\frac{(+)}{(-)} = (-) \qquad \qquad \frac{(-)}{(+)} = (-)$$

EXAMPLE 28 Divide.
(a) $(-8) \div (+1)$
(b) $(-6) \div (-2)$
(c) $(+9)/(-3)$

Solution   (a) $(-8) \div (+1) = \dfrac{-8}{+1} = -\dfrac{8}{1} = -8$ ✓

(b) $(-6) \div (-2) = \dfrac{-6}{-2} = +\dfrac{6}{2} = +3$ ✓

(c) $(+9)/(-3) = \dfrac{+9}{-3} = -\dfrac{9}{3} = -3$ ✓

EXAMPLE 29 Simplify $\dfrac{+15}{-1} + \dfrac{-8}{-2} + \dfrac{+4}{+4}$.

Solution
$$\frac{+15}{-1} + \frac{-8}{-2} + \frac{+4}{+4} = -\frac{15}{1} + \left(+\frac{8}{2}\right) + \left(+\frac{4}{4}\right)$$
$$= -15 + (+4) + (+1)$$
$$= -15 + 4 + 1$$
$$= -10 \ \checkmark$$

Be careful *not* to interpret a problem like this as addition of fractions. Although there may appear to be fractions here, when the division is done the results in all cases are integers. There will be no fractions. So *do not* seek a common denominator in order to add the "fractions." It's easy to make a mistake that way.

EXAMPLE 30 Simplify $\dfrac{-6}{-3} + \dfrac{-14}{+2} - \dfrac{+10}{-5}$.

Solution
$$\frac{-6}{-3} + \frac{-14}{+2} - \frac{+10}{-5} = +\frac{6}{3} + \left(-\frac{14}{2}\right) - \left(-\frac{10}{5}\right)$$
$$= +2 + (-7) - (-2)$$
$$= +2 - 7 + 2$$
$$= -3 \quad \checkmark$$

EXAMPLE 31   Simplify $\dfrac{(+6)(-2)}{(+4)} + (-2)(+3) - 5$.

Solution
$$\frac{(+6)(-2)}{(+4)} + (-2)(+3) - 5 = \frac{(-12)}{(+4)} + (-6) - 5$$
$$= -3 - 6 - 5$$
$$= -14 \quad \checkmark$$

EXAMPLE 32   Simplify $\dfrac{(+10)(-2)}{(-4)(-5)} + \dfrac{(+9)(+4)}{(-1)(+1)}$.

Solution
$$\frac{(+10)(-2)}{(-4)(-5)} + \frac{(+9)(+4)}{(-1)(+1)} = \frac{-20}{+20} + \frac{+36}{-1}$$
$$= -\frac{20}{20} + \left(-\frac{36}{1}\right)$$
$$= -1 + (-36)$$
$$= -1 - 36$$
$$= -37 \quad \checkmark$$

EXAMPLE 33   Simplify $\dfrac{(+4)(-3) + (-2)(-5)}{-7 + 5 + 2 - 6 - 9}$.

Solution   Note that
$$(+4)(-3) = -12$$
$$(-2)(-5) = +10$$
$$-7 + 5 + 2 - 6 - 9 = -15$$

So the fraction simplifies to

$$\frac{-12 + (+10)}{-15} = \frac{-12 + 10}{-15} = \frac{-2}{-15} = +\frac{2}{15} \quad \checkmark$$

Example 33 points out that even though you do not actually carry out the division of 2 by 15, the sign of the fraction 2/15 should be determined. Similarly,

$$\frac{+4}{-7} = -\frac{4}{7}$$

$$\frac{-4}{+7} = -\frac{4}{7}$$

$$\frac{+4}{+7} = +\frac{4}{7}$$

The sign of the fraction should be determined and placed in front of it, as shown above.

As a final note, you should recall from arithmetic (Section 1.7) that *division by zero is not defined; you cannot divide by zero.*

> You cannot divide by 0.
>
> $\dfrac{x}{0}$ is undefined for any number x.

Thus $\dfrac{5}{0}$ is undefined and $\dfrac{-8}{0}$ is undefined. On the other hand,

> $\dfrac{0}{x} = 0$ for any nonzero number x.

For example, $\dfrac{0}{5} = 0$ and $\dfrac{0}{-8} = 0$.

EXERCISES 10.6

*1. Divide.

(a) $(+10) \div (+2)$ (b) $(-9) \div (+3)$ (c) $(+6) \div (-2)$

(d) $(-28) \div (-7)$ (e) $(-8)/(-2)$ (f) $(-18)/(+3)$

(g) $(+4)/(-1)$ (h) $(-100)/(+20)$ (i) $\dfrac{(-84)}{(-7)}$

(j) $\dfrac{(-3)}{(+3)}$ (k) $\dfrac{(+42)}{(-7)}$ (l) $\dfrac{-9}{-3}$

(m) $\dfrac{-81}{-3}$ (n) $\dfrac{54}{-9}$ (o) $\dfrac{0}{+3}$

(p) $\dfrac{0}{-5}$ (q) $\dfrac{-1}{+1}$ (r) $\dfrac{-4}{-1}$

*2. Simplify each expression. Your answer should take the form of a fraction with a sign in front of it, as explained in the note after Example 33.

(a) $\dfrac{-3}{+5}$ (b) $\dfrac{-4}{+7}$ (c) $\dfrac{-7}{-9}$

(d) $\dfrac{+6}{-7}$ (e) $\dfrac{+5}{+8}$ (f) $\dfrac{+1}{-3}$

(g) $\dfrac{-5}{+6}$ (h) $\dfrac{-3}{-4}$ (i) $\dfrac{+6}{+7}$

3. Simplify.

*(a) $\dfrac{+9}{-3} + \dfrac{+4}{+1}$

(b) $\dfrac{-15}{-3} + \dfrac{-20}{+4}$

*(c) $\dfrac{-10}{+2} - \dfrac{+6}{-1}$

(d) $\dfrac{+12}{-4} - \dfrac{-8}{-2}$

*(e) $\dfrac{-25}{+5} + \dfrac{+12}{-4} - \dfrac{-6}{-2}$

(f) $\dfrac{+8}{-4} - \dfrac{+30}{+10} + \dfrac{-2}{+1}$

*(g) $\dfrac{-16}{-4} + \dfrac{+18}{-2} + \dfrac{+10}{+2}$

(h) $\dfrac{+7}{+1} - \dfrac{+20}{-2} - \dfrac{-32}{-4}$

*4. Perform each division, if possible. If the division cannot be done, then indicate "not defined."

(a) $\dfrac{+6}{-2}$ (b) $\dfrac{0}{-7}$ (c) $\dfrac{-2}{0}$

(d) $\dfrac{0}{-4}$ (e) $\dfrac{-1}{+1}$ (f) $\dfrac{+3}{-3}$

(g) $\dfrac{0}{0}$ (h) $\dfrac{+3}{0}$ (i) $\dfrac{0}{+2}$

(j) $\dfrac{-8}{+8}$ (k) $\dfrac{0}{+4}$ (l) $\dfrac{-4}{0}$

5. Determine the value of each expression.

*(a) $\dfrac{+10}{-2} + \dfrac{-8}{-4} + (-9)$

*(b) $\dfrac{(+2)(-5) - (+2)(-3)}{(+2)}$

*(c) $\dfrac{(-2)^2 - 8 + 6 - 5 - 3}{2 - 3 + 4 - 5}$

*(d) $\left(\dfrac{+7 - 2 + 5 - 3}{-5 - 12 + 8 + 2}\right)^9$

(e) $(-2)^3(-3) + \dfrac{+3 - 2 - 1}{(-3)^5(+2)^4}$

(f) $\dfrac{(+12)(-10)}{(+5)(-2)} - \dfrac{(+8)(-1)}{(-2)(+4)}$

*(g) $\dfrac{(+4)(-7) - (+8)(-11)}{+5 - 7 + 8 - 1 - 3}$

*(h) $\dfrac{(+3)(-2) - (-1)(-5)}{(-1)(-2)(-3)(-4)}$

(i) $\dfrac{(-1)(-7) - (-5)(+2)}{(-1)(+2) + (+4)(-2)}$

*(j) $\dfrac{(-3)(+2) + (-8)(-2)}{(-3)(-4) - (+5)(+2)}$

10.7 EVALUATION OF EXPRESSIONS

When a specific value is used for x in an expression such as $x^2 - 5x + 2$, it becomes an arithmetic expression very similar to those that we have been evaluating in this chapter.

EXAMPLE 34   Evaluate the expression $x^2 - 5x + 2$ (a) when x is 6, (b) when $x = 2$, (c) when $x = 0$, and (d) when $x = -3$.

Solution   (a) If x is 6, then

$$x^2 - 5x + 2 = (6)^2 - 5(6) + 2 \qquad \underline{5x \text{ means } 5 \cdot x.}$$
$$= 36 - 30 + 2$$
$$= 8 \quad \checkmark$$

(b) If $x = 2$, then

$$x^2 - 5x + 2 = (2)^2 - 5(2) + 2$$
$$= 4 - 10 + 2$$
$$= -4 \quad \checkmark$$

(c) If $x = 0$, then

$$x^2 - 5x + 2 = (0)^2 - 5(0) + 2$$
$$= 0 - 0 + 2$$
$$= 2 \quad \checkmark$$

(d) If $x = -3$, then

$$x^2 - 5x + 2 = (-3)^2 - 5(-3) + 2$$
$$= +9 + 15 + 2$$
$$= 26 \quad \checkmark$$

EXAMPLE 35   Evaluate the expression

$$\frac{6a + b - c}{2d}$$

for $a = -2$, $b = 3$, $c = -1$, and $d = -2$.

Solution   If $a = -2$, $b = 3$, $c = -1$, and $d = -2$, then

$$\frac{6a + b - c}{2d} = \frac{6(-2) + (3) - (-1)}{2(-2)}$$
$$= \frac{-12 + 3 + 1}{-4}$$
$$= \frac{-8}{-4}$$
$$= 2 \quad \checkmark$$

EXERCISES 10.7

*1. Determine the value of the expression $x^2 - 7x + 3$ for the following values of x.
   (a) 9   (b) 5   (c) 0   (d) −4

2. Determine the value of the expression $x^2 + 3x - 8$ for the following values of x.
   (a) 1   (b) −1   (c) −3   (d) 0

*3. Determine the value of the expression $x^2 + 5x - 1$ for the following values of x.
   (a) −1   (b) 1   (c) 0   (d) −2

4. Determine the value of the expression $x^2 - 4x + 2$ for the following values of x.
    (a) 0   (b) -3 (c) 3   (d) -1

*5. Determine the value of the expression $-x^2 + 7x + 2$ for the following x.
    (a) 3   (b) -3 (c) 2   (d) -2

6. Determine the value of the expression $-x^2 + 4x + 1$ for the following x.
    (a) -1 (b) 1   (c) -4 (d) 4

7. Evaluate each expression for $x = -2$.
    *(a) $x^2 - 10x$ (b) $x^2 + 7x$
    *(c) $x^2 + 6x$ (d) $x^2 - 8x$
    *(e) $x^2 - x - 7$ (f) $x^2 + x - 3$
    *(g) $15 - 3x$ (h) $10 - 8x$
    *(i) $1 + 10x + x^2$ (j) $1 - 3x + x^2$
    *(k) $-x^2 + 3x + 1$ (l) $-x^2 + 5x - 2$
    *(m) $-x^2 - 7x - 3$ (n) $-x^2 - 6x + 8$

8. Evaluate the expression

$$\frac{2a + 4b}{a - b}$$

for the a and b values given.
    *(a) $a = 3, b = 1$ (b) $a = 5, b = 2$
    *(c) $a = 1, b = -1$ (d) $a = 2, b = -2$
    *(e) $a = -4, b = 0$ (f) $a = 0, b = -3$
    *(g) $a = -2, b = -1$ (h) $a = -1, b = -2$
    *(i) $a = 0, b = -7$ (j) $a = -4, b = 0$

9. Determine the value of the expression $b^2 - 4ac$ for the values of a, b, and c given below. Note that $4ac$ means $4 \cdot a \cdot c$, the product of 4, a, and c.
    *(a) $a = 2, b = 7, c = 1$ (b) $a = 1, b = 5, c = 3$
    *(c) $a = 1, b = 2, c = -3$ (d) $a = 1, b = 4, c = -2$
    *(e) $a = 3, b = 2, c = 4$ (f) $a = 5, b = 1, c = 2$
    *(g) $a = 1, b = -5, c = -1$ (h) $a = 8, b = -1, c = 6$
    *(i) $a = 1, b = -3, c = 0$ (j) $a = 2, b = 0, c = -1$

10.8 INEQUALITY AND ABSOLUTE VALUE

The expressions *magnitude* and **absolute value** mean the same thing—the value (or size) of the number—without sign. On the number line magnitude can be considered the undirected distance (number of units) from 0.

The magnitude of $+9$ is 9.

The magnitude of -9 is 9.

The absolute value of $+9$ is 9.

The absolute value of -9 is 9.

We use the notation | | for absolute value. Thus

$$|+9| = 9$$
$$|-9| = 9$$

Also,

$$|(+6) + (+4)| = |+10| = 10$$
$$|(+6) + (-4)| = |+2| = 2$$
$$|(-6) + (-4)| = |-10| = 10$$
$$|(-6) + (+4)| = |-2| = 2$$

Note that the absolute value indicator | | acts like parentheses by grouping elements inside. The absolute value is taken *after* the sum inside is computed.

The symbol > means **greater than.** Read 5 > 2 as "five is greater than two." On a number line the larger of two numbers is always farther right.

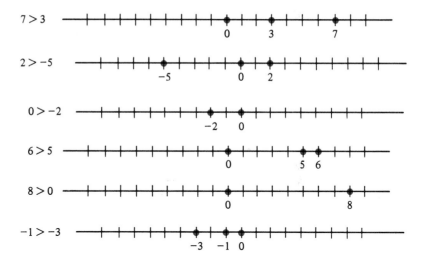

$|-8| > -8$, since $|-8| = 8$ and $8 > -8$.

The symbol < means **less than.** Read 2 < 5 as "two is less than five." The relationships above can be written by using < instead of >.

$$3 < 7$$
$$-5 < 2$$
$$-2 < 0$$
$$5 < 6$$
$$0 < 8$$
$$-3 < -1$$
$$-8 < |-8|, \text{ since } |-8| = 8 \text{ and } -8 < 8$$

The symbol ≥ means **greater than or equal to.** For example, $5 \geq 2$, $2 \geq 2$. $5 \geq 2$ is true, since 5 is greater than 2 (thus it is greater than or equal to 2). $2 \geq 2$ is true, since 2 is equal to 2 (thus it is greater than or equal to 2).

The symbol ≤ means **less than or equal to.** For example, $6 \leq 8$, $8 \leq 8$.

The symbol ≠ means **not equal to.** For example, $5 \neq 3$.

EXERCISES 10.8

*1. Determine the value of each expression.
   (a) $|-15| + |+15|$
   (b) $|(-15) + (+15)|$
   (c) $|(+3) + (-2) + (-3)|$
   (d) $|(-9) + (-5) + (-4) + (+1)|$
   (e) $|(+6) + (-4) + (-3) + (-7) + (+1)|$
   (f) $|(+5) + (-8) + (+8) + (-5)|$
   (g) $|(-6) + (+2) + (-9) + (+3) + (-7) + (+5)|$
   (h) $|(+1) + (-3) + (+7) + (-9) + (+1)|$

*2. Indicate whether each statement is true or false.
   (a) $6 > 2$ (b) $5 > -3$
   (c) $0 > -7$ (d) $-4 > -2$
   (e) $-4 > -8$ (f) $-7 < 0$
   (g) $0 < 0$ (h) $12 < -13$
   (i) $-6 \geq 6$ (j) $0 \leq -2$
   (k) $|-8| = 8$ (l) $|-3| \neq -3$
   (m) $|(-5) + (+2)| = 7$ (n) $|(-5) + (-7)| \geq 12$
   (o) $|(8) + (-12)| \neq |(0) + (+4)|$ (p) $7 + |-7| = 0$
   (q) $-7 + |-7| = 0$ (r) $|(-7) + (7)| > 0$

CHAPTER 10 REVIEW EXERCISES

*1. Carry out each operation.
   (a) $(+6) + (-3)$ (b) $(-8) + (-7)$
   (c) $(-2) + (+1)$ (d) $(+6) + (+4)$
   (e) $(-7) - (-4)$ (f) $(-9) - (+7)$
   (g) $(+5) + (-1)$ (h) $(-6) - (-10)$
   (i) $(-9)(+5)$ (j) $(-7)(-6)$
   (k) $(-6) - (-7)$ (l) $(-9) \div (+9)$
   (m) $(-54) \div (-9)$ (n) $(-7)^2$
   (o) $(-17) + (+54)$ (p) $(-17) - (+54)$
   (q) $(+100) \div (-5)$ (r) $(-1)^{50}$
   (s) $(-2)^5$ (t) $(-5) + (+2)$
   (u) $(-9)(+3)$ (v) $(-19) - (-14)$

*2. Simplify each expression.

(a) $(-9)(+2)(-1)$

(b) $+9 + (-7) - (-2)$

(c) $(-3)(-4)(+1)(-1)$

(d) $+7 - 9 + 3 + 2 - 18$

(e) $(-2)(-9) + (-2)(+1)$

(f) $(+7)(-6) + (-4)(+5)$

(g) $(+5)(-6) - (-7)(+3)$

(h) $(-1)(-1)(-1) - (-1)(-2)(-4)$

(i) $-6^2 + (-1)$

(j) $(-1)(-2)^2(+3)^2(-1)^3$

(k) $(-3)^4 + (-4)^3 + 2^3(-1)$

(l) $(-3)^2 + (-2)^3 + 3^2(-2)$

(m) $(-6)(-5) - (+4)(-6)$

(n) $-9 + (-4)(-9) - (-7)^2$

(o) $(-8)(+4) - (-7) + 1$

(p) $1 - (-3)(-2)(-1) - (-9)(-1)$

(q) $\dfrac{(-5)(-10)(-3)}{(+15)}$

(r) $\dfrac{(-3)(+8)}{(-2)} + \dfrac{(-1)}{(+1)}$

*3. Determine the value of each expression.

(a) $-9 + (-3)(+1) - (-1)^3 + (-1)^5 + (-4)$

(b) $-5 + (-2)^3 + (-5)^2 - (+8)(-1)$

(c) $(-2 + 3 - 6)^2 - (+4 - 2)^3$

(d) $(-5)(-3 + 2 - 1) + (2 - 5 + 6)(-1)^2 + (-3)$

(e) $-2 + 6 - 3 - 15 + 2 - 0 + 6 + 4 - 9 + 2 - 12$

(f) $+7 - 3 + 2 - (-3) + (-4) + (-2) - (-2)^3 + (-2)(+1)$

11

ALGEBRAIC EXPRESSIONS

11.1 INTRODUCTION

Throughout arithmetic you have worked with **constants,** numbers such as 2, -8, 13.569, 0, and $\frac{3}{4}$. In algebra we use letters to represent numbers whose values we do not know. For example, in the equation $x + 5 = 9$, the letter x is an **unknown.** The equation can be solved to determine that x is 4 in this example. Letters are also used to represent **variables.** For example, in $x + y = 10$, x and y are variables; their values can change or vary. Specifically, if x is 7 and y is 3, $x + y = 10$. Also, if x is 4 and y is 6, $x + y$ is 10. Many other values are also possible for the variables x and y in this example.

An **algebraic expression** is a combination of constants and variables created by using arithmetic operations.

Algebraic Expression	Comment
$3x$	$3x$ is $3 \cdot x$, 3 times x
$9 + x^2$	Uses exponentiation and addition
$\dfrac{8x}{c}$	Uses multiplication and division
t	A variable alone is considered to be an expression
-6	A constant alone is considered to be an expression

Technically speaking, even a constant alone or a variable alone is considered an expression. Thus 7 is an expression and x is an expression.

An expression can be thought of as a sum of one or more quantities. A **term** of an expression is any constant, variable, product, or quotient that is added to form the expression. Here are some examples.

Expression	Terms of This Expression
$3x + 8$	$3x$ and 8
$9xy^2 - 17x + \dfrac{a}{b}$	$9xy^2$ and $-17x$ and $\dfrac{a}{b}$
$\dfrac{7abc}{5x}$	$\dfrac{7abc}{5x}$

An expression consisting of just one term is called a **monomial.** An expression having two terms is called a **binomial.** A three-term expression is known as a **trinomial.** Here are some examples.

Monomials	Binomials	Trinomials
a	$a + b$	$a + b + c$
$5x^2$	$x^2 + 7$	$3x^2 - 9x + 2$
$\dfrac{2mn}{3}$	$2x^3 - y^2$	$m^2 + mn + n^2$

If we consider terms containing x, then in each term the **coefficient** of x is the number(s) multiplied by x.

Term	Coefficient of x	
$5x$	5	
$-7x$	-7	
$7xy$	$7y$	
x	1	(x is $1 \cdot x$)
$-x$	-1	($-x$ is $-1 \cdot x$)

Before continuing, we note two key observations made in the table above.

$$\boxed{\begin{array}{rll} x & \text{means} & 1 \cdot x \\ -x & \text{means} & -1 \cdot x \end{array}}$$

Coefficients play an important role when like terms of an expression are combined. Terms having the same variables and the same exponents are called **like terms.**

Like Terms	Unlike Terms
$6x$ and $7x$	$3x$ and $2x^2$
$3y^2$ and $-9y^2$	$5x$ and $4y$

Like terms can be combined by using the distributive property:

$$ba + ca = (b + c)a$$

You can see that if two terms contain the same factor, that factor can be removed from each term and the coefficients can be added.

EXAMPLE 1　　Combine $3x + 8x$.

Solution　　By the distributive property,

$$3x + 8x = (3 + 8)x$$
$$= 11x \quad \checkmark$$

EXAMPLE 2　　Combine $3x - 9x$.

Solution
$$3x - 9x = (3 - 9)x$$
$$= -6x \quad \checkmark$$

As you can see, *like terms can be added simply by adding the coefficients.* More than two like terms are combined in the same manner.

EXAMPLE 3　　Combine $5x + 8x - 12x + 2x$.

Solution
$$5x + 8x - 12x + 2x = (5 + 8 - 12 + 2)x$$
$$= 3x \quad \checkmark$$

Only like terms can be combined. *Unlike terms cannot be combined.*

EXAMPLE 4　　Combine $3x + 5y$.

Solution　　$3x$ and $5y$ are not like terms; they cannot be combined. $3x + 5y$ is the simplest form of this expression.

Sometimes when we are asked to combine terms, not all the terms are like terms. Here are two examples.

EXAMPLE 5　　Combine $5x + 8x - 7x + 4y + x + 6y$.

Solution　　The expression contains x terms and y terms. All the x terms can be combined ($5x + 8x - 7x + x = 7x$) and all the y terms can be combined ($4y + 6y = 10y$). Thus

$$5x + 8x - 7x + 4y + x + 6y = 7x + 10y \quad \checkmark$$

Note, however, that $7x$ cannot be combined with $10y$; they are not like terms.

EXAMPLE 6 Combine like terms: $3x + 5y + 7 + 4x - 3$.

Solution $3x + 5y + 7 + 4x - 3$

$$= \underline{3x + 4x} + \underline{5y} + \underline{7 - 3} \qquad \left\{ \begin{array}{l} 3x \text{ and } 4x \text{ are like terms.} \\ 7 \text{ and } -3 \text{ are like terms.} \end{array} \right.$$

$$= 7x + 5y + 4 \quad \checkmark$$

EXAMPLE 7 Combine like terms: $7x^2 + x^2 - 9x^2 + 5x^2$.

Solution All four of the terms are like terms. They are all x^2 terms. This means that all of the terms can be combined simply by adding their coefficients.

$$7x^2 + x^2 - 9x^2 + 5x^2 = 4x^2$$

EXAMPLE 8 Combine like terms: $5x + 2x - 7x + 5x^2 + x$.

Solution All are like terms except $5x^2$, which is an x^2 term rather than an x term. Combine the x terms only.

$$5x + 2x - 7x + 5x^2 + x = \underline{5x + 2x - 7x + x} + \underline{5x^2}$$
$$= (5 + 2 - 7 + 1)x + 5x^2$$
$$= 1x + 5x^2$$
$$= x + 5x^2 \quad \checkmark$$

EXERCISES 11.1

Answers to starred exercises are given in the back of the book.

*1. Combine like terms.

(a) $5x + 3x$
(b) $8y + 9y$
(c) $3m + 4m + 7m$
(d) $6y + 3y - y$
(e) $8x - 4x + x$
(f) $4x - 7x + x$
(g) $6x - x + 2x$
(h) $-8t + t + 3t - 5t$
(i) $x + x + x + x + x$
(j) $y - y + y - y + y - y$
(k) $n - n + n - n + n$
(l) $7x - 8x + 2x - 6x - x + 3x$
(m) $-2t + t + 5t - 10t$
(n) $7a - 10a + a - 4a + a$
(o) $2y - 8y + y + 3y$
(p) $t - 7t + 5t + 12t$
(q) $b - b + b - b + b$
(r) $y^2 + 7y^2 - 10y^2 - y^2$
(s) $4x^2 - 2x^2 - 6x^2 + x^2$
(t) $-3t^2 - 12t^2 - t^2 + 5t^2$

2. Combine like terms, when possible.

*(a) $3x + 5x + 4x + 7$
(b) $7x + 3 + 2x + 9$
*(c) $9x + 3x - 6y - 12x + 4y$
(d) $6y - 15 + y$
*(e) $5mn + 2mn - mn + 5m$
(f) $5x - 8x + x - y + 7y + 3$
*(g) $5x^2 + 6x - x + x^2$
(h) $6x - 13x + 1 - 4x$

*(i) $2x + 3y + 9z$ (j) $9xy - xy + 2xy - 19xy$
*(k) $5 + x + 2 - x$ (l) $3 - y + x - 3 + x - y$
*(m) $3x^2 - 7x^2 + 2x^2 - x^2$ (n) $10x^2 - x + 3x^2 + 5x$
*(o) $7y^2 - 8y + 2y^2 - y^2$ (p) $5t - 2t^2 + 6t - 15$
*(q) $1 - t^2 + t - 10t + t^2$ *(r) $ab + 6ab - 10ab + 5$
*(s) $-4a^2 + b - 6a^2 + 15b$ (t) $a^2 - b^2 + 5b^2 - 3a^2 + a^2$

*3. Indicate the coefficient of x in each term.

(a) $2x$ (b) x (c) $-5x$
(d) $-x$ (e) xy (f) $3xy$
(g) $-5xy$ (h) $-xy$ (i) $25mxy^2$
(j) wx (k) $-abx$ (l) $-tx$

4. Combine like terms.

*(a) $1.2x + 3.5x + 8.5x$ (b) $3.2y + 5.1y + 6.0y$
*(c) $9.4x^2 - 1.8x^2 + x^2$ (d) $1.2t + 4.1t + 9.9t$

*(e) $\dfrac{2}{3}x + \dfrac{3}{4}x$ (f) $\dfrac{2}{5}x + \dfrac{3}{10}x$

*(g) $\dfrac{1}{2}x + \dfrac{1}{8}x$ (h) $\dfrac{3}{4}x + \dfrac{1}{8}x$

*(i) $\dfrac{3}{5}t - \dfrac{1}{10}t$ (j) $\dfrac{7}{12}y - \dfrac{1}{3}y$

11.2 EXPONENTS, MONOMIALS, AND PARENTHESES

The concept of exponent was introduced in Chapter 1. Recall, for example, that in 4^3, 4 is the **base** and 3 is the **exponent** or **power**. The exponent specifies the number of times the base will appear in the multiplication. Here are a few examples to serve as a review.

$$4^3 = 4 \cdot 4 \cdot 4 = 64$$
$$3^2 = 3 \cdot 3 = 9$$
$$2^5 = 2 \cdot 2 \cdot 2 \cdot 2 \cdot 2 = 32$$
$$x^5 = x \cdot x \cdot x \cdot x \cdot x$$
$$x^4 = x \cdot x \cdot x \cdot x$$
$$x^1 = x$$

To continue, let's consider $x^5 \cdot x^4$, a product of two powers of x.

$$x^5 \cdot x^4 = \underline{x \cdot x \cdot x \cdot x \cdot x} \cdot \underline{x \cdot x \cdot x \cdot x}$$
$$= x \cdot x \cdot x \cdot x \cdot x \cdot x \cdot x \cdot x \cdot x$$
$$= x^9$$

Thus we see that $x^5 \cdot x^4 = x^9$.

Similarly, $m^3 \cdot m^7 = \underline{m \cdot m \cdot m} \cdot \underline{m \cdot m \cdot m \cdot m \cdot m \cdot m \cdot m} = m^{10}$.

It should be apparent that if two numbers having the *same base* are *multiplied*, then the result has the same base and the exponent is the *sum* of the two exponents. In symbols,

$$\boxed{x^a \cdot x^b = x^{a+b}}$$

EXAMPLE 9 Simplify $x^3 \cdot x^4$.

Solution
$$x^3 \cdot x^4 = x^{3+4} = x^7 \quad \checkmark$$

EXAMPLE 10 Simplify $y^2 \cdot y^9$.

Solution
$$y^2 \cdot y^9 = y^{2+9} = y^{11} \quad \checkmark$$

EXAMPLE 11 Simplify $x^5 \cdot y^9$.

Solution This term cannot be simplified because the bases x and y are different.

EXAMPLE 12 Simplify $x^4 + x^6$.

Solution The property of exponents deals with *multiplication*, not addition, and so it cannot be used to simplify this expression.

Some knowledge of exponents is needed when multiplying monomials. Such multiplication is easy, although sometimes you must rearrange the factors involved. This is illustrated in the next examples.

EXAMPLE 13 Multiply $3x \cdot 4x$.

Solution
$$\begin{aligned} 3x \cdot 4x &= 3 \cdot x \cdot 4 \cdot x \\ &= 3 \cdot 4 \cdot x \cdot x \\ &= 12 \cdot x^2 \\ &= 12x^2 \quad \checkmark \end{aligned}$$

EXAMPLE 14 Multiply $(-5y^3)(-7y^7)$.

Solution
$$\begin{aligned} (-5y^3)(-7y^7) &= (-5)(-7) \cdot y^3 \cdot y^7 \\ &= 35y^{10} \quad \checkmark \end{aligned}$$

EXAMPLE 15 Simplify $5x^2 \cdot x^{10} \cdot x^8$.

Solution
$$\begin{aligned} 5x^2 \cdot x^{10} \cdot x^8 &= 5 \cdot x^2 \cdot x^{10} \cdot x^8 \\ &= 5 \cdot x^{2+10+8} \\ &= 5 \cdot x^{20} \quad \text{or} \quad 5x^{20} \quad \checkmark \end{aligned}$$

EXAMPLE 16 Simplify $(-2a^3b^2)(5ab^7)$.

Solution
$$\begin{aligned} (-2a^3b^2)(5ab^7) &= (-2)(5) \cdot a^3 \cdot a^1 \cdot b^2 \cdot b^7 \\ &= -10a^{3+1}b^{2+7} \\ &= -10a^4b^9 \quad \checkmark \end{aligned}$$

NOTE
Note the *form* of the final result in Example 16. The constant comes first, followed by the variables in *alphabetical order*. This form is considered to be standard and will be used throughout the text. It is an easy matter for people to compare results if everyone uses the same form.

Multiplication of monomials can also occur when an expression to be simplified contains parentheses. If some of the terms you wish to combine are contained within parentheses, the parentheses must be removed before all the like terms can be combined. Note how the distributive property is used in the example that follows. Here there are two multiplications of monomial by monomial.

EXAMPLE 17   Simplify the expression $7x + 3(4 + 2x) - 9$.

Solution   Here parentheses contain a binomial consisting of a constant term (4) and an x term ($2x$). Use the distributive property to multiply out $3(4 + 2x)$.

$$7x + 3(4 + 2x) - 9 = 7x + 3 \cdot 4 + 3 \cdot 2x - 9$$
$$= 7x + 12 + 6x - 9$$
$$= 7x + 6x + 12 - 9 \qquad \text{when rearranged}$$
$$= 13x + 3 \quad \checkmark$$

after like terms are combined.

EXAMPLE 18   Simplify the expression $5x - 2(3 - 8x) + (4 - x)6$.

Solution
$$5x - 2(3 - 8x) + (4 - x)6 = 5x - 6 + 16x + 24 - 6x$$
$$= 15x + 18 \quad \checkmark$$

The expression $15x + 18$ can also be written as

$$18 + 15x \quad \checkmark$$

The order of the two terms does not matter. However, expressions of this type are *ordinarily* written with the x-term first, and *then* the constant.

There is one special case that should be considered carefully in order to avoid misunderstanding. In the next example an algebraic expression like $(7x + 5) - (4x - 3)$ will be simplified. To do this, you must know how to simplify the expression $-(4x - 3)$. Consider the minus in front of the expression in parentheses to mean "opposite of." Then since the opposite of any term is the same magnitude term with opposite sign, we have

$$-(4x - 3) = -4x + 3$$

If you prefer, $-(4x - 3)$ can be written as $-1(4x - 3)$ and multiplied out. The result is the same as obtained above.

$$-(4x - 3) = -1(4x - 3) = -4x + 3$$

In general,

> A minus sign in front of an expression in parentheses has the effect of changing the sign of each term within parentheses.

Examples:

$$-(4x - 3) = -4x + 3$$
$$-(1 - y) = -1 + y$$
$$-(a + b + c) = -a - b - c$$
$$-(-x^2 + 4x - 5) = x^2 - 4x + 5$$

EXAMPLE 19   Simplify the expression $(2x + 5) - (6x - 4)$.

Solution   A minus in front of parentheses has the effect of changing the signs of all terms within parentheses. That is, $-(6x - 4) = -6x + 4$. Thus

$$(2x + 5) - (6x - 4) = 2x + 5 - 6x + 4$$
$$= -4x + 9 \quad \checkmark \quad \text{or} \quad 9 - 4x \quad \checkmark$$

NOTE
A two-term expression such as $-4x + 9$ is often written as $9 - 4x$ in order to avoid leading with a minus sign. The form $9 - 4x$ is considered neater and it also "protects" the minus, which could be lost or misinterpreted when kept in front as in $-4x + 9$.

EXAMPLE 20   Simplify $(8x^2 + 7x - 1) - (3x^2 - 2x - 19)$.

Solution

$$(8x^2 + 7x - 1) - (3x^2 - 2x - 19)$$
$$= 8x^2 + 7x - 1 - 3x^2 + 2x + 19 \qquad \text{by boxed rule}$$
$$= 8x^2 - 3x^2 + 7x + 2x - 1 + 19 \qquad \text{regrouping}$$
$$= 5x^2 + 9x + 18 \quad \checkmark \qquad \text{combining like terms}$$

EXAMPLE 21   Simplify $(5x^2 - 8x + 1) + (2x^2 + x - 9)$.

Solution

$$(5x^2 - 8x + 1) + (2x^2 + x - 9) = 5x^2 - 8x + 1 + 2x^2 + x - 9$$
$$= 5x^2 + 2x^2 - 8x + x + 1 - 9$$
$$= 7x^2 - 7x - 8 \quad \checkmark$$

The next example involves brackets and parentheses. To avoid errors, work *from the inside* out, as explained in the example.

EXAMPLE 22   Simplify the expression $5 - 3[7 + (2x - 6)8 - 4x] + 19$.

Solution   Working from the inside out means simplifying the expression inside the brackets before multiplying by the -3. And you cannot combine the terms within the brackets until you multiply out $(2x - 6)8$. So the first step is to multiply out $(2x - 6)8$.

$$5 - 3[7 + (2x - 6)8 - 4x] + 19 = 5 - 3[7 + 16x - 48 - 4x] + 19$$

Now we can combine the like terms within the brackets. The result is

$$5 - 3[12x - 41] + 19$$

Next, multiply out by distributing the -3 to the terms within the brackets.

$$5 - 3[12x - 41] + 19 = 5 - 36x + 123 + 19$$

Finally, we can combine like terms to produce the simplest form of the expression, namely,

$$-36x + 147 \quad \checkmark$$

EXAMPLE 23   Simplify the expression $-7x - 2[3x - 4(y - 3) + 7] - y$.

Solution
$$\begin{aligned}
&-7x - 2[3x - 4(y - 3) + 7] - y \\
&= -7x - 2[3x - 4y + 12 + 7] - y && \text{after distributing } -4 \\
&= -7x - 2[3x - 4y + 19] - y && \text{after combining } +12 \text{ and } +7 \\
&= -7x - 6x + 8y - 38 - y && \text{after distributing } -2 \\
&= -13x + 7y - 38 \quad \checkmark
\end{aligned}$$

EXERCISES 11.2

*1. Simplify, if possible.
(a) $x^5 \cdot x^7$ (b) $y^9 \cdot y^4$
(c) $m^6 \cdot m^8$ (d) $t^2 \cdot t^3$
(e) $x^7 + x^4$ (f) $x^{17} \cdot x^4$
(g) $a^{20} \cdot a^{20}$ (h) $m^{10} + m^{15}$
(i) $y^{40} \cdot y^{80}$ (j) $x^{13} \cdot x^8$

2. Simplify.
*(a) $8y \cdot 3y$ (b) $2x \cdot 9x$
*(c) $5m^2 \cdot m^8$ (d) $3x^8 x^4$
*(e) $3x^2 \cdot x^5 \cdot x^4$ (f) $7x^4 \cdot x^3 \cdot x^6$
*(g) $8x^7 \cdot x^3 \cdot x$ (h) $2x \cdot x^9 \cdot x^{15}$
*(i) $(-3a^2b^3)(2ab^4)$ (j) $(-7m^7u^4)(4m^5u^3)$
*(k) $(-5x^2y^5)(-2x^8y^2)$ (l) $(-3t^4w)(-8tw^5)$
*(m) $(10u^2v)(-3u^8)$ (n) $(-a^3b^5)(-7ab^6)$

3. Simplify each expression.
*(a) $4x + 2(x + 5) - 3$ *(b) $7 + 14(7 - 3x) + 9$
*(c) $(2y - 9) - (3y + 1)$ (d) $(9x + 2) - (5x - 3)$
*(e) $(4x + 1) - (1 - 2x)$ (f) $-(2x - 7) - (x - 3)$
*(g) $3x - 4(x + 1) + x$ *(h) $9 - 3(1 - 5t) + t$

(i) $5x + 2(4x - 9) - 3$ (j) $x - (8 - x) - x - 8$
*(k) $3x + 5(6x + 1) - 2$ (l) $7x - 4(1 + 5x) - 1$
*(m) $8x - (1 - 9x) + 7$ (n) $6x - (3 - 2x) + 4$
*(o) $3(2 + x) - (3x - 2)$ *(p) $5(4 - x) - (7 - 4x)$
*(q) $5(3 - x) + 2x - 4(2x + 1)$
*(r) $4(y - 5) - 3y + 3(2y + 6)$

4. Simplify.
*(a) $(4x^2 + 5x - 2) + (3x^2 + 7x + 1)$
(b) $(6x^2 - 8x + 1) + (x^2 + 3x - 9)$
*(c) $(x^2 - 19x + 5) + (5x^2 - 3x - 4)$
*(d) $(3x^2 - 8x + 14) - (5x^2 - 2x - 6)$
*(e) $(x^2 - 13x + 7) - (x^2 + 2x - 4)$
(f) $(4x^2 + 13x - 6) - (2x^2 - 3x - 4)$

5. Simplify each expression.
*(a) $-5[x - (8 - x) + 4] - 5x$
*(b) $7 - 2[6 + (3x - 2)9 - 5x] + 23$
*(c) $9 + 4[7 + (4 - 2x) - 9x] + x$
(d) $8 - 3[3x - (4 - 2x) + 13x] - x$
*(e) $9 - 7[4x - 2(3x - 5) + 2] - 9x$
*(f) $3 - 4[3 - (x + 2)] + 4(x + 2) - 3$
(g) $6(4 - 3x) + 5 - 2[3 + 2(x + 5)]$
*(h) $4x + 7 + 3[5 + (3x + 2) - 7]$
*(i) $5x - 4 - [2 + (7x - 3) - 9]$
*(j) $7 - 8x - 3[1 - (4x + 8) + 2]$
(k) $8 + 7x - 2[7x + (1 + 5x) - 6]$
*(l) $5 - 3[3x + (6 + 4x)5 - x]$
(m) $2 - 8[9x + (7x + 2)4 - 3x]$

6. Simplify each expression.
*(a) $5x - 3[2x - 5(y - 4) + 6] - y$
(b) $7x + 4[5x - 3(4 - 3y) - 9] + 6x$
*(c) $5y - 7[2(3 - 2x) + 4y - 6] - 3y$
(d) $3y + 5[(5x - 3)7 - 3y + 2] + 2x$
*(e) $4x + 3[7(1 - 2y) + 6x - y]$
(f) $2y - 4[9(2x - 3) - 3y + x]$
*(g) $15 - 2[x - 4(3y + 7) - 12]$
(h) $12 - 5[y - 7(8 - 3x) - 2x]$

11.3 MULTIPLICATION

In this section the concept of multiplication will be extended to involve binomials and trinomials. The distributive property will be used along with the property of exponents presented in the preceding section.

EXAMPLE 24   Multiply $3x^2(7x^3 + 4)$.

Solution   Apply the distributive property.

$$3x^2(7x^3 + 4) = 3x^2 \cdot 7x^3 + 3x^2 \cdot 4$$
$$= 3 \cdot 7 \cdot x^2 \cdot x^3 + 3 \cdot 4 \cdot x^2$$
$$= 21x^5 + 12x^2 \quad \checkmark$$

EXAMPLE 25   Multiply.
(a) $(x^2 + 3)8$ (b) $(2x + 3)4x$.

Solution   The distributive property can be applied from the right as well as from the left.

(a)
$$(x^2 + 3)8 = x^2 \cdot 8 + 3 \cdot 8$$
$$= 8x^2 + 24 \quad \checkmark$$

(b)
$$(2x + 3)4x = 2x \cdot 4x + 3 \cdot 4x$$
$$= 8x^2 + 12x \quad \checkmark$$

We can handle $(x + 2)(y + 3)$ similarly if we consider $(x + 2)$ as one number and distribute it to each of y and 3. Thus

$$\boxed{(x + 2)}\,(y + 3) = (x + 2)y + (x + 2)3$$

Now distribute the y to both x and 2 and distribute the 3 to both x and 2.

$$(x + 2)y + (x + 2)3$$

$$x \cdot y + 2 \cdot y + x \cdot 3 + 2 \cdot 3$$

This can be rewritten as

$$xy + 2y + 3x + 6$$

Thus

$$(x + 2)(y + 3) = xy + 2y + 3x + 6 \quad \checkmark$$

In effect, all that we have done is multiplied x by each of y and 3 and multiplied 2 by each of y and 3.

$$(x + 2)(y + 3) = xy + 3x + 2y + 6$$

EXAMPLE 26   Multiply $(a + b)(c + d)$.

Solution
$$(a + b)(c + d) = a \cdot c + a \cdot d + b \cdot c + b \cdot d$$
$$= ac + ad + bc + bd \quad \checkmark$$

EXAMPLE 27   Multiply $(x + 2)(x + 5)$.

Solution
$$(x + 2)(x + 5) = x \cdot x + x \cdot 5 + 2 \cdot x + 2 \cdot 5$$
$$= x^2 + 5x + 2x + 10$$
$$= x^2 + 7x + 10 \quad \checkmark$$

EXAMPLE 28   Multiply $(x + 3)(x - 3)$.

Solution
$$(x + 3)(x - 3) = x \cdot x + x(-3) + 3 \cdot x + 3(-3)$$
$$= x^2 - 3x + 3x - 9$$
$$= x^2 + 0x - 9$$
$$= x^2 - 9 \quad \checkmark$$

EXAMPLE 29   Multiply $(x + y)^2$ to show that $(x + y)^2 \neq x^2 + y^2$.

Solution
$$(x + y)^2 = (x + y)(x + y)$$
$$= x \cdot x + x \cdot y + y \cdot x + y \cdot y$$
$$= x^2 + xy + yx + y^2$$
$$= x^2 + xy + xy + y^2 \qquad \text{since } yx = xy$$
$$= x^2 + 2xy + y^2 \quad \checkmark \qquad \text{since } xy + xy = 2xy$$

So you can see that

$$\boxed{(x + y)^2 = x^2 + 2xy + y^2}$$

This means that

$$\boxed{(x + y)^2 \neq x^2 + y^2}$$

A similar result, which you can show by multiplying out $(x - y)^2$ is

$$\boxed{(x - y)^2 = x^2 - 2xy + y^2}$$

As an arithmetic example, we show that $(3 + 2)^2 \neq 3^2 + 2^2$.

$$(3 + 2)^2 \stackrel{?}{=} 3^2 + 2^2$$
$$(5)^2 \stackrel{?}{=} 9 + 4$$
$$25 \neq 13$$

EXAMPLE 30   Multiply $(2x + 5)(7x - 4)$.

Solution
$$(2x + 5)(7x - 4) = 2x(7x) + 2x(-4) + 5(7x) + 5(-4)$$
$$= 14x^2 - 8x + 35x - 20$$
$$= 14x^2 + 27x - 20 \quad \checkmark$$

EXAMPLE 31 Multiply $(3x^2 + 5x + 2)(4x + 1)$.

Solution $(3x^2 + 5x + 2)(4x + 1)$

$$= \underline{3x^2}(4x + 1) + \underline{5x}(4x + 1) + \underline{2}(4x + 1)$$

$$= \underline{3x^2} \cdot 4x + \underline{3x^2} \cdot 1 + \underline{5x} \cdot 4x + \underline{5x} \cdot 1 + \underline{2} \cdot 4x + \underline{2} \cdot 1$$

$$= 12x^3 + 3x^2 + 20x^2 + 5x + 8x + 2$$

$$= 12x^3 + 23x^2 + 13x + 2 \quad \checkmark$$

EXAMPLE 32 Multiply $(5x^2 + 3x + 1)(2x^2 + 7x - 4)$.

Solution $(5x^2 + 3x + 1)(2x^2 + 7x - 4)$

$$= \underline{5x^2}(2x^2 + 7x - 4) + \underline{3x}(2x^2 + 7x - 4) + \underline{1}(2x^2 + 7x - 4)$$

$$= 10x^4 + 35x^3 - 20x^2 + 6x^3 + 21x^2 - 12x + 2x^2 + 7x - 4$$

$$= 10x^4 + 41x^3 + 3x^2 - 5x - 4 \quad \checkmark$$

EXERCISES 11.3

1. Multiply.
*(a) $5x^2(3x^4 + 6)$ (b) $7y^4(2y^6 + 1)$
*(c) $8y(7 + 3y^5)$ (d) $2m^6(m^3 + 4m)$
*(e) $x^6(5x^3 + 8x)$ (f) $n^3(n^6 + p^3)$
*(g) $6m^7(8m^9 - 16)$ (h) $6x^4(9x^7 - 15x)$
*(i) $(x^2 + 5)6$ (j) $(3x^2 + 1)4$
*(k) $(9x + 2)3x$ (l) $(5 + 6x)7x$
*(m) $(7 + 8x)x^2$ (n) $(9x + 5)x^2$
*(o) $(5 - 2y)5y$ (p) $(15 - 2t)6t$

2. Multiply; then simplify, if possible.
*(a) $(x + 3)(x + 2)$ (b) $(x + 1)(x + 4)$
*(c) $(x + 7)(x - 5)$ (d) $(x - 5)(x + 4)$
*(e) $(y - 8)(y - 2)$ (f) $(m - 4)(m - 3)$
*(g) $(c + 5)(d - 6)$ (h) $(u - 9)(v + 4)$
*(i) $(a + 2)(b + 6)$ (j) $(n + 5)(p + 3)$
*(k) $(m + 1)(m - 1)$ (l) $(y + 9)(y - 9)$
*(m) $(3 + x)(3 - x)$ (n) $(5 + m)(8 + m)$
*(o) $(2 + x)(7 - x)$ (p) $(y - 3)(y + 5)$

3. Multiply; then simplify, if possible.
*(a) $(5x + 3)(4x + 1)$ (b) $(5x + 3)(2x + 4)$
*(c) $(4x - 1)(2x + 1)$ (d) $(3x - 2)(x - 5)$
*(e) $(3x - 5)(x - 2)$ (f) $(2x + 7)(3x + 9)$
*(g) $(m + 5)(7m - 5)$ (h) $(5m + 2)(m - 6)$
*(i) $(2m + 7)(3n + 5)$ (j) $(7a - 2)(3b + 1)$
*(k) $(5 + 9x)(2 + x)$ (l) $(1 + y)(1 - 2y)$
*(m) $(6 + 3x)(2x - 4)$ (n) $(5t + 4)(t - 6)$
*(o) $(x + y)(2x + y)$ (p) $(x - y)(2x - y)$

4. Multiply, then simplify.
   *(a) $(m + n)^2$ (b) $(r + v)^2$ *(c) $(a - b)^2$ (d) $(t - w)^2$
   *(e) $(x + 7)^2$ (f) $(y + 4)^2$ *(g) $(y - 5)^2$ (h) $(x - 6)^2$

5. Multiply; then simplify, if possible.
   *(a) $(x^2 + 2x + 4)(x + 3)$ (b) $(y^2 - 6y + 5)(y - 2)$
   *(c) $(x - 6)(x^2 - 3x + 2)$ (d) $(m - 4)(m^2 - 7m - 8)$
   *(e) $(x + 2)(x^2 + 4x - 5)$ (f) $(a + b + c)(d + e)$
   *(g) $(2x^2 - 3x + 5)(x^2 - 4x - 1)$ (h) $(x^2 + 5x + 4)(x^2 + 7x + 3)$
   *(i) $(3x - 4y + 7)(5x - 8y + 2)$ (j) $(2m + 3n + 4)(m - 5n - 7)$

6. (a) Verify that $x^3 - y^3 = (x - y)(x^2 + xy + y^2)$ by multiplying out and simplifying the product on the right-hand side of the equation.
   (b) Verify that $x^3 + y^3 = (x + y)(x^2 - xy + y^2)$ by multiplying out and simplifying the product on the right-hand side of the equation.

11.4 SHORT DIVISION, EXPONENTS, AND LONG DIVISION

Short division in algebra is division of an algebraic expression by a mono-mial. As a simple example, consider the division of $10x^3$ by $2x$, which can appear in either of two ways.

$$2x\overline{)10x^3} \qquad \text{or} \qquad \frac{10x^3}{2x}$$

The second form looks like a fraction to reduce, and in fact reduction of fractions like this and others is presented in Chapter 18. In this section we will use only the first form. A division such as

$$\frac{8x^3 + 9x^2 - 7x}{x}$$

will appear as

$$x\overline{)8x^3 + 9x^2 - 7x}$$

Let us return now to consider the division $2x\overline{)10x^3}$.

EXAMPLE 33   Divide $2x\overline{)10x^3}$.

Solution   The quotient is $5x^2$, since when $5x^2$ is multiplied by $2x$ the result is $10x^3$.

$$2x\overline{)10x^3}^{\,5x^2 \;\checkmark}$$

since $5x^2 \cdot 2x = 10x^3$.

EXAMPLE 34   Divide $2x\overline{)10x^3 + 16x^2 - 6x}$.

Solution   This division can be done as three individual divisions. Divide each of the terms of the trinomial $10x^3 + 16x^2 - 6x$ by $2x$. From Example 33 we already know the first term of the quotient.

$$\frac{5x^2}{2x)\overline{10x^3 + 16x^2 - 6x}}$$

Next, divide $16x^2$ by $2x$. The result is $8x$, since $8x \cdot 2x = 16x^2$.

$$\frac{5x^2 +\ \ 8x}{2x)\overline{10x^3 + 16x^2 - 6x}}$$

Finally, divide $-6x$ by $2x$, to get -3, since $-3 \cdot 2x = -6x$.

$$\frac{5x^2 +\ \ 8x\ -\ 3}{2x)\overline{10x^3 + 16x^2 - 6x}} \quad \checkmark$$

As a check, multiply $5x^2 + 8x - 3$ by $2x$ to produce $10x^3 + 16x^2 - 6x$.

$$(2x)(5x^2 + 8x - 3) = 10x^3 + 16x^2 - 6x$$

This type of division can be used to determine another property of exponents. Consider, for example,

$$x^3)\overline{x^{10}}$$

The quotient must be x^7, since $x^7 \cdot x^3 = x^{10}$. Thus

$$\frac{x^{10}}{x^3} = x^7$$

Similarly,

$$\frac{x^9}{x^5} = x^4$$

since $x^4 \cdot x^5 = x^9$.

In each of these examples the exponent in the result could have been obtained by subtracting the denominator exponent from the numerator exponent.

$$\boxed{\frac{x^a}{x^b} = x^{a-b}}$$

For now, we must insist that a is greater than b.

EXAMPLE 35   Simplify $\dfrac{x^{15}}{x^4}$.

Solution

$$\frac{x^{15}}{x^4} = x^{15-4} = x^{11} \quad \checkmark$$

Much more will be said about exponents in Chapter 20.

When the divisor consists of two or more terms, then the division process becomes **long division.** The nature and sequence of the steps in algebraic long division are similar to those of the arithmetic long division presented in Chapter 1.

EXAMPLE 36   Divide.

$$x + 3 \overline{)\, x^3 + 2x^2 - 5x - 6}$$

Solution   Begin by dividing the *first term* of $x + 3$ into the *first term* of expression $x^3 + 2x^2 - 5x - 6$. The result of dividing x into x^3 is x^2, of course.

$$\begin{array}{r} x^2 \\ x + 3 \overline{)\, x^3 + 2x^2 - 5x - 6} \end{array}$$

Next, *multiply back* the x^2 by the divisor $x + 3$.

$$\begin{array}{r} x^2 \\ x + 3 \overline{)\, x^3 + 2x^2 - 5x - 6} \\ x^3 + 3x^2 \end{array}$$

Next, *subtract* $x^3 + 3x^2$ from $x^3 + 2x^2$. This is done by changing the signs of $x^3 + 3x^2$ (to $-x^3 - 3x^2$) and adding.

$$\begin{array}{r} x^2 \\ x + 3 \overline{)\, x^3 + 2x^2 - 5x - 6} \\ \underline{-x^3 - 3x^2} \\ -x^2 \end{array}$$

Now that the subtraction has been completed, *bring down* the next term $(-5x)$.

$$\begin{array}{r} x^2 \\ x + 3 \overline{)\, x^3 + 2x^2 - 5x - 6} \\ \underline{-x^3 - 3x^2} \\ -x^2 - 5x \end{array}$$

Continue the process by dividing $x + 3$ into the new two-term expression $-x^2 - 5x$. This is done by dividing x (the first term of $x + 3$) into $-x^2$ (the first term of $-x^2 - 5x$). The result is $-x$.

$$\begin{array}{r} x^2 - x \\ x + 3 \overline{)\, x^3 + 2x^2 - 5x - 6} \\ \underline{-x^3 - 3x^2} \\ -x^2 - 5x \end{array}$$

Now, multiply back $-x$ by $x + 3$. The result is $-x^2 - 3x$ and is shown below.

$$\begin{array}{r} x^2 - x \\ x + 3 \overline{)\, x^3 + 2x^2 - 5x - 6} \\ \underline{-x^3 - 3x^2} \\ -x^2 - 5x \\ -x^2 - 3x \end{array}$$

Next, subtract $-x^2 - 3x$ by changing its signs ($-x^2 - 3x$ becomes $x^2 + 3x$) and adding it. Then bring down the next term (-6).

$$
\begin{array}{r}
x^2 - x \\
x + 3 \overline{) \ x^3 + 2x^2 - 5x - 6} \\
\underline{-x^3 - 3x^2} \\
-x^2 - 5x \\
\underline{x^2 + 3x} \\
-2x - 6
\end{array}
$$

Next, divide the x (of $x + 3$) into the $-2x$ (of $-2x - 6$) and get -2. Multiply it back and note that there is no remainder in the subtraction. The result of the division is $x^2 - x - 2$ exactly.

$$
\begin{array}{r}
x^2 - x - 2 \\
x + 3 \overline{) \ x^3 + 2x^2 - 5x - 6} \\
\underline{-x^3 - 3x^2} \\
-x^2 - 5x \\
\underline{x^2 + 3x} \\
-2x - 6 \\
\underline{-2x - 6} \\
0
\end{array}
$$

As a check, you could multiply out $(x + 3)(x^2 - x - 2)$ to show that it is indeed equal to $x^3 + 2x^2 - 5x - 6$.

EXAMPLE 37   Divide.

$$ x - 2 \overline{) \ 2x^4 + x^3 - 23x + 9} $$

Solution   Note that the x^2 term is missing. To avoid confusion later, insert the term $0x^2$ where the x^2 term would be. It will serve as a placeholder.

$$ x - 2 \overline{) \ 2x^4 + x^3 + 0x^2 - 23x + 9} $$

Now perform the division.

$$
\begin{array}{r}
2x^3 + 5x^2 + 10x - 3 \\
x - 2 \overline{) \ 2x^4 + \ x^3 + \ 0x^2 - 23x + 9} \\
\underline{2x^4 - 4x^3} \\
5x^3 + \ 0x^2 \\
\underline{5x^3 - 10x^2} \\
10x^2 - 23x \\
\underline{10x^2 - 20x} \\
-3x + 9 \\
\underline{-3x + 6} \\
3
\end{array}
$$

The remainder is 3. The result of the division can be written

$$2x^3 + 5x^2 + 10x - 3, \text{ R: } 3 \quad \checkmark$$

or

$$2x^3 + 5x^2 + 10x - 3 + \frac{3}{x - 2} \quad \checkmark$$

As a check, you should show that

$$(x - 2)(2x^3 + 5x^2 + 10x - 3) + 3 = 2x^4 + x^3 - 23x + 9$$

In Example 37, $0x^2$ was inserted as a placeholder because no x^2 term was present in the dividend expression $2x^4 + x^3 - 23x + 9$. Here are two more examples of how placeholders should be inserted in dividends.

$$\overline{)x^3 + 3x^2 - 2} \quad \longrightarrow \quad \overline{)x^3 + 3x^2 + 0x - 2}$$

$$\overline{)x^3 + 5} \quad \longrightarrow \quad \overline{)x^3 + 0x^2 + 0x + 5}$$

EXERCISES 11.4

1. Divide (short division).

   *(a) $x\overline{)x^3 + 6x^2 + x}$

   (b) $m\overline{)m^3 - 7m^2 + 3m}$

   *(c) $y\overline{)3y^3 - 2y^2 - y}$

   (d) $t\overline{)7t^4 + 5t^3 - 2t^2 - t}$

   *(e) $3x\overline{)12x^4 + 3x^2 + 6x}$

   (f) $2y\overline{)8y^3 + 4y^2 + 2y}$

   *(g) $5t\overline{)20t^5 - 10t^4 - 5t^2 + 15t}$

   (h) $3x\overline{)9x^5 - 3x^4 - 6x^2 + 15x}$

   *(i) $2x^2\overline{)6x^4 + 18x^3 - 20x^2}$

   (j) $3x^2\overline{)12x^4 - 9x^3 + 3x^2}$

*2. Simplify.

   (a) $\dfrac{x^{20}}{x^7}$
   (b) $\dfrac{x^{14}}{x^8}$
   (c) $\dfrac{y^6}{y^4}$
   (d) $\dfrac{m^{17}}{m^9}$
   (e) $\dfrac{a^{12}}{a^5}$

   (f) $\dfrac{x^{11}}{x^8}$
   (g) $\dfrac{b^8}{b^3}$
   (h) $\dfrac{u^{16}}{u^7}$
   (i) $\dfrac{n^{19}}{n^{11}}$
   (j) $\dfrac{t^{18}}{t^{10}}$

3. Divide (long division).

   *(a) $x + 1\overline{)\, x^2 + 9x + 8}$

   (b) $y + 3\overline{)\, y^2 - y - 12}$

   *(c) $x - 3\overline{)\, x^2 + 2x - 15}$

   (d) $x - 4\overline{)\, x^2 - 13x + 36}$

   *(e) $x - 1\overline{)\, x^3 + 4x^2 - 8x + 3}$

   (f) $x - 2\overline{)\, x^3 - 3x^2 + 4x - 4}$

   *(g) $y + 5\overline{)\, 7y^3 + 35y^2 - 19y - 95}$

   (h) $x + 2\overline{)\, x^3 + 7x^2 + 7x - 6}$

*(i) $2x + 1\overline{)\ 8x^2 - 2x - 3}$

(j) $3n - 2\overline{)\ 6n^2 - 13n + 6}$

4. Divide (long division). Each result will have a *remainder*.

*(a) $x - 2\overline{)\ x^3 + 5x^2 - 6x + 5}$

(b) $x - 1\overline{)\ x^3 + 2x^2 + 6x - 7}$

*(c) $x + 3\overline{)\ x^3 + 7x^2 + 2x - 4}$

(d) $x + 4\overline{)\ x^3 + 5x^2 - x + 3}$

*(e) $x - 5\overline{)\ x^3 - 3x^2 + 7x + 2}$

(f) $x - 3\overline{)\ x^3 - 4x^2 - 8x + 4}$

5. Divide. Be sure to use placeholders for missing terms (long division).

*(a) $x - 2\overline{)\ x^3 - x^2 - 4}$ (b) $x - 1\overline{)\ x^3 + 5x + 17}$

*(c) $x + 1\overline{)\ 3x^4 + x^3 + 7x + 7}$ (d) $x - 2\overline{)\ 5x^4 - x^3 - 36x + 7}$

*(e) $x + 1\overline{)\ x^3 + 1}$ (f) $x - 1\overline{)\ x^3 - 1}$

*(g) $x - 1\overline{)\ x^4 - 1}$ (h) $x - 1\overline{)\ x^5 - 1}$

6. Perform each long division.
   *(a) Divide $x^2 + 3x + 4$ by $x - 2$.
   (b) Divide $x^2 + 17x + 60$ by $x + 5$.
   *(c) Divide $x^3 + 5x^2 - 7x + 24$ by $x - 5$.
   (d) Divide $x^3 + 8x^2 - 4x - 32$ by $x + 8$.
   *(e) Divide $5x^3 - 19x^2 - 4x$ by $x - 4$.
   (f) Divide $x^3 - 4x^2 - 31x + 70$ by $x - 7$.
   *(g) Divide $x^3 + 3x^2 - 13x + 1$ by $x - 2$.

CHAPTER 11 REVIEW EXERCISES

1. Combine like terms, when possible.
   *(a) $7x - x + 4x - 9x + x - 3x$
   *(b) $3x + 2y - 8x - 9y + x - y + 4y$
   *(c) $9x - 1 + x - 15 + 3x + 4$
   (d) $7r + 6 - 4w + r - w + 18 - 64$

2. Simplify.

   *(a) $x^{10} \cdot x^7$ (b) $m^{16} \cdot m^7$ *(c) $\dfrac{x^{15}}{x^8}$ (d) $\dfrac{n^9}{n^4}$

   *(e) $7x^8 \cdot x^4 \cdot x$ *(f) $3u^5 \cdot u^6 \cdot u^3$

*3. Simplify each expression.
   (a) $3x - 4(x + 2) - 9$
   (b) $8 - (1 - 4x) + 5x$
   (c) $(x^2 + 6x + 6) - (x^2 - 5)$
   (d) $6 - 2[2 + (3x - 1)7 + x] - 3x$

4. Multiply; then simplify when possible.

$*$(a) $3x^2(2x^3 + 7)$ $*$(b) $4x(x^5 + 9x)$
$*$(c) $(m + 3)(m + 7)$ (d) $(m - 2)(m - 5)$
$*$(e) $(x + 10)^2$ (f) $(p + q)^2$
$*$(g) $(3y - 5)(3y - 1)$ (h) $(5m - 3)(m + 2)$
$*$(i) $(x - 3)(x^2 + 8x - 9)$ $*$(j) $(5x^2 - 3x - 1)(2x + 7)$

5. Perform each short division.

$*$(a) $4x\overline{)8x^3 - 12x^2 - 4x}$ (b) $3m\overline{)15m^3 + 9m^2 - 6m}$

6. Perform each long division.

$*$(a) $x - 1\overline{)\, x^2 - 11x + 10}$

$*$(b) $x - 4\overline{)\, 3x^3 - 10x^2 - 3x - 23}$

$*$(c) $x - 2\overline{)\, x^3 - 8}$

(d) $x + 2\overline{)\, x^3 - 8}$
$*$(e) Divide $x + 3$ into $x^3 + 5x^2 + 4x + 6$.
(f) Divide $x - 2$ into $x^3 - x^2 + 3x - 4$.

CUMULATIVE REVIEW EXERCISES (CHAPTERS 10–11)

$*$**1.** Simplify each expression completely.
(a) $(-7) - (+4)$ (b) $-8 + 3 - 17 + 3 + 1 - 12$
(c) $(-5)(-2) - (-3)(+5)$ (d) $(-2)(-1)(-2)(+4)$
(e) $(-2)^3 - (-2)^4$ (f) $\dfrac{-15}{+15} - \dfrac{18}{-3}$

$*$**2.** Evaluate the expression $x^2 - 3x + 10$ for the given value of x.
(a) $x = 5$ (b) $x = -5$ (c) $x = 0$

$*$**3.** Combine like terms, when possible.
(a) $3x - 19x + x + 15x - x$ (b) $xy + 10xy - 2xy + 4x + 2y$

$*$**4.** Indicate the coefficient of x in each term.

(a) $-7x$ (b) x (c) $\dfrac{2x}{7}$

$*$**5.** Simplify, if possible.
(a) $4x \cdot 3x \cdot x$ (b) $5x^7 \cdot 6x^9$
(c) $3x^2 + 7x^3$ (d) $(-3x^4y)(-xy)$

$*$**6.** Simplify each expression.
(a) $5x - (2 - 7x) + 5$ (b) $3(x + 4) + 8(3x - 4)$

$*$**7.** Multiply; then simplify, if possible.
(a) $(2y + 1)(y - 6)$ (b) $(x - 9)^2$

$*$**8.** Simplify.
(a) $\dfrac{x^{15}}{x^{12}}$ (b) $\dfrac{y^{19}}{y^6}$

$*$**9.** Divide.

$$x - 4\overline{)\, x^3 - 3x^2 - 10x + 19}$$

12

LINEAR EQUATIONS

12.1 INTRODUCTION

An **equation** is a statement of equality between two quantities. The statement $4 = 3 + 1$ is a very simple equation. In algebra, equations usually contain variables or unknowns as well as constants. In this chapter we shall be concerned with **linear equations,** that is, equations in which the highest power of the variable or unknown is the first power. Here are some examples of linear equations in x:

$$x + 5 = 17$$
$$9x = 14$$
$$15x - 7 = 3x - 2$$
$$4(x - 3) + 15 = 31$$

A **solution** of an equation is a number which when substituted for the unknown will make the equation a true statement. For example, in the equation $x + 5 = 12$, the number 7 is a solution, since when 7 is substituted for x the equation becomes the true statement $12 = 12$. Sometimes a solution is referred to as a **root** of the equation.

To solve linear equations, you must make use of your knowledge of signed numbers (Chapter 10) and simplification of expressions (Chapter 11). Our approach to solving linear equations begins with the use of opposites.

12.2 ADDING OPPOSITES

The equation $x + 2 = 7$ can be solved just by looking at it to determine the value of x that will make $x + 2$ equal to 7. The solution is 5, since when x is replaced by 5 the two sides of the equation will be the same. Both sides will equal 7.

Although you can sometimes "see" the solution, as in $x + 2 = 7$, very often this is not the case, especially in more complicated equations presented later in the chapter. Accordingly, a plan for solving equations is desirable. The method presented next is based on the following idea.

> If the same number is added to both sides
> of an equation, the equality is maintained.

For example, in the simple equation $35 = 35$, if 6 is added to each side, the equality is maintained.

$$\begin{array}{rcr} 35 &=& 35 \\ +6 && +6 \\ \hline 41 &=& 41 \end{array}$$

Of course, if you add a number (other than zero) to only one side of an equation, the equality will not be maintained.

The equation $x + 2 = 7$ can be solved by adding -2 to both sides, as shown next.

$$\begin{array}{rcr} x + 2 &=& 7 \\ -2 && -2 \\ \hline x + 0 &=& 5 \end{array}$$

or $$x = 5 \ \checkmark$$

It was decided to add -2 to both sides because -2 is the opposite of the $+2$ that appeared next to (added to) the x. Because $(+2) + (-2) = 0$, the $+2$ no longer appears on the left side, thus x is left by itself.

> To eliminate a term from one side
> of an equation, always add the opposite
> of the term to both sides of the equation.

EXAMPLE 1   Solve for m: $m - 13 = 6$.

Solution   Add $+13$ to both sides in order to leave m by itself. We chose $+13$ because $+13$ is the opposite of -13; that is, $+13$ and -13 add to zero, thereby leaving m alone.

$$\begin{array}{rcr} m - 13 &=& 6 \\ +13 && +13 \\ \hline m &=& 19 \ \checkmark \end{array}$$

The solution is 19.

EXERCISES 12.2

Answers to starred exercises are given in the back of the book.

*1. Solve each linear equation.

    (a) $x + 1 = 6$ (b) $x + 5 = 8$ (c) $x - 7 = 2$
    (d) $x - 2 = 15$ (e) $m + 5 = 2$ (f) $y + 8 = 1$
    (g) $r + 2 = -7$ (h) $x - 2 = 158$ (i) $x - 1 = -1$
    (j) $m - 7 = -5$ (k) $c + 6 = 0$ (l) $x - 3 = -9$
    (m) $x - 9 = 0$ (n) $x + 9 = -17$ (o) $y + 4 = 4$
    (p) $y - 5 = 0$

2. Solve each linear equation.

    *(a) $x - \dfrac{1}{8} = \dfrac{1}{4}$ *(b) $y + \dfrac{2}{3} = \dfrac{3}{4}$

    *(c) $x + 9.6 = 15.2$ *(d) $t - 7.4 = -5.8$

    *(e) $x - \dfrac{3}{8} = \dfrac{3}{2}$ (f) $x - \dfrac{2}{3} = \dfrac{11}{12}$

    *(g) $x + \dfrac{1}{3} = \dfrac{1}{3}$ (h) $x + \dfrac{5}{8} = \dfrac{5}{8}$

    *(i) $x - 1.4 = -2.6$ (j) $n - 9.2 = -15.7$
    *(k) $y + 8.4 = -5.7$ (l) $y + 17.6 = -11.9$

12.3 DIVIDING BY THE COEFFICIENT

The equation $3x = 60$ can be solved simply by looking at it to determine the value of x that will make $3x$ equal to 60. The solution is 20, since when x is replaced by 20 the two sides of the equation will be the same. Both sides will equal 60.

You cannot always "see" the solution, as you can in $3x = 60$. This is especially true with more complicated equations. In view of this, a method for solving equations of this type is presented here. The method is based on the following idea.

> If both sides of an equation are divided
> by the same number, the equality is maintained.

For example, in the simple equation $40 = 40$, if both sides are divided by 8, the equality is maintained.

$$\frac{40}{8} = \frac{40}{8}$$
$$5 = 5$$

Of course, if you divide only one side of an equation by a number (other than 1), the equality will not be maintained.

The equation $3x = 60$ can be solved by dividing both sides by 3, as shown next.

$$3x = 60$$

$$\frac{3x}{3} = \frac{60}{3}$$

$$1x = 20$$

$$x = 20$$

The idea is to change the coefficient of the unknown to 1. In this case the coefficient of x was 3, so we divided both sides by 3. Dividing $3x$ by 3 produces $1x$, or x, as desired.

> To change the coefficient of the unknown to 1, divide both sides of the equation by the coefficient of the unknown.

EXAMPLE 2   Solve for x: $4x = 20$.

Solution   The coefficient of x is 4, so divide both sides by 4. The result is

$$4x = 20$$

$$\frac{4x}{4} = \frac{20}{4}$$

$$x = 5 \quad \checkmark$$

The solution is 5.

EXAMPLE 3   Solve for x: $-x = 16$.

Solution $-x$ is the same as $-1x$, so the coefficient of x is -1 in this case. This means we will divide both sides of the equation by -1.

$$-x = 16$$

becomes
$$\frac{-x}{-1} = \frac{16}{-1}$$

or
$$x = -16 \quad \checkmark$$

EXAMPLE 4   Solve $-3y = 7$.

Solution   The coefficient of y is -3, so divide both sides by -3. The result is

$$\frac{-3y}{-3} = \frac{7}{-3}$$

$$y = -\frac{7}{3} \quad \checkmark$$

NOTE
In algebra it is usually acceptable to leave the solution as an improper fraction, and this will be done throughout. However, if the fraction can be reduced, it should be. For example, 8/6 should be reduced to 4/3, and 4/3 is an acceptable form.

EXAMPLE 5   Solve $9x = 0$.

Solution   Divide both sides by 9, the coefficient of x.

$$\frac{9x}{9} = \frac{0}{9}$$
$$x = 0 \quad \checkmark$$

EXAMPLE 6   Solve $2.6x = 88.4$.

Solution   Divide both sides of the equation by 2.6, the coefficient of x. Then

$$2.6x = 88.4$$

becomes

$$\frac{2.6x}{2.6} = \frac{88.4}{2.6}$$

or

$$x = 34 \quad \checkmark$$

EXERCISES 12.3

*1. Solve each linear equation.
   (a) $7x = 21$ (b) $5x = 30$
   (c) $3m = 144$ (d) $9y = 189$
   (e) $4x = 17$ (f) $8x = 3$
   (g) $5t = 2$ (h) $3x = 19$
   (i) $7x = -35$ (j) $4p = -56$
   (k) $-3x = 36$ (l) $-7x = 42$
   (m) $-5c = 17$ (n) $-8x = 5$
   (o) $-6x = -60$ (p) $-2n = -184$
   (q) $-5x = -9$ (r) $-6x = -1$
   (s) $-x = 19$ (t) $-y = -7$
   (u) $7x = 0$ (v) $-4w = 0$
   (w) $1.9x = 38$ (x) $3.2t = 72.32$
   (y) $-2.1y = 1050$ (z) $-9.1x = -59.15$

2. Solve each linear equation.
   *(a) $4x = 65$ (b) $3h = -17$ *(c) $-x = 16$
   (d) $19x = 0$ *(e) $8x = -63$ (f) $-5x = -120$
   *(g) $-65y = -17$ (h) $-m = -93$ *(i) $8x = 0$
   (j) $10t = -160$

3. Solve each linear equation.
   *(a) $x + 13 = -7$ (b) $13x = -7$

*(c) $-4t = 1$ (d) $m - 5 = -1$
*(e) $2m = 9$ (f) $y - 8 = 0$
*(g) $y - 3 = -3$ (h) $2y = 0$
*(i) $-6p = -37$ (j) $p - 6 = -37$

12.4 MULTIPLYING BY RECIPROCALS

As you have seen in Section 12.3, equations such as $3x = 19$, $1.2x = 72$, and $-5x = -16$ are solved by dividing both sides by the coefficient of x. However, when the coefficient of the unknown is a fraction or if the constant on the other side is a fraction, then another approach is recommended.

To prepare for the alternative approach, we need to know about reciprocals. The **reciprocal** of a fraction is the fraction created by interchanging the numerator and denominator. Here is a table showing several examples.

Number	Reciprocal
$\dfrac{3}{7}$	$\dfrac{7}{3}$
$\dfrac{7}{3}$	$\dfrac{3}{7}$
$\dfrac{1}{9}$	9
4	$\dfrac{1}{4}$
-2	$-\dfrac{1}{2}$

The result of multiplying a number by its reciprocal is always 1. Here are some examples.

$$\frac{3}{7} \cdot \frac{7}{3} = 1$$

$$\frac{1}{9} \cdot 9 = 1$$

$$(-2)\left(-\frac{1}{2}\right) = 1$$

In general:

$$\boxed{(\text{number}) \cdot (\text{its reciprocal}) = 1}$$

The method presented next is based on the use of reciprocals and the following idea.

> If both sides of an equation are multiplied
> by the same number, the equality is maintained.

Specifically, we will multiply both sides of the equation by the reciprocal of the coefficient of the unknown. As a result, the coefficient of the unknown will become 1.

> When fractions are present, to change the
> coefficient of the unknown to 1, multiply
> both sides of the equation by the reciprocal
> of the coefficient of the unknown.

EXAMPLE 7   Solve $\frac{1}{3}x = 7$.

Solution   Multiply both sides of the equation by 3, the reciprocal of the coefficient of x. This produces

$$3 \cdot \frac{1}{3}x = 3 \cdot 7$$

$$x = 21 \quad \checkmark$$

In some cases it is more difficult to determine the coefficient of the unknown. Here are some examples.

Term	Coefficient of x	
$\dfrac{3x}{8}$	$\dfrac{3}{8}$	$\left(\text{since } \dfrac{3x}{8} \text{ is } \dfrac{3}{8}x\right)$
$\dfrac{x}{4}$	$\dfrac{1}{4}$	$\left(\text{since } \dfrac{x}{4} \text{ is } \dfrac{1x}{4} \text{ or } \dfrac{1}{4}x\right)$
$-\dfrac{7x}{3}$	$-\dfrac{7}{3}$	

EXAMPLE 8   Solve $-\dfrac{x}{5} = 13$.

Solution   The coefficient of the unknown is $-\frac{1}{5}$, so multiply both sides of the equation by -5, the reciprocal of the coefficient. The result is

$$(-5)\left(-\frac{x}{5}\right) = (-5)(13)$$

or
$$x = -65 \quad \checkmark$$

EXAMPLE 9 Solve $\dfrac{4x}{7} = 9$.

Solution Multiply both sides by $\frac{7}{4}$, the reciprocal of the coefficient of x. This produces

$$\frac{7}{4} \cdot \frac{4x}{7} = \frac{7}{4} \cdot 9$$

or

$$x = \frac{63}{4} \quad \checkmark$$

EXAMPLE 10 Solve $3y = \dfrac{4}{11}$.

Solution Since there is a fraction in the equation, multiply both sides by the reciprocal of the coefficient of y, namely, $\frac{1}{3}$.

$$\frac{1}{3} \cdot 3y = \frac{1}{3} \cdot \frac{4}{11}$$

$$y = \frac{4}{33} \quad \checkmark$$

EXERCISES 12.4

*1. Solve each linear equation.

(a) $\dfrac{1}{2}x = 7$ (b) $\dfrac{3}{4}y = 7$ (c) $\dfrac{n}{3} = 5$

(d) $-\dfrac{x}{6} = 4$ (e) $-\dfrac{5}{7}x = \dfrac{9}{14}$ (f) $\dfrac{5x}{4} = 2$

(g) $\dfrac{2m}{7} = -8$ (h) $-\dfrac{2m}{9} = -\dfrac{11}{5}$ (i) $3p = \dfrac{4}{5}$

(j) $-7t = -\dfrac{11}{13}$ (k) $\dfrac{3x}{7} = \dfrac{11}{4}$ (l) $\dfrac{5x}{3} = \dfrac{2}{7}$

2. Solve each linear equation.

*(a) $\dfrac{2x}{9} = 5$ (b) $7t = \dfrac{11}{4}$ *(c) $-\dfrac{m}{7} = -2$

(d) $\dfrac{5y}{9} = \dfrac{3}{2}$ *(e) $6y = -\dfrac{13}{7}$ (f) $\dfrac{x}{2} = -9$

*(g) $-\dfrac{3}{5}x = \dfrac{7}{4}$ (h) $-x = \dfrac{2}{7}$ *(i) $\dfrac{x}{3} = \dfrac{2}{5}$

3. Solve each linear equation.

*(a) $7x = -3$ (b) $-3x = 17$ *(c) $\dfrac{m}{9} = -6$

(d) $5x = 0$ *(e) $-4x = \dfrac{27}{11}$ (f) $\dfrac{3y}{14} = 5$

$$*(g) \quad -t = \frac{6}{5} \qquad (h) \quad -f = -13 \qquad *(i) \quad -\frac{1}{3}x = \frac{9}{7}$$

$$(j) \quad -\frac{y}{3} = -17$$

12.5 COMBINING THE METHODS

To solve the equation $3x - 1 = -8$ requires addition of an opposite (add $+1$ to both sides) and then division by the coefficient of x (divide both sides by 3).

$$
\begin{array}{rl}
3x - 1 = -8 \\
\underline{+1 \quad +1} \\
3x \quad\;\; = -7
\end{array}
\qquad
\left\{
\begin{array}{l}
\text{First, add } +1 \\
\text{to both sides.}
\end{array}
\right.
$$

$$
\frac{3x}{3} = \frac{-7}{3}
\qquad
\left\{
\begin{array}{l}
\text{Second, divide} \\
\text{both sides by 3.}
\end{array}
\right.
$$

$$x = -\frac{7}{3} \quad \checkmark$$

In general,

> To solve an equation in which you must both
> *add an opposite* and *divide by the coefficient,*
> it is advisable to add the opposite *first* and
> then divide by the coefficient.

By the way, if you happen to divide first (before adding the opposite) you will usually introduce fractions immediately and then have to add fractions before obtaining a solution. Consequently, we will follow the boxed suggestion above.

EXAMPLE 11   Solve $5n + 3 = 12$.

Solution

$$
\begin{array}{rl}
5n + 3 = 12 \\
\underline{-3 \quad -3} \\
5n \quad\;\; = 9
\end{array}
\qquad
\left\{
\begin{array}{l}
\text{adding } -3 \text{ to both sides} \\
\text{to get } 5n = \cdots
\end{array}
\right.
$$

$$
\frac{5n}{5} = \frac{9}{5}
\qquad
\left\{
\begin{array}{l}
\text{dividing both sides} \\
\text{by 5 to get } n = \cdots
\end{array}
\right.
$$

$$n = \frac{9}{5} \quad \checkmark$$

EXAMPLE 12   Solve $-2x + 9 = -5$.

Solution

$$-2x + 9 = -5$$

$$\underline{ -9 -9}$$

$$-2x = -14$$

$\begin{cases} \text{adding } -9 \text{ to both sides} \\ \text{to get } -2x = \cdots \end{cases}$

$$\frac{-2x}{-2} = \frac{-14}{-2}$$

$\begin{cases} \text{dividing both sides by} \\ -2 \text{ to get } x = \cdots \end{cases}$

$$x = 7 \quad \checkmark$$

EXAMPLE 13   Solve $5 - 3y = 0$.

Solution   First, add -5 to both sides, in order to get the y term alone.

$$5 - 3y = 0$$

$$\underline{-5 -5}$$

$$-3y = -5$$

Next, divide both sides by -3, the coefficient of y.

$$\frac{-3y}{-3} = \frac{-5}{-3}$$

$$y = \frac{5}{3} \quad \checkmark$$

Remember that when fractions are present, the coefficient of the unknown can be changed to 1 by multiplying both sides by the reciprocal of the coefficient. This means that with an equation such as $\frac{4}{3}t - 1 = -12$ we first add $+1$ to both sides and then multiply both sides by $\frac{3}{4}$, the reciprocal of $\frac{4}{3}$.

EXAMPLE 14   Solve $\frac{4}{3}t - 1 = -12$.

Solution

$$\frac{4}{3}t - 1 = -12$$

$$\underline{\phantom{\frac{4}{3}t} +1 = +1}$$

$$\frac{4}{3}t = -11$$

adding $+1$ to both sides

$$\frac{3}{4} \cdot \frac{4}{3}t = \frac{3}{4}(-11)$$

multiplying both sides by $\frac{3}{4}$

$$t = -\frac{33}{4} \quad \checkmark$$

In a linear equation, *the unknown can appear on the right side rather than on the left side of the equals sign.* While you may prefer to have the unknown on the left side, there is nothing wrong with it being on the right side instead. In reality it does happen, so you should get used to it. Here are some examples of equations where the unknown is on the right side.

$$19 = 2x + 3$$
$$15 = 4x$$
$$8 = x - 5$$

As an example, $19 = 2x + 3$ is solved next.

EXAMPLE 15   Solve $19 = 2x + 3$.

Solution   First, add -3 to both sides to get the x term by itself.

$$\begin{array}{r} 19 = 2x + 3 \\ -3 -3 \\ \hline 16 = 2x \end{array}$$

Next, divide both sides by 2, the coefficient of x.

$$\frac{16}{2} = \frac{2x}{2}$$
$$8 = x \quad \checkmark$$

EXERCISES 12.5

*1. Solve each linear equation.
   (a) $3x - 4 = 6$ (b) $2x + 1 = 5$
   (c) $7m - 3 = -18$ (d) $5x + 2 = -3$
   (e) $7x + 3 = 0$ (f) $4y + 1 = 1$
   (g) $-2x - 7 = -5$ (h) $-7x - 3 = 41$

   (i) $\frac{1}{2}x + 6 = 9$ (j) $-\frac{x}{5} + 3 = 18$

   (k) $5 + 2x = 17$ (l) $9 - 7x = 4$

   (m) $\frac{17}{2}n + 1 = 35$ (n) $\frac{3m}{5} - 2 = 1$

   (o) $-\frac{3}{5}x - 5 = 16$ (p) $-\frac{2x}{3} + 6 = -19$

*2. Solve each linear equation.
   (a) $25 = 3x$ (b) $13 = t - 7$ (c) $15 = 3 + 7x$
   (d) $-8 = 3x - 10$ (e) $0 = 1 - 15y$ (f) $7 = 8x + 7$
   (g) $0 = 10t + 3$ (h) $2 = 15x + 6$ (i) $-3 = -4y - 11$

 3. Solve each linear equation.
   *(a) $\frac{x}{3} = 13$ (b) $5x = -\frac{1}{7}$

   *(c) $5 - 2m = 17$ (d) $9 + 6x = -14$

$$*(e)\ 3 = 1 - \frac{1}{2}x \qquad\qquad (f)\ \frac{4}{5}x + 3 = -22$$

$$*(g)\ 6y - \frac{1}{3} = \frac{1}{6} \qquad\qquad (h)\ 5w + \frac{1}{2} = \frac{3}{4}$$

$$*(i)\ 1.6x + 3 = 9.4 \qquad (j)\ 1.3y - 8.6 = -3.4$$

$$*(k)\ -7.8p + 9.5 = -6.1 \qquad (l)\ -2.3t + 1.9 = -2.7$$

12.6 OTHER LINEAR EQUATIONS

This section presents several examples of more complicated linear equations along with step-by-step solutions. In each instance the equations are simplified to resemble those of previous sections. Here are guidelines we will use in the seven examples presented in this section.

To solve complicated linear equations, follow these steps—in sequence.

1. If the unknown is within parentheses, multiply out to free the unknown (see Examples 20–22).
2. Combine like terms (Examples 16, 17, 19, 21, 22).
3. If the unknown appears on both sides of the equation, use an opposite to eliminate it from one of the sides (Examples 18, 19, 21).
4. At this point the equation has been simplified to resemble those solved in earlier sections, so proceed now as you have in previous sections (Examples 16–22).

EXAMPLE 16   Solve $3x + 4x - 5x = 6 + 2 - 3$.

Solution   There are no parentheses, so begin by combining like terms. The terms on the left combine to $2x$; the terms on the right combine to 5. Thus, the equation becomes

$$2x = 5$$

Next,

$$x = \frac{5}{2}\ \checkmark$$

after dividing both sides by 2.

EXAMPLE 17   Solve $7n + 2 - 3n + 4 = 6 - 4$.

Solution   Begin by combining like terms. The result is

$$4n + 6 = 2$$

This type of equation was solved in Section 12.5. First, add -6 to both sides.

$$4n + 6 = 2$$
$$\underline{ -6 \quad -6}$$
$$4n = -4$$

Next, divide both sides by 4, the coefficient of the unknown.

$$\frac{4n}{4} = \frac{-4}{4}$$

or
$$n = -1 \quad \checkmark$$

EXAMPLE 18   Solve $8x + 1 = x - 4$.

Solution   There are no parentheses and there are no like terms to be combined. However, the unknown does appear on both sides. So (step 3) add $-x$ to both sides to eliminate the x term from the right side.

$$8x + 1 = x - 4$$
$$\underline{-x -x }$$
$$7x + 1 = -4$$

Continuing, we add -1 to both sides to get the x term alone.

$$7x + 1 = -4$$
$$\underline{ -1 \quad -1}$$
$$7x = -5$$

Finally, after dividing both sides by 7,

$$x = -\frac{5}{7} \quad \checkmark$$

EXAMPLE 19   Solve $8y + 2 - 3 + y = 5 + 15y - 4$.

Solution   First, combine like terms, so that

$$8y + 2 - 3 + y = 5 + 15y - 4$$

becomes

$$9y - 1 = 15y + 1$$

Notice that the unknown, y, appears on both sides of the equation. So apply step 3 and eliminate the $15y$ from the right side by adding $-15y$ to both sides.

$$9y - 1 = 15y + 1$$
$$\underline{-15y -15y }$$
$$-6y - 1 = +1$$

Add $+1$ to both sides to get the y term alone.

$$-6y - 1 = +1$$
$$\underline{ +1 \quad +1}$$
$$-6y = 2$$

Finally, divide both sides by -6.

$$\frac{-6y}{-6} = \frac{2}{-6}$$
$$y = -\frac{2}{6} = -\frac{1}{3} \quad \checkmark$$

EXAMPLE 20 Solve $2(t + 1) = 17$.

Solution The unknown, t, is within parentheses, so multiply out the $2(t + 1)$ in order to free the unknown. Thus

$$2(t + 1) = 17$$

becomes

$$2t + 2 = 17$$

and then

$$2t = 15 \qquad \text{by adding } -2 \text{ to both sides}$$

Finally,

$$t = \frac{15}{2} \quad \checkmark$$

EXAMPLE 21 Solve $5(3 - x) + 6 = 2 + 3x$.

Solution The unknown, x, is within parentheses, so multiply out the $5(3 - x)$ in order to free the unknown.

$$5(3 - x) + 6 = 2 + 3x$$

becomes

$$15 - 5x + 6 = 2 + 3x$$

Now combine the like terms 15 and 6 that appear on the left side.

$$21 - 5x = 2 + 3x$$

Next, add $-3x$ to both sides to eliminate the x term from the right side.

$$21 - 8x = 2$$

Add -21 to both sides to get the x term alone on the left.

$$-8x = -19$$

Divide both sides by -8 to obtain

$$x = \frac{19}{8} \quad \checkmark$$

EXAMPLE 22   Solve $5x - (3 + 2x) = 7$.

Solution   Remove the x from parentheses by multiplying out $-(3 + 2x)$ to get $-3 - 2x$. So

$$5x - (3 + 2x) = 7$$

first becomes

$$5x - 3 - 2x = 7$$

On combining the x terms,

$$3x - 3 = 7$$

Adding $+3$ to both sides yields

$$3x = 10$$

Finally,

$$x = \frac{10}{3} \quad \checkmark$$

EXERCISES 12.6

*__1.__ Solve each linear equation.
   (a) $5x + 7x = 12 + 8$
   (b) $3x - 7x + x = -3 - 5 + 19$
   (c) $7p + 5 - 3p + 1 = 0$
   (d) $n + 3n = 2n + 8$
   (e) $3(m + 2) = 13$
   (f) $y - (2y + 1) = 7$
   (g) $15x - 17 = 2 - 3x$
   (h) $8 - 3x = 5 - 4x + 1$
   (i) $2(x - 1) + 5x = 3$
   (j) $-7(x + 2) = 4x$
   (k) $15 = 4(1 + 3x)$
   (l) $t - (3t + 1) = 17$
   (m) $9y - 14 = 15 - 2y$
   (n) $7 - 5x = 2 - 6x + 3$
   (o) $10(t - 2) + 3t = 1$
   (p) $5 - 3(x + 1) = 11$
   (q) $7 - 2(1 - x) = 0$
   (r) $8 + 5(4 - 7y) = 0$

 2. Solve each linear equation.
   *(a) $5(2 + 3x) = 13$
   (b) $n + n + 1 = n + 9$
   *(c) $7x + 8 = 4x + 3$
   (d) $3n + 7 - n = 4n + 3$
   *(e) $7(x + 2) - 4(x - 1) = 0$
   (f) $7(x + 5) - 3(x + 1) = 0$
   *(g) $5(1 - 9x) + 3(15x + 4) = x$

(h) $8 - 3y = 5 - (4y + 1)$

*(i) $4 - 2(2 - y) + 3 = 10$

(j) $2m + 3 = m + 3$

*(k) $2(x + 1) = 3(1 - x)$

(l) $2x + 3x + 4x + 2 = 5 - x$

*(m) $5(t + 1) = 5(2t + 9)$

(n) $7x - 1 = 4(2x + 1)$

*(o) $13 - 3(1 - p) + 2 = 17$

(p) $2(3m + 1) + 5 = 4 + m$

*(q) $5x + 2(x - 3) = 19$

(r) $5(x - 3) - 4x = 0$

*(s) $1 - 4(1 - 3x) = 10x$

(t) $5(y + 1) = 6(2y - 7)$

*(u) $(7x - 6)3 = 1 - 4x$

(v) $1 - 9x = 5(3x - 2)$

*(w) $2t - (1 - 3t) + 5t = 0$

(x) $13y - (2y + 7) - y = 0$

*3. Solve each linear equation.

(a) $\dfrac{1}{4}x + 6 = 5 - \dfrac{3}{4}x$

(b) $\dfrac{3x}{4} + 7 = \dfrac{5x}{8} + 3$

(c) $\dfrac{2x}{3} - 1 = 5 - \dfrac{3x}{4}$

(d) $7.4x + 5.2 = 1.3x + 11.3$

(e) $9.5y - 1.6 = 2.3y + 7.4$

(f) $1.9t + 4.6 = 5.8t - 3.2$

12.7 CHECKING YOUR SOLUTION

To solve an equation like $5x + 2 = 17$ is to determine the value of x that "works" in the equation—in other words, determine the number that can be substituted for x to make the left-hand side equal to the right-hand side.

If you solve $5x + 2 = 17$, you get 3 as the solution. This means that if x is replaced by 3 in the original equation, the two sides will have the same value.

Take the equation

$$5x + 2 = 17$$

and substitute 3 for x.

$$5(\underline{3}) + 2 \overset{?}{=} 17$$

$$15 + 2 \overset{?}{=} 17$$

$$17 = 17 \quad \checkmark$$

The two sides have the same value, 17.

As you can see, the foregoing process is a **check.** We can check "solutions" to equations to determine whether they are indeed solutions. Only a solution will check. For instance, if we "solved" $5x + 2 = 17$ and obtained 2 as a "solution," the check would fail, meaning that 2 is not a solution. This situation is shown below.

$$5x + 2 = 17$$

$$5(\underline{2}) + 2 \stackrel{?}{=} 17 \qquad \text{substituting 2 for } x$$

$$10 + 2 \stackrel{?}{=} 17$$

$$12 \neq 17 \qquad \text{The symbol} \neq \text{means } \textbf{not equal to.}$$

The check fails, since the two sides have different values.

EXAMPLE 23 Is 4 a solution of $2(3 - x) + 5 = x - 1$?

Solution

$$2(3 - \underline{x}) + 5 = \underline{x} - 1$$

$$2(3 - \underline{4}) + 5 \stackrel{?}{=} \underline{4} - 1 \qquad \left\{ \begin{array}{l} \text{substituting 4 for } x \\ \text{throughout the equation} \end{array} \right.$$

$$2(-1) + 5 \stackrel{?}{=} 3$$

$$-2 + 5 \stackrel{?}{=} 3$$

$$3 = 3 \quad \checkmark$$

Yes, 4 is a solution of given equation.

EXAMPLE 24 Is 1 a solution of $5y + 2y - 3 = 5(y + 1)$?

Solution

$$5y + 2y - 3 = 5(y + 1)$$

$$5(\underline{1}) + 2(\underline{1}) - 3 \stackrel{?}{=} 5(\underline{1} + 1) \qquad \left\{ \begin{array}{l} \text{substituting 1 for } y \\ \text{throughout the equation} \end{array} \right.$$

$$5 + 2 - 3 \stackrel{?}{=} 5(2)$$

$$4 \neq 10 \qquad \text{The check fails.}$$

No, 1 fails to check. It is not a solution of the given equation.

EXAMPLE 25 Solve and check $3x + 1 = 9$.

Solution

$$\begin{array}{rcl} 3x + 1 &=& 9 \\ -1 & & -1 \\ \hline 3x & = & 8 \end{array}$$

$$x = \frac{8}{3} \quad \checkmark$$

$$\textit{Check:} \quad 3\left(\frac{8}{3}\right) + 1 \stackrel{?}{=} 9$$

$$8 + 1 \stackrel{?}{=} 9$$

$$9 = 9 \quad \checkmark$$

EXERCISES 12.7

*1. For which of the following equations is -2 a solution?
(a) $3x = -6$ (b) $5x - 3x = 4$
(c) $2x + 3 = 1$ (d) $2(x + 1) + 2 = 0$
(e) $x + 9 = 2x + 11$ (f) $x + 2 = 0$
(g) $3 - (x + 1) = 8$ (h) $2(x + 3) + 5 = 7 - (1 - x)$
(i) $7x + 10 + x = x - 4$ (j) $5(1 - x) - 11 = x + 6$

2. For which of the following equations is -3 a solution?
(a) $4x = 12$ (b) $4x = -12$
(c) $3(x + 1) + 6 = 0$ (d) $-x = 3$
(e) $1 - 4(1 - x) = -15$ (f) $2(x + 3) + x = -3$
(g) $x + 2 - 4x = 3x$ (h) $5(x + 4) = 4(x + 2)$

3. Solve and check each equation.
*(a) $5x - 2 = 18$ *(b) $3x + 1 = 10$
*(c) $x + 2 = 2$ *(d) $5m = 12$
*(e) $3(1 - x) + 5 = -1$ *(f) $7x + 2 + 5x = 2x + 12$
(g) $4x + 1 = 6$ (h) $6x - 2 + x = 4 - x$
*(i) $3(1 + 5x) - 1 = x$ (j) $7(2x - 1) + x = 3$

12.8 HOW MANY SOLUTIONS? (OPTIONAL)

Thus far each linear equation in one unknown has had just one solution. We've solved dozens of such equations, and always there has been exactly one solution. However, not all linear equations have one solution as you will see.

EXAMPLE 26 Solve $x + 3 = x + 2$.

Solution First, add -3 to both sides.

$$\begin{array}{r} x + 3 = x + 2 \\ -3 \qquad -3 \\ \hline x \qquad = x - 1 \end{array}$$

Next, add $-x$ to both sides.

$$\begin{array}{r} x = \quad x - 1 \\ -x \quad -x \\ \hline 0 = -1 \end{array}$$

But 0 is not equal to -1. Moreover, the x has disappeared. The inconsistent result $0 = -1$ means that the equation has *no solution*. Look back at the original equation $x + 3 = x + 2$ and think. We seek a number x such that $x + 2$ and $x + 3$ are equal. This cannot be because $x + 3$ is always 1 larger than $x + 2$.

EXAMPLE 27 Solve $5(2x + 3) + 8 = 10x + 23$.

Solution First, multiply out the $5(2x + 3)$.

$$5(2x + 3) + 8 = 10x + 23$$

becomes

$$10x + 15 + 8 = 10x + 23$$

Next, combine the $+15$ and the $+8$.

$$10x + 23 = 10x + 23$$

At this point you can see that the two sides of the equation are identical. In other words, $10x + 23$ is equal to $10x + 23$ regardless of the value of x. This means that *any number is a solution* of the equation $10x + 23 = 10x + 23$ or of the original equation $5(2x + 3) + 8 = 10x + 23$. The equation has an *infinite number of solutions*. Had we attempted to combine like terms and proceed from $10x + 23 = 10x + 23$, the result would have been the equation $0 = 0$, which is true regardless of the value of x.

EXERCISES 12.8

1. Determine the number of solutions of each equation: none, one, infinite.
   *(a) $x + 9 = x + 5$
   *(b) $3x + 7 = 3x + 7$
   *(c) $2x + 10 = x + 3$
   (d) $4(x + 8) = 2x - 9$
   *(e) $5x - 3 + x = 5 + 6x$
   (f) $5x + 5 + x = 5 + 6x$
   *(g) $3(1 - x) + 5x = 1 + 2(1 + x)$
   (h) $3(x + 2) - x = 2(3x + 7)$

CHAPTER 12 REVIEW EXERCISES

1. Solve each linear equation.
   *(a) $m - 19 = 24$ *(b) $-5t = -17$
   *(c) $-6y = 19$ *(d) $7r = 4$
   *(e) $\dfrac{1}{2}x = -16$ (f) $\dfrac{1}{4}x = 11$
   *(g) $9x - 2 = -13$ (h) $6x + 5 = 22$
   *(i) $17 - 4x = 28$ (j) $8 + 4y = -11$
   *(k) $9y - 12 = 51$ (l) $\dfrac{3}{7}x + 8 = -9$
   *(m) $\dfrac{4}{5}x + 3 = 10$ *(n) $\dfrac{5}{3}x + 1 = -2$
   *(o) $8x + 5 - x = 4 + 3x$ (p) $14 - 5x + 2 = -3x - 17$

*(q) $5(2x - 3) + 4 = 3x + 1$ (r) $8(x + 4) - 50 = 18 - x$
*(s) $3(1 - 2t) - 2 = 16 + 3t$ (t) $3 - 2(x + 1) = 9x + 4$
*(u) $7 - (5 - x) = x - (2 - x)$ (v) $1 - 3(x - 2) = 7 + 3x$

CUMULATIVE REVIEW EXERCISES (CHAPTERS 10–12)

*1. Simplify each expression completely.
    (a) $-5 + 8 - 7 + 2 - 1 + 6$ (b) -6^2
    (c) $1 - (-4)(+2)$ (d) $5(-3)^2 + 6(-3) - 11$

*2. Evaluate the expression $b^2 - 4ac$ when $a = -1$, $b = 3$, and $c = 6$.

*3. Combine like terms, when possible.
    (a) $8x^2 + 3x - x^2 + 5x^2 + x$ (b) $10x + 2 - 17x - x$

*4. Simplify, if possible.
    (a) $5x^3 \cdot 4x^7$ (b) $3x^2 - 2x^5$

*5. Simplify each expression.
    (a) $x - (3 - x) + (x - 1)4$ (b) $50 + 10(3x - 7) - 17x$

*6. Multiply; then simplify, if possible.
    (a) $(x - 7)(x - 8)$ (b) $(x + 5)(x - 5)$

*7. Divide.
$$x - 2\overline{)\ x^3 - x^2 + 7x - 18}$$

*8. Solve each linear equation.
    (a) $x - 10 = 2$ (b) $-7x = 12$
    (c) $5(2x + 1) = 13$ (d) $8x + 3 = 12x - 19$
    (e) $\frac{x}{3} + 17 = 2$ (f) $5 + 6(1 + 6x) = 3$
    (g) $7 - 4(1 + 2x) = 0$ (h) $3x + 5(1 + x) = 13$

13

WORD PROBLEMS

13.1 INTRODUCTION

This chapter consists of a variety of examples intended to show how algebra is used in solving problems that are stated in words. In the first section you'll see how to change word phrases to algebraic expressions. The other sections present entire problems stated in sentences. Such sentences must be examined in order to set up equations that can be solved.

The abstract nature of word problems can make them seem difficult to solve. In view of this, when you begin the study of word phrases in the next section, you will find arithmetic examples presented first. This is done to point out that many of those problems are easy to solve when you think in terms of arithmetic first. For example, if you know how to work with 4 dimes, then x dimes should not be impossible to work with. Or, if you must work with y nickels and you find this difficult, consider something more concrete, like 7 nickels, in order to see what's going on.

13.2 WORD PHRASES

Here are several examples to introduce you to the concept of changing words to algebra.

EXAMPLE 1 How many cents are there in 4 dimes?

Solution There are 40 cents in 4 dimes. Since each dime is worth 10 cents, if you have 4 dimes, you have $10 \cdot 4 = 40$ cents.

EXAMPLE 2 How many cents are there in x dimes?

Solution Since each dime is worth 10 cents, there are $10 \cdot x$ or $10x$ cents—that is, 10 cents each *times* the number of dimes, x.

EXAMPLE 3 How many feet are there in 72 inches?

Solution Since there are 12 inches in every foot, there are $72/12 = 6$ feet in 72 inches. We divide the number of inches by 12 in order to get the number of feet.

EXAMPLE 4 How many feet are there in $5m$ inches?

Solution Change inches to feet by dividing by 12 as in the last example. Thus

$$5m \text{ inches} = \frac{5m}{12} \text{ feet} \quad \checkmark$$

> **NOTE**
> Some people get confused about when to multiply and when to divide. When changing dimes to cents (Examples 1 and 2) we multiplied because there must be more cents than dimes in the same amount of money. When changing inches to feet (Examples 3 and 4) we divided because there must be fewer feet than inches in the same measurement.

EXAMPLE 5 If Bob is x years old today, how old will he be in 7 years?

Solution Today he is x years old. In 7 years he will have increased his age by 7 years. He will be $x + 7$ years old then, seven years older.

EXAMPLE 6 Write expressions to represent:
(a) Seven more than n.
(b) Nine less than twice x.
(c) The next even number after the even number x.

Solution (a) Seven more than n is simply $n + 7$, or $7 + n$ if you prefer. Both represent a number that is seven more than n.

(b) Twice x is $2x$—that is, two times x. So nine less than twice x is nine less than $2x$—namely, $2x - 9$.

(c) Even numbers are 0, 2, 4, 6, 8, 10, 12, ... You add two to any even number to get the next one in sequence. If x is an even number, then the next one must be $x + 2$, two more than x.

EXAMPLE 7   Write an algebraic expression to represent the total number of cents in n nickels and q quarters.

Solution   Each nickel is worth 5 cents, and so n nickels are worth $5n$ cents (5 cents each for each of n nickels). Similarly, each quarter is worth 25 cents; so q quarters are worth $25q$ cents. Thus the total cents value of n nickels and q quarters is $5n + 25q$.

The next example makes use of the formulas for perimeter and area of a rectangle. Using the letters P for perimeter, A for area, l for length, and w for width, we have the formulas shown below.

$$P = 2l + 2w$$
$$A = l \cdot w$$

EXAMPLE 8   Suppose that the width of a rectangle is x units and the length is 3 more than the width. Determine (a) the perimeter and (b) the area.

Solution   (a) You are given that the width is x. Since the length is 3 more than the width, the length is $x + 3$. The perimeter is two lengths plus two widths, so the perimeter is

$$2(x + 3) + 2(x)$$

which simplifies to

$$2x + 6 + 2x$$

and finally becomes

$$4x + 6 \quad \checkmark$$

(b) Since area is length times width, the area is

$$(x + 3)x$$

which, when multiplied out, is

$$x^2 + 3x \quad \checkmark$$

The next example makes use of the formula for the simple interest (i) on an investment of p dollars at r percent per year for t years. The percent r must be expressed as a fraction or decimal for calculation purposes.

Simple Interest

$$i = prt \qquad \begin{cases} p = \text{amount invested } (\textit{principal}) \\ r = \text{rate of interest (fraction or decimal)} \\ t = \text{time (in years)} \\ i = \text{amount of interest obtained} \end{cases}$$

EXAMPLE 9 Determine the interest on each investment.
 (a) $1200 at 8% per year for 2 years.
 (b) $1000 at 6.5% per year for 10 years.
 (c) d dollars at 7% per year for 3 years.
 (d) $137 at m% per year for x years.

Solution (a) Here p is $1200, r is .08, and t is 2. So

$$i = (\$1200)(.08)(2) = \$192 \quad \checkmark$$

 (b) Here p is $1000, r is .065, and t is 10. So

$$i = (\$1000)(.065)(10) = \$650 \quad \checkmark$$

 (c) Here p is d, r is .07, and t is 3. So

$$i = (\$d)(.07)(3) = .21d \text{ dollars, or } \$.21d \quad \checkmark$$

 (d) Here p is $137, r is $\dfrac{m}{100}$ or .01m, and t is x. So

$$i = (\$137)\left(\frac{m}{100}\right)(x) \quad \text{or} \quad (\$137)(.01m)(x)$$

$$= \frac{\$137mx}{100} \quad \checkmark \quad \text{or} \quad \$1.37mx \quad \checkmark$$

Either form is acceptable as a final answer. Note that percent means hundredths, and this m% can be expressed as hundredths by using either a fraction or decimal form.

EXERCISES 13.2

Answers to starred exercises are given in the back of the book.

*1. Write an algebraic expression to represent each phrase.
 (a) Four more than y.
 (b) Seventeen less than w.
 (c) Ten more than 5x.
 (d) Half of m.
 (e) Twice t.
 (f) Three more than twice x.

(g) Five less than three times y.

(h) m more than n.

(i) x less than y.

(j) The product of x and y.

(k) Seven times $6c$.

(l) 23% of x.

(m) $k\%$.

(n) $k\%$ of x.

2. Write an algebraic expression to represent each phrase.

    *(a) The number of cents in t nickels.

    (b) The number of cents in x half dollars.

    *(c) The number of dollars in v cents.

    (d) The number of dollars in y dimes.

    *(e) The total number of cents in x nickels and y half dollars.

    (f) The total number of cents in p pennies and q quarters.

    *(g) The total cost in cents of x tickets at $1.50 each and y tickets at $3 each.

    (h) The total cost in cents of m tickets at $2 each and n tickets at $3.25 each.

3. Write an algebraic expression to represent each phrase.

    *(a) The number of shoes in p pairs.

    (b) The number of legs in $3x$ spiders. (A spider has eight legs.)

    *(c) The number of minutes in H hours.

    (d) The number of hours in D days.

    *(e) The number of days in T hours.

    (f) The number of hours in M minutes.

    *(g) The number of feet in $5y$ yards.

    (h) The number of inches in $7x$ feet.

    *(i) The number of feet in p inches.

    (j) The number of yards in $2k$ feet.

    *(k) The cost in cents of m articles at x cents each.

    (l) The cost in francs of n items at y francs each.

    *(m) The total cost in lira of a items at x lira each and b items at y lira each.

    (n) The number of miles in f feet. (1 mile = 5280 feet.)

    *(o) The profit on a book that is bought for y dollars and sold for x dollars.

    *(p) The amount of money that remains if x dollars out of $5000 is spent.

    (q) The number of cattle remaining in a herd of h cattle if s cattle are sold.

4. Write an algebraic expression to represent each phrase.

    *(a) The interest on d dollars invested at 6% per year for 1 year.

    *(b) The interest on d dollars invested at $p\%$ per year for 2 years.

    (c) The interest on x dollars invested at 5% per year for 3 years.

(d) The interest on m dollars invested at 9% per year for y years.

*(e) The interest on m dollars invested at n% per year for w years.

*(f) 23% more than x.

(g) 17% more than y.

*5. (metric) Write an algebraic expression to represent each phrase.

(a) The number of meters in x kilometers. (1 kilometer = 1000 meters.)

(b) The number of meters in y millimeters. (1 meter = 1000 millimeters.)

(c) The number of centimeters in $3x$ meters. (1 meter = 100 centimeters.)

(d) The number of millimeters in $2y$ meters. (1 meter = 1000 millimeters.)

(e) The number of liters in m milliliters. (1 liter = 1000 milliliters.)

(f) The number of grams in x kilograms. (1 kilogram = 1000 grams.)

(g) The number of kilograms in y grams. (1 kilogram = 1000 grams.)

6. Write an algebraic expression to represent each phrase.

*(a) The total distance walked by Madeline if she walks x yards from the apartment to the pool, then y yards from the pool to her car, and finally z yards from her car to the apartment. Make a drawing if you do not see the answer immediately.

*(b) The perimeter of a rectangle with length x and width y.

*(c) The area of a rectangle with length x and width y.

*(d) The perimeter of a square with sides of length x.

(e) The area of a square with sides of length x.

*(f) The area of a rectangle with length $x + 2$ and width 7.

(g) The perimeter of a rectangle with length $x + 2$ and width 7.

*(h) The area of a rectangle whose width is x and whose length is twice the width.

(i) The perimeter of a rectangle whose width is x and whose length is twice the width.

*(j) The perimeter of a rectangle whose width is x and whose length is four more than its width.

(k) The area of a rectangle whose width is x and whose length is four more than its width.

*(l) The perimeter of a triangle with sides of lengths a, b, and c.

13.3 INTRODUCTION TO WORD PROBLEMS

The experience you have gained from working with the word phrases of Section 13.2 should enable you to pursue the study of word problems. The presentation of this section consists of a series of examples that will demonstrate the use of algebra to solve problems.

EXAMPLE 10   The sum of four consecutive numbers is 234. What are the numbers?

Solution   Consecutive numbers are those that come one right after the other in the counting sequence; for example, 7, 8, 9, 10 or 21, 22, 23, 24.

If we let x represent the first (or smallest) of the four consecutive numbers, then $x + 1$ is the second number, since the second number must be 1 larger than the first. Similarly, the third number is $x + 2$ and the fourth number is $x + 3$. The representation, then:

$$x = \text{smallest number (first)}$$
$$x + 1 = \text{second number in sequence}$$
$$x + 2 = \text{third number in sequence}$$
$$x + 3 = \text{fourth number in sequence}$$

Since the sum of the four numbers is 234, we can form the equation

$$(x) + (x + 1) + (x + 2) + (x + 3) = 234$$

We next proceed to solve this equation. The parentheses can be removed, since they were used only to make it easier to see the four separate numbers $x, x + 1, x + 2,$ and $x + 3$. Thus we have

$$x + x + 1 + x + 2 + x + 3 = 234$$

Combining like terms yields the simplified equation

$$4x + 6 = 234$$

Then
$$4x = 228$$
or
$$x = 57$$
and
$$x + 1 = 58$$
$$x + 2 = 59$$
$$x + 3 = 60$$

The numbers are 57, 58, 59, and 60.

As a check, we return to the statement of the problem. First, note that the numbers are in fact consecutive. Second, check to be sure that their sum is indeed 234. It is, since $57 + 58 + 59 + 60$ does add up to 234.

The step-by-step procedure demonstrated in this first example is recommended for success in solving word problems. To summarize, the following steps should be followed, in order, when solving word problems.

1. Set up the representation of unknowns.
2. Form an equation.
3. Solve the equation.
4. Answer the question asked by the problem.
5. Check the solution by referring to the statement of the problem.

EXAMPLE 11   If the sum of two numbers is 94 and the larger number is five less than two times the smaller, what are the numbers?

Solution   Let x represent the smaller number, since the larger one is described by comparing it with the smaller.

The larger is five less than two times the smaller. Since x is the smaller, the larger is five less than two times x. This means that the larger is five less than $2x$, which means that it is $2x - 5$. So we have the representation

$$x = \text{smaller number}$$
$$2x - 5 = \text{larger number}$$

The statement of the problem indicates that the sum of the smaller and the larger is 94; that is,

$$\text{smaller} + \text{larger} = 94$$

or

$$(x) + (2x - 5) = 94$$

We now solve this equation.

$$x + 2x - 5 = 94$$
$$3x - 5 = 94$$
$$3x = 99$$
$$x = 33 \qquad \text{smaller}$$

Also,

$$2x - 5 = 2(33) - 5$$
$$= 66 - 5$$
$$= 61 \qquad \text{larger}$$

The smaller number is 33. The larger number is 61.

Check the answers by returning to the statement of the problem to determine whether your answers check with the words of the problem. *Do not* merely see if they check in your equation. Your equation could be wrong!

Check:

"If the sum of two numbers $33 + 61 \overset{?}{=} 94$
is 94" $94 = 94$

"Larger number is five less $61 \overset{?}{=} 2 \cdot 33 - 5$
than two times the smaller" $61 \overset{?}{=} 66 - 5$
$61 = 61$

EXAMPLE 12   15% of what number is 27?

Solution   Let x be the number. Then 15% of x is 27. This statement can be written as the equation shown below. Keep in mind that 15% of x means .15 times x.

$$.15x = 27$$

The equation is solved by dividing both sides by .15, the coefficient of x.

$$\frac{.15x}{.15} = \frac{27}{.15}$$
$$x = 180$$

The number is 180. That is, 15% of 180 is 27. Check this answer by actually computing 15% of 180 in order to show that it is indeed 27.

EXAMPLE 13 A dress on sale at 40% off is now selling for the reduced price of $51. What was the original price of the dress?

Solution *One approach:* Let x = original price of the dress. Then, using the fact that

original price − 40% of original price = sale price

we obtain the equation

$$x - .40x = 51$$

or

$$1.00x - .40x = 51$$
$$.60x = 51$$
$$x = \frac{51}{.60} = 85$$

The original price of the dress was $85.

Another approach: If $51 is 40% *off* the original price, then $51 is 60% *of* the original price. Thus, if x is the original price, then we have immediately

$$.60x = 51$$

which can be continued as done in the first approach.

The check is left for you to do.

EXAMPLE 14 Dorys and Adolfo went to a nice restaurant for their anniversary and spent $68.20, including 4% tax and a 20% tip. What was the amount of the bill before tax and tip?

Solution Let x be the amount of the bill before tax and tip. Then we can establish the following representation.

$$x = \text{amount of the bill before tax and tip}$$
$$.20x = \text{tip} \quad \text{(a 20\% tip on the amount of the bill)}$$
$$.04x = \text{tax} \quad \text{(a 4\% tax on the amount of the bill)}$$

Since the sum of these three numbers must be the total amount spent, $68.20, we have the following equation.

$$x + .20x + .04x = 68.20$$

or

$$1.24x = 68.20$$

which is readily solved by dividing both sides by 1.24, the coefficient of x.

$$x = \frac{68.20}{1.24} = 55$$

Thus *the amount of the bill before tax and tip was $55.*
        To check this answer, see if $55 plus 4% of $55 plus 20% of $55 is indeed equal to $68.20.

bill:	$55.00
4% of $55:	2.20
20% of $55:	11.00
Total:	$68.20

Thus, our answer checks.

EXAMPLE 15   A 972-meter length of fence will be used to enclose a rectangular field that is twice as long as it is wide. Determine the length and width of the field to be enclosed.

Solution   Let x = width

   $2x$ = length (since the length is 2 times, or twice, the width).

Here is a drawing of the rectangle.

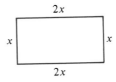

The perimeter is the sum of the four sides, $x + 2x + x + 2x$, and must be equal to 972 meters, which is the amount of fencing to be used. Thus

$$x + 2x + x + 2x = 972$$

or $$6x = 972$$

or $$x = \frac{972}{6} = 162$$

and $$2x = 324 \quad \checkmark$$

The length of the field will be 324 meters and the width will be 162 meters.

Check:
   "Twice as long as it is wide" $324 \stackrel{?}{=} 2(162)$
                                      $324 = 324 \quad \checkmark$

   "972-meter length" $324 + 162 + 324 + 162 \stackrel{?}{=} 972$
                          $972 = 972 \quad \checkmark$

EXAMPLE 16   If you drive at the rate of 55 miles per hour, how long will it take you to drive from Riverside to Sacramento, which is a distance of 440 miles?

Solution   The relationship involving rate of speed, time, and distance is:

$$\boxed{\begin{array}{c} \text{distance} = \text{rate} \times \text{time} \\ d = rt \end{array}}$$

Since we do not know how long the drive will take, let x be that time.

$$x = \text{time it takes to complete the drive}$$

Now we can use $d = rt$, with $d = 440$, $r = 55$, and $t = x$. The resulting equation is

$$440 = 55x$$

This equation is readily solved by dividing both sides by 55.

$$\frac{440}{55} = \frac{55x}{55}$$

or $8 = x$

Thus we find that *it will take 8 hours to drive from Riverside to Sacramento.* The check is left for you to do.

EXERCISES 13.3

Use algebra to solve each word problem.

*1. The sum of three consecutive numbers is 114. Determine the numbers.

2. The sum of five consecutive numbers is 420. Determine the numbers.

*3. The sum of two numbers is 342. One number is twice the size of the other. Determine the two numbers.

4. The sum of two numbers is 252. The larger number is twice the smaller. What are the two numbers?

*5. The sum of two numbers is 153. The larger number is 7 more than the smaller. What are the two numbers?

6. The sum of two numbers is 195. The larger is 3 more than twice the smaller. What are the two numbers?

*7. A board 17 meters long will be sawed into two pieces. One piece will be 2 meters longer than the other. Determine the length of each piece.

8. A wire 85 centimeters long will be cut into two pieces. One piece will be 9 centimeters shorter than the other. Determine the length of each piece.

*9. 16 percent of what number is 36?

10. 24 percent of what number is 54?

*11. 35% of what number is 91?

12. 15% of what number is 21?

*13. What is the cost of a lamp if 3% sales tax on its cost amounts to $1.65?

14. What is the cost of a TV if 4% sales tax is $29.40?

*15. How much was the restaurant bill if a 12% tip is $5.25?

16. How much was the beauty salon bill if a 15% tip is $3.60?

*17. Edward earns a 4% commission on all compact disc (CD) players he sells. If his commission this week was $95, how many dollars worth of CD players did he sell?

18. Judy makes a commission of 5% on all jewelry she sells. If her commission this month was $240, how many dollars worth of jewelry did she sell?

*19. A suit on sale at 30% off is now selling for the reduced price of $196. What was the original price of the suit?

20. The Jones family paid $750 for a sofa on sale at 40% off. What was the full price of the sofa before the sale?

*21. The Prince family spent $594 for a dinette set on sale at 45% off. What was the full price of the dinette set before the sale?

*22. A popular car has been marked up by a dealer and is now selling for 10% more than its sticker price. If the marked up price is $18,920, what is the sticker price of the car?

23. Professor Cogniz just sold a painting for $12,600. He is bragging that he got 40% more than he paid for it two years ago. How much did he pay for it?

*24. A woman opens an account with x dollars. At the end of one year, interest equal to 8% of the deposit is credited to the account. If the total in the account after one year is $351, how much did the woman deposit when the account was opened?

25. A man opens a bank account with y dollars. At the end of one year, interest equal to 5% of the original deposit is credited to his account. If the total in the account after one year is $882, how much did the man deposit originally?

*26. A personal computer (PC) salesperson earns $150 a week in salary plus $80 for every PC she sells. Last week she earned $470. How many PCs did she sell?

27. An encyclopedia salesman earns $100 per week plus $65 for each set of encyclopedias he sells. Last week he made a total of $295. How many sets of encyclopedias did he sell?

*28. What was the amount of the bill before tax and tip if the bill plus a 15% tip plus 5% tax totals $18.48?

29. What was the amount of the bill before tax and tip if the bill plus a 10% tip plus 5% tax totals $27.60?

*30. An 832-meter length of fence will be used to enclose a garden that is three times as long as it is wide. Determine the length and width of the field to be enclosed.

31. The length of a rectangle is 4 feet more than three times the width. If the perimeter is 64 feet, find the length and width.

*32. A square piece of paper has a perimeter of 364 centimeters. How long is each side?

33. A wire 84 inches long is bent to form a square. How long is each side of the square?

*34. The sum of the angles of a triangle is 180°. The largest angle is twice the smallest. The third angle is 16° more than the smallest. Determine the number of degrees in each angle.

35. Two angles of a triangle are equal and the third angle is 24°. Determine the size of the two equal angles.

*36. Two sides of a triangle are equal. The third side is 34 inches. If the perimeter is 156 inches, how long are the two equal sides?

37. The longest side of a triangle is three times the length of the shortest side. The remaining side is 4 inches longer than the shortest side. If the perimeter of the triangle is 424 centimeters, how long is each side?

*38. Two angles are *complementary* if their sum is 90°. For example, angles of 35° and 55° are complementary, since 35° + 55° = 90°. If there are two complementary angles such that one angle is four times the other, determine the number of degrees in each angle.

39. Find two complementary angles such that one angle is 6° less than twice the other angle.

*40. Two angles are *supplementary* if their sum is 180°. If there are two supplementary angles such that the larger is 10° less than four times the smaller, determine the angles.

41. One of two supplementary angles is 73° more than the other. Determine the angles.

*42. If the area of the rectangle shown here is 117 square inches, determine the length of the unknown side.

43. If the perimeter of the rectangle shown here is 226 centimeters, determine the length of each side.

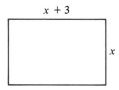

*44. The area of a triangle is given by $A = \frac{1}{2}bh$, where b is the base and h is the height. If the base of a triangle is 24 centimeters and the height is $x + 2$ centimeters, find the height of the triangle if its area is 102 square centimeters.

*45. If you drive at the rate of 60 miles per hour, how long will it take to drive from Dallas to San Antonio, which is a distance of 270 miles?

46. If you drive at the rate of 50 miles per hour, how long will it take to drive from Chicago to St. Louis, which is a distance of 290 miles?

*47. If you can walk at an average of 4 miles per hour, how long will it take to walk 22 miles?

48. If you can walk at an average 8 kilometers per hour, how long will it take to walk 38 kilometers?

*49. If you walked 15 miles in 4 hours, what was your average rate of speed for the walk?

50. If you biked 22 miles in $2\frac{1}{2}$ hours, what was your average rate of speed for the journey?

*51. Suppose that Ann bought x items at $5 each and then sold them all for a total of $120. If her total profit was $45, how many items did she buy?

52. Allison sold x items at $7 each. She had purchased all x of them for a total of $160. If her profit from the sale was $106, how many items did she sell?

*53. A rental car company offers two rates: (1) $43 per day with no charge for miles driven and (2) $31 per day *plus* 20¢ per mile driven. How many miles (x) would you need to drive in a day for the two rates to result in the same amount due in payment to the rental car company?

54. Same as Exercise 53, except assume the rates are (1) $58 per day and (2) $39 per day *plus* 10¢ per mile driven.

13.4 MORE WORD PROBLEMS

This section is intended to provide additional experience with word problems. The examples offer a variety of situations that are different from those presented in Section 13.3.

EXAMPLE 17   Dennis goes to the bank with $4.80 to exchange for nickels and dimes. He wants to get four times as many nickels as dimes. How many of each will he get?

Solution   To begin, let

$$x = \text{number of dimes}$$
$$4x = \text{number of nickels, since he wants four times}$$
$$\text{as many nickels as dimes}$$

The total value of the nickels and dimes is $4.80, which is 480 cents. The number of cents in x dimes and in $4x$ nickels is readily determined by using the concept considered in Section 13.2.

$$x \text{ dimes} = \underline{10} \cdot x \text{ or } 10x \text{ cents } (\underline{10} \text{ cents in each dime})$$

$$4x \text{ nickels} = \underline{5} \cdot 4x \text{ or } 20x \text{ cents } (\underline{5} \text{ cents in each nickel})$$

Clearly,

     value of dimes (in cents) + value of nickels (in cents) = total cents

Since the dimes are worth $10x$ cents and the nickels are worth $20x$ cents, the equation becomes

$$10x + 20x = 480$$

which simplifies to

$$30x = 480$$

Thus

$$x = 16 \quad \checkmark \quad \text{(dimes)}$$

and
$$4x = 64 \quad \checkmark \quad \text{(nickels)}$$

Dennis will have 16 dimes and 64 nickels.

Check:
(a) *Is the total value of 16 dimes and 64 nickels equal to $4.80?*

$$\underline{\begin{array}{l} 16 \text{ dimes } = \$1.60 \\ 64 \text{ nickels} = \$3.20 \end{array}}$$
$$\text{total} = \$4.80 \quad \checkmark$$

(b) *Are there four times as many nickels as dimes?*
     Yes, 64 is four times 16. $\checkmark$

EXAMPLE 18   Dan is 4 years older than Sylvia. 32 years ago he was twice as old as she was then. How old is each today?

Solution   Let us begin by considering their ages today. We will let x be Sylvia's age and then compare Dan's age with x.

Today:
$$x = \text{Sylvia's age}$$
$$x + 4 = \text{Dan's age} \quad \text{(he's 4 years older)}$$

32 years ago:
$$\underline{x - 32} = \text{Sylvia's age}$$
$$x + 4 \underline{- 32} = x - 28 = \text{Dan's age}$$

The statement "32 years ago he was twice as old as she was then" means that Dan's age then $(x - 28)$ was twice Sylvia's age then $(x - 32)$. The equation that follows from this is

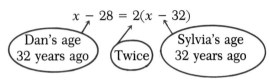

Now let us solve this equation,
$$x - 28 = 2(x - 32)$$

To begin, multiply out the right side to eliminate the parentheses.
$$x - 28 = 2x - 64$$

Next, add 64 to both sides of the equation. The result is
$$x + 36 = 2x$$

Adding $-x$ to both sides yields
$$36 = x$$

This means that
$$x = 36 \quad \checkmark$$
$$x + 4 = 40 \quad \checkmark$$

We conclude that *Sylvia is 36 years old and Dan is 40 years old.*

As a check, note that Dan (age 40) is 4 years older than Sylvia (age 36). 32 years ago he was 8 and she was 4, which means that he was indeed twice her age then.

EXAMPLE 19   John invests a total of $9500, part at 6% (per year) and the rest at 7%. If his (1 year) return on the investments is $622, how much of the $9500 was invested at 6% and how much was invested at 7%?

Solution   If we let x be the amount invested at 6%, then how can we represent the amount invested at 7%? If he invested $2000 at 6%, it is obvious that $7500 is invested at 7%; this is simply $9500 − $2000. *Similarly,* if x dollars are invested at 6%, then $9500 − x$ dollars are left for investment at 7%. Thus

$$x = \text{amount invested at } 6\%$$
$$9500 - x = \text{amount invested at } 7\%$$

$\left\{\vphantom{\begin{matrix}a\\b\\c\end{matrix}}\right.$ Note that the two amounts add up to be 9500.

$$\begin{array}{r} x \\ \underline{9500 - x} \\ 9500 \end{array}$$

We note that

$$\text{interest at } 6\% + \text{interest at } 7\% = \text{total interest (\$622)}$$

Since x dollars are invested at 6%, the interest on it is 6% of x; that is,

$$\text{interest at } 6\% = .06 \cdot x = .06x$$

Similarly, the interest on $(9500 − x)$ dollars at 7% is 7% of $(9500 − x)$; that is,

$$\text{interest at } 7\% = .07\,(9500 - x)$$

So we have the equation

$$.06x + .07\,(9500 - x) = 622$$

It will be easier to work with the equation if we multiply each term by 100 in order to eliminate decimals. The result is

$$6x + 7\,(9500 - x) = 62{,}200$$

After distributing the 7, we have

$$6x + 66{,}500 - 7x = 62{,}200$$

The like terms $6x$ and $-7x$ can be combined.

$$-x + 66{,}500 = 62{,}200$$

Next, add $-66{,}500$ to both sides in order to isolate the x term.

$$-x = -4300$$

Finally, multiply both sides by -1 in order to solve for x.

$$x = 4300 \quad \checkmark \quad \text{(amount at 6\%)}$$

and $\qquad\qquad 9500 - x = 5200 \quad \checkmark \quad \text{(amount at 7\%)}$

Thus *John invests $4300 at 6% and $5200 at 7%.*

Check:
(a) Does the investment return $622?

$$.06 \ (\$4300) = \$258.00$$
$$.07 \ (\$5200) = \$364.00$$
$$\text{total} = \$622.00 \ \checkmark$$

(b) Is the total investment equal to $9500?

$$\$4300 + \$5200 = \$9500 \ \checkmark$$

EXAMPLE 20   Two planes leave an airport at the same time and travel in opposite directions. One plane travels 50 mph faster than the other. How fast is each traveling if they are 1000 miles apart after 2 hours?

Solution   Let's make a drawing of the situation. It will help us to see just what is happening.

Since one plane is 50 mph faster than the other, let

$$x = \text{speed of the slower plane}$$
$$x + 50 = \text{speed of the faster plane}$$

These speeds can be placed in the drawing. Use x for the slower plane and $x + 50$ for the faster plane.

The fact that the planes will be 1000 miles apart after 2 hours can be added to the drawing as shown next.

Now we can use the standard formula that involves distance, rate of speed, and time.

$$\boxed{\begin{array}{c} \text{distance} = \text{rate} \times \text{time} \\ d = r \cdot t \end{array}}$$

The total distance (d) is 1000 miles. The time (t) is 2 hours. The rates are x and $x + 50$. Also,

$$\left(\begin{array}{c}\text{distance of} \\ \text{faster plane}\end{array}\right) + \left(\begin{array}{c}\text{distance of} \\ \text{slower plane}\end{array}\right) = \left(\begin{array}{c}\text{total distance the} \\ \text{planes are apart}\end{array}\right)$$

or

$$\left(\begin{array}{c}\text{rate} \times \text{time} \\ \text{(faster)}\end{array}\right) + \left(\begin{array}{c}\text{rate} \times \text{time} \\ \text{(slower)}\end{array}\right) = \left(\text{distance apart}\right)$$

or

$$(x + 50)2 \quad + \quad x \cdot 2 \quad = \quad 1000$$

Note that the time (2 hours) is the same for each plane.

Since they are traveling in opposite directions, the total distance (1000) represents the sum of the distances of the two planes.

Now solve the equation formed,

$$(x + 50)2 + x \cdot 2 = 1000$$

To begin, multiply out $(x + 50)2$.

$$2x + 100 + 2x = 1000$$

Combine like terms.

$$4x + 100 = 1000$$

Next,

$$4x = 900$$
$$x = 225 \quad \checkmark \quad \text{(slower)}$$
$$x + 50 = 275 \quad \checkmark \quad \text{(faster)}$$

The slower plane travels 225 mph; the faster travels at 275 mph.

Check:

(a) Is one plane traveling 50 mph faster than the other?
   Yes, the speeds are 275 mph and 225 mph, and $275 - 225 = 50$.

(b) Will they be 1000 miles apart in 2 hours?
   Yes, since $2(225) + 2(275) = 1000$.

EXERCISES 13.4

*1. Judy has $9.75 in nickels and dimes. She has twice as many dimes as nickels. How many of each does she have?

2. Carroll has $7.50 in nickels and dimes. If he has 15 more nickels than dimes, how many of each does he have?

*3. Carla goes to the bank to exchange $6.05 for dimes and quarters. If she gets three times as many dimes as quarters, how many of each will she have?

*4. Carol has $3.68 in pennies, nickels, and dimes. She has the same number of each. How many of each does she have?

*5. Pat buys stamps in 25¢ and 15¢ denominations. She spends a total of $8.20 and receives a total of 42 stamps. How many of each did she buy?

6. Jose buys stamps in 25¢ and 20¢ denominations. For $31.40 he receives 136 stamps. How many 25¢ stamps did he buy?

*7. School 66 raises money by selling 102 tickets for a benefit concert. They sell $4 and $5 tickets and obtain a total of $435. How many of each did they sell?

*8. The college has been selling tickets for its variety show. So far they have collected $1224. They have sold twice as many student tickets ($5 each) as general public tickets ($7 each). How many of each have they sold?

9. The PTA is selling concert tickets in order to raise money. They will offer an equal number of tickets at $10, $12, and $15. The total value of all tickets is $1591. How many tickets are there altogether?

*10. Amy has $2.95 in nickels, dimes, and quarters. If she has twice as many dimes as quarters and four more nickels than dimes, how many of each does she have?

11. Randy has $4 in nickels, dimes, and quarters. He has twice as many nickels as quarters. He has four more nickels than dimes. How many of each does he have?

*12. Mike has 15 coins—nickels and dimes. Their value is $1.15. How many of each does he have?

*13. Charles is twice as old as Liz. If the sum of their ages is 48, how old is each of them?

14. In 16 years Jennifer will be three times as old as she is now. How old is she today?

*15. Ginger is 6 years older than Yung. Eleven years ago she was twice as old as Yung was then. How old is each person today?

16. Troy is 5 years older than Rick. Sixteen years ago he was twice as old as Rick was then. How old is Troy today?

*17. Bill is 57 years old and Martha is 53. How many years ago was he three times as old as she was then?

18. Tomas is 36 and Boris is 31. How many years ago was Tomas twice as old as Boris was then?

*19. Charlotte is twice as old as Kay. Five years from now the sum of their ages will be 88. How old is Kay now?

20. Travis is 13 years older than Matt. In 3 years Travis will be twice as old as Matt. How old is Travis now?

*21. Ken plans to invest $4500, part at 5% and the rest at 6% annual interest. If his income from the two investments must be $248 for the year, determine how much he should invest at each rate.

22. Anita will invest $2300, part at 6% and the remainder at 7%. If her total interest from the two investments is to be $153, how much will she invest at each rate?

*23. Naomi invests twice as much money in corporate bonds paying 9% annually as she does in municipal bonds paying 6% annually. If her total annual return from the two investments is $864, how much did she invest in each?

24. Betsy invests in an income fund that pays 7% annually. At the same time, she also buys shares of a utility stock that pays a 6% annual dividend. The amount invested in the income fund is $800 more than that invested in the utility stock. If the total annual yield from the two investments is $576, determine how much she has invested in each.

*25. Arlene invests $5400 in bonds paying 8% and a fund paying 5% interest. After a year, she notices that the 8% investment returns a dividend that is $68 more than that returned by the 5% investment. How much is invested at each rate?

26. Margaret invests $6700, part at 6% and the rest at 5%. If the return from the 5% investment is $60 more than that of the 6% investment, how much money was invested at each rate?

*27. Two trains leave a terminal at the same time and travel in opposite directions. One train travels 30 mph faster than the other. How fast is each train moving if they are 800 miles apart after 4 hours of traveling?

28. Two trains leave a terminal at the same time and travel in opposite directions. After 3 hours the trains are 465 miles apart. If one train traveled at the average rate of 70 miles per hour, how fast did the other train travel?

*29. Jack and Wilbur are 190 miles apart. They begin to drive toward each other. Wilbur travels 15 mph faster than Jack. If they meet in 2 hours, how fast will they be going?

30. Two cyclists ride toward each other. At the beginning they are 51 miles apart. One cyclist goes 3 miles per hour faster than the other. They will meet in 3 hours. Determine the rate of speed of each cyclist.

*31. Two planes leave an airport at the same time and travel in opposite directions. Both planes travel at the same speed. If the planes are 1540 miles apart after 2 hours, how fast are the planes traveling?

*32. Two planes leave an airport at the same time and travel in opposite directions. One plane travels at 320 miles per hour. The other plane travels at 340 miles per hour. When will the planes be 2145 miles apart?

*33. A car is traveling at a speed that is 34 mph slower than that of a train. In 3 hours the train can cover the same distance that the car covers in 5 hours. Find the speed of each one.

34. A bicyclist is traveling at a speed that is 27 mph slower than that of a car. In 2 hours the car can go the same distance that the bicycle goes in 5 hours. Find the speed of each one.

CHAPTER 13 REVIEW EXERCISES

*1. Write an algebraic expression to represent each phrase.
   (a) 15 less than the product of a and b.
   (b) The total number of cents in x dimes and $2y$ half dollars.
   (c) The number of yards in $17x$ feet.
   (d) The perimeter of a rectangle with length x and width $x - 4$.
   (e) The interest on \$213 invested at $c\%$ per year for x years.
   (f) 37% more than y.

*2. Solve each word problem.
   (a) The sum of two numbers is 168. One number is three times as big as the other. Determine the two numbers.
   (b) 36% of what number is 37.8?

*3. Solve each word problem.
   (a) Flo has \$4.60 in nickels and quarters. If she has two more nickels than quarters, how many of each does she have?
   (b) Jane invests \$10,000, part at 9% and the rest at 8%. The total interest from the two investments is \$865. How much did she invest at each rate?

CUMULATIVE REVIEW EXERCISES (CHAPTERS 10–13)

*1. Determine the value of each expression.
   (a) $(-4)(+2) - (-5)(-1)$ (b) $-(-4)(+7)^2$
   (c) $-6 + 2 - 5 + 11 - 7$ (d) $(+6) - (-7) + (-2)^3$

*2. Combine like terms, when possible.
   (a) $8x - 9x - x + 6x - 15x$ (b) $y - 2y + 3 - 4y + 5$

*3. Simplify.
   (a) $5m^3 \cdot 3m^6$ (b) $(-5x^2y^3)(4xy^9)$
   (c) $7x - 3(5x + 2) + x$ (d) $(y^2 + 3y - 8) - (2y^2 - 7y + 1)$

*4. Multiply. Then simplify, if possible.
   (a) $(x + 5)(x - 6)$ (b) $(u + 5)^2$

*5. Divide.
$$x + 2 \overline{)\ x^3 + 5x^2 - 3x + 8}$$

*6. Solve each linear equation.
   (a) $9x - 1 = 15$ (b) $5(1 + 2x) = 13$
   (c) $2x - 3 = 7x + 4$ (d) $5 - 2(1 - 3x) = 0$

*7. Write an algebraic expression to represent each phrase.
   (a) The number of tires on $3t$ bicycles.
   (b) 47% of y.
   (c) The perimeter of a rectangle with length $x + 9$ and width $2x - 1$.

*8. Solve each word problem.
   (a) How much was the bakery bill, if the 5% tax on the bill is $1.23?
   (b) A wire 144 inches long is bent to form an equilateral triangle (that is, a triangle with three equal sides). How long is each side of the triangle?
   (c) Andy is 11 years older than Chip. Nine years ago he was twice Chip's age then. How old is Chip today?

14

GRAPHING
STRAIGHT LINES

14.1 INTRODUCTION

Let us take the two number lines

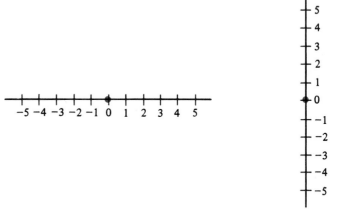

and join them so that they are perpendicular at their zeros. The result is

The point where the lines cross is called the **origin.** The diagram below shows some points in this *plane.* Look at point *A.* How can we get from the origin to the point *A*? One way is to move 5 units to the right and then 1 unit up. There are other ways, of course.

We can get to *B* (from the origin) by moving 1 unit to the right and then up 2 units. Similarly, we get to *C* by moving right 3 and down 1. We get to *D* by moving left 3 and up 4. *E* is left 1 and down 2.

We use the notation (5, 1) to mean 5 units to the right (of the origin) and 1 unit up. Similarly, (−3, 4) means 3 units to the left and 4 units up.

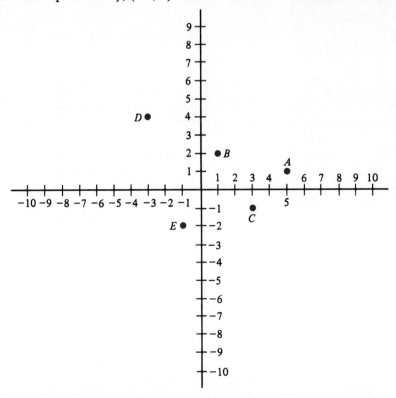

Right is positive and left is negative; up is positive and down is negative.

We tabulate the five points: A, B, C, D, E.

Point	Left or Right	Up or Down	Notation
A	right 5	up 1	(5, 1)
B	right 1	up 2	(1, 2)
C	right 3	down 1	(3, −1)
D	left 3	up 4	(−3, 4)
E	left 1	down 2	(−1, −2)

We have represented a point as a pair of numbers—the **coordinates** *of the point.* The first coordinate of the pair indicates the left-right position of the point. The second coordinate indicates the up-down position. So you can see that the order of the numbers is important, and you can also see why the pair of numbers is called an **ordered pair.**

The system that we have drawn and described is called the **rectangular** *coordinate system.* It is also called the **Cartesian** *coordinate system* after the French mathematician René Descartes, who developed the scheme in 1637.

EXAMPLE 1 The points (5, 1) and (1, 5) are not the same. The order of the numbers 5 and 1 is important in describing a point.

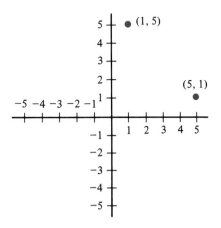

EXAMPLE 2 Plot the points (3, 4), (−2, −5), (4, 0), and (0, 3).

Solution To plot (3, 4), begin at the origin and move 3 to the right and up 4. To plot (−2, −5), begin at the origin and move 2 to the left and down 5. To plot (4, 0), begin at the origin and move 4 to the right and up 0. To plot (0, 3), begin at the origin and move 0 to the right (or left) and up 3.

The graph on which we plot points is divided into four distinct regions. Each is called a **quadrant.** The quadrants are numbered I, II, III, IV, as shown in the diagram. Points on the number lines are not considered as being in a quadrant.

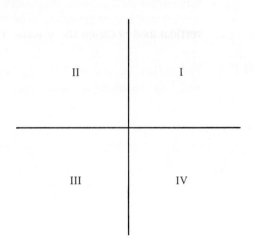

EXERCISES 14.1

Answers to starred exercises are given in the back of the book.

1. Plot each point and indicate which quadrant it is in.

   (a) (3, 2)   (b) (2, 3)   (c) (4, 1)   (d) (1, 4)
   (e) (1, −5)   (f) (1, 5)   (g) (3, −2)   (h) (−4, 3)
   (i) (−5, −2)   (j) (0, 6)   (k) (−2, −4)   (l) (−3, 1)
   (m) (3, 0)   (n) (4, 4)   (o) (5, −3)   (p) (7, −1)
   (q) (0, 0)   (r) (−2, 0)   (s) (0, 2)   (t) (0, −1)

*2. Determine the coordinates of each point.

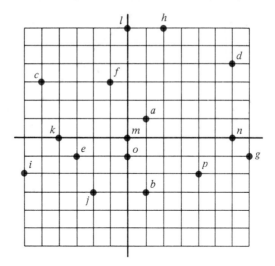

14.2 GRAPHING STRAIGHT LINES

The number lines that we joined to form a plane are called **axes,** and they are usually labeled x and y. The horizontal axis is called the **x axis.** The vertical axis is called the **y axis.** The plane is then called the xy plane.

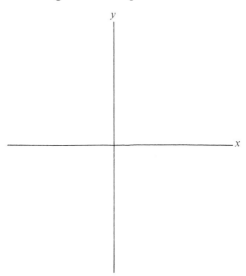

Distances left or right are values for x; distances up or down are values for y. The point $(1, 2)$ means right 1 and up 2. It is also the point (x, y), where x is 1 and y is 2.†

_____

† The x coordinate is sometimes referred to as the **abscissa** and the y coordinate as the **ordinate.**

Next, let's plot the points $(0, 0)$, $(1, 2)$, $(2, 4)$, $(3, 6)$, and $(-1, -2)$ in the plane.

If you connect the points, you'll have a *straight line segment*. In other words, a straight line can be passed through the points. Also, if you chart the x coordinates and y coordinates as

		x	y
$(0, 0)$	$\longrightarrow$	0	0
$(1, 2)$	$\longrightarrow$	1	2
$(2, 4)$	$\longrightarrow$	2	4
$(3, 6)$	$\longrightarrow$	3	6
$(-1, -2)$	$\longrightarrow$	-1	-2

you might see the relationship between x and y. Each y value is *exactly twice* the corresponding x value. That is,

$$y = \text{twice } x$$

or

$$y = 2x$$

By plotting points and then passing a straight line through them, we would be sketching the **graph** of the relationship $y = 2x$. Later in the book we will sketch graphs that are not straight lines.

EXAMPLE 3   Sketch the graph of $y = 3x$.

Solution   We choose some values for x and then determine the corresponding y values. This procedure will yield some points of the form (x, y). Let us choose the values $-1, 0, 2$, and 3 for x and obtain points. Keep in mind that you can choose any values; they need not be $-1, 0, 2$, and 3.

- If $x = -1$,   then $y = 3x$ becomes $y = 3(-1) = -3$.
- If $x = 0$,   then $y = 3x$ becomes $y = 3(0) = 0$.

- If $x = 2$, then $y = 3x$ becomes $y = 3(2) = 6$.
- If $x = 3$, then $y = 3x$ becomes $y = 3(3) = 9$.

So we now have the points shown in the chart below.

x	$y = 3x$	Point
-1	-3	$(-1, -3)$
0	0	$(0, 0)$
2	6	$(2, 6)$
3	9	$(3, 9)$

The graph is

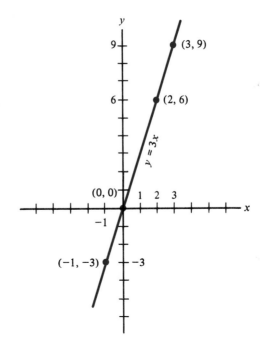

EXAMPLE 4   Sketch the graph of $y = 2x + 1$.

Solution   If we choose x values of 0, 1, 2, 3, and 4, we get the points shown below.

x	$y = 2x + 1$	Point
0	1	$(0, 1)$
1	3	$(1, 3)$
2	5	$(2, 5)$
3	7	$(3, 7)$
4	9	$(4, 9)$

Of course, you can choose *any* values for x, but keep in mind that you are the one who must compute the y values and then plot the points. So numbers like -20 or 30 would not be good choices.

The graph is

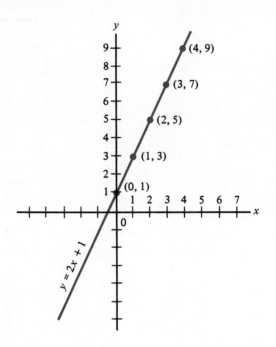

EXAMPLE 5   Sketch the graph of $y = -3x + 2$.

Solution   Again y is given in terms of x, so select values for x and use them to determine the corresponding y values. We have chosen 0, 1, 2, and -1 as the x values. Other values could be used.

x	$y = -3x + 2$	Point
0	2	(0, 2)
1	-1	(1, -1)
2	-4	(2, -4)
-1	5	(-1, 5)

The graph is

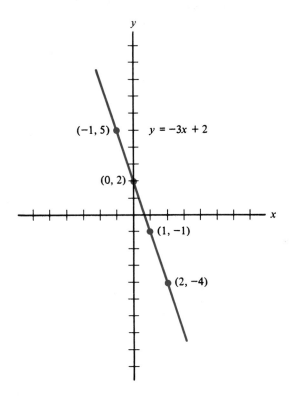

EXAMPLE 6   Sketch the graph of $x = 2y + 1$.

Solution   Since x is given in terms of y, choose values for y and then determine the corresponding x values.

y	$x = 2y + 1$	(x, y)
0	1	(1, 0)
1	3	(3, 1)
2	5	(5, 2)
3	7	(7, 3)

The graph is

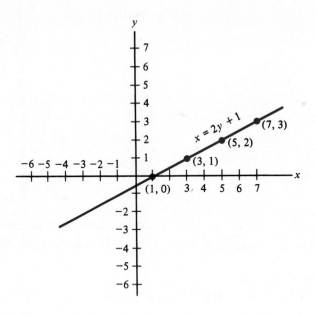

EXAMPLE 7   Sketch the graph of the line whose equation is $3x + 2y = 6$.

Solution   The equation $3x + 2y = 6$ does not give y in terms of x or x in terms of y. The easiest way to obtain points for such a line is by letting x be 0 to get y and then letting y be 0 to get x. This will give you two points, which is technically all you need to obtain the line. But be careful, since you have no third or fourth points to serve as a check in the event you make a mistake.

First, let's use 0 for x.

$$3(0) + 2y = 6$$
$$0 + 2y = 6$$
$$2y = 6$$
$$y = 3$$

This yields the point $(0, 3)$.

Similarly, if y is 0, we have

$$3x = 6$$
$$x = 2$$

This yields the point $(2, 0)$. The graph is shown on the next page.

x	y
0	3
2	0

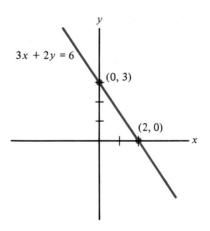

The next two examples illustrate rather unusual lines—one is the graph of an equation having no x in it, the other is the graph of an equation having no y in it.

EXAMPLE 8 Sketch the graph of the line $y = 2$.

Solution The equation states that y equals 2, regardless of what value x takes on. If x is 1, y is 2. If x is 3, y is 2. If x is 0, y is 2. And so on. The points: $(1, 2)$, $(3, 2)$, $(0, 2)$. The graph:

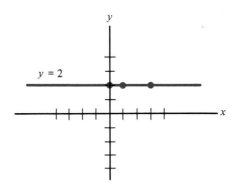

So the graph of $y = 2$ is a horizontal line 2 units above the x axis.

EXAMPLE 9 Sketch the graph of the line $x = 3$.

Solution The equation states that x equals 3, regardless of what value y takes on. If y is 0, x is 3. If y is 2, x is 3. If y is 3, x is 3. And so on. The points: $(3, 0)$, $(3, 2)$, $(3, 3)$. The graph:

So the graph of $x = 3$ is a vertical line 3 units to the right of the y axis.

The graphs of two lines can be sketched on the same axes in order to determine their approximate point of intersection. Here is an example.

EXAMPLE 10   Sketch the graphs of $y = x + 2$ and $5x + 2y = 10$ and estimate where they meet.

Solution   The points:

$y = x + 2$			$5x + 2y = 10$	
x	y		x	y
0	2		0	5
1	3		2	0
2	4			
3	5			

The graph:

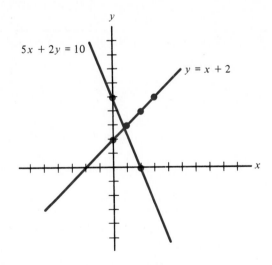

It appears that the lines meet at about $(1, 3)$, but we cannot tell *exactly* by using a graph. Actually, these lines meet at the point $(\frac{6}{7}, 2\frac{6}{7})$. Methods for obtaining such exact solutions are presented in the next chapter.

If two lines are parallel, then they will not meet when graphed. Here is an example.

EXAMPLE 11   Sketch the graphs of the lines whose equations are $y = 2x + 3$ and $y = 2x - 1$. You will see that the lines are parallel.

Solution   The points:

$y = 2x + 3$			$y = 2x - 1$	
x	y		x	y
0	3		0	-1
1	5		1	1
2	7		2	3
-1	1		3	5

The graph:

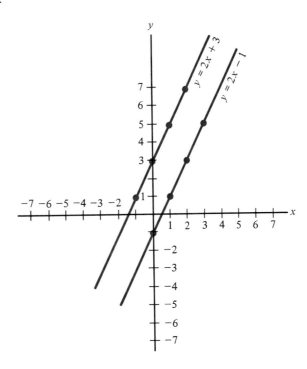

The lines appear to be parallel, and in fact, they are parallel. The next section will present an algebraic method for determining whether two lines are parallel.

NOTE

In this section we have graphed equations such as $y = 3x$, $x = 2y + 1$, and $3x + 2y = 6$. These equations, as well as the other equations presented in the section, are called *linear* equations. The graph of a linear equation in two variables (say x and y) is a straight line.

EXERCISES 14.2

1. Sketch the graph of each line.

   *(a) $y = 2x$ (b) $y = 4x$ *(c) $y = x + 3$
   (d) $y = x + 4$ *(e) $y = x - 2$ (f) $y = x - 3$
   *(g) $y = x$ (h) $y = x - 1$ *(i) $y = 2x + 2$
   (j) $y = 2x + 4$ *(k) $y = 3x - 2$ (l) $y = 4x - 3$
   *(m) $y = -2x$ (n) $y = -3x$ *(o) $y = -3x + 4$
   (p) $y = -2x + 1$ *(q) $y = -2x - 3$ (r) $y = -3x - 5$

   *(s) $y = \frac{1}{2}x$ (t) $y = -\frac{1}{2}x$ *(u) $y = \frac{1}{2}x + 3$

   (v) $y = \frac{1}{2}x - 1$ *(w) $y = -x$ (x) $y = -x + 2$

2. Sketch the graph of each line.

   *(a) $x = y + 3$ (b) $x = y - 4$ *(c) $x = 2y - 3$
   (d) $x = 3y + 1$ *(e) $x = -y$ (f) $x = -2y$
   *(g) $x = -2y + 3$ (h) $x = -y + 4$

3. Sketch the graph of each line.

   *(a) $x + 2y = 8$ (b) $2x + y = 6$ *(c) $x - y = 3$
   (d) $x + y = 4$ *(e) $2x + 3y = 12$ (f) $2x + 5y = 10$
   *(g) $2x - 3y = 12$ (h) $2x - 5y = 10$ *(i) $3y - 2x = 6$
   (j) $2y - 3x = 12$ *(k) $x + y = -4$ (l) $x - y = -4$
   *(m) $x + y = 0$ (*Hint:* Use values of x or y other than zero.)
   (n) $x + 2y = 0$

4. Sketch the graph of each line. Use Example 8 as a guide.
   *(a) $y = 3$ (b) $y = 1$ *(c) $y = -5$ (d) $y = 0$

5. Sketch the graph of each line. Use Example 9 as a guide.
   *(a) $x = 2$ (b) $x = 4$ *(c) $x = -3$ (d) $x = 0$

6. Graph each pair of lines on the same axes and determine approximately where the lines meet, if they do meet.

   *(a) $\begin{cases} y = x + 1 \\ y = 3x - 1 \end{cases}$ *(b) $\begin{cases} y = x - 1 \\ y = -x + 3 \end{cases}$ *(c) $\begin{cases} y = x \\ y = 2x - 4 \end{cases}$

   (d) $\begin{cases} y = -4x + 2 \\ y = 2x - 5 \end{cases}$ *(e) $\begin{cases} y = x \\ y = -x \end{cases}$ (f) $\begin{cases} y = x + 4 \\ y = -x + 4 \end{cases}$

   *(g) $\begin{cases} y = x + 2 \\ y = x - 2 \end{cases}$ (h) $\begin{cases} y = -x + 3 \\ y = 2x - 1 \end{cases}$ *(i) $\begin{cases} y = 0 \\ x = 0 \end{cases}$

   (j) $\begin{cases} y = 2x - 3 \\ y = 2x + 2 \end{cases}$ *(k) $\begin{cases} y = 2x + 1 \\ y = 4 \end{cases}$ (l) $\begin{cases} y = x - 4 \\ x = 3 \end{cases}$

$$*(m) \begin{cases} x + y = 6 \\ y = -x + 1 \end{cases} \qquad (n) \begin{cases} y = 1 - 3x \\ y = 3x - 1 \end{cases}$$

14.3 THE CONCEPT OF SLOPE

This section will introduce the concept of slope. The slope of a line is a measure of the steepness or inclination of the line with respect to the horizontal (x) axis. Consider the two lines graphed below.

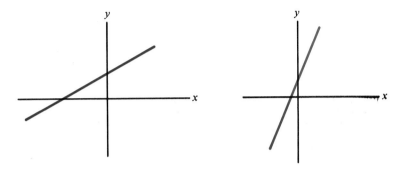

The line on the right is steeper. Intuitively, we might say that it has the greater slope. The following definition of slope takes this intuition into account and, in fact, assigns numerical values as slopes. The definition can be used to determine the slope of a line, if you know two points that are on the line. Informally,

$$\boxed{\text{slope} = \frac{\text{vertical change}}{\text{horizontal change}} = \frac{\text{rise}}{\text{run}}}$$

Clearly, the steeper the line the greater will be the vertical change compared with the horizontal change.

A more formal definition can be given by calling one point (x_1, y_1) and the other point (x_2, y_2). The slope (m) of the line through the two points is

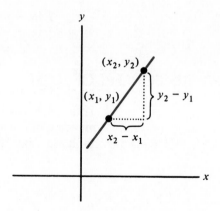

$$\boxed{\text{slope } (m) = \frac{y_2 - y_1}{x_2 - x_1}}$$

EXAMPLE 12   Compute the slope of the line passing through the points $(3, -2)$ and $(5, 4)$.

Solution   Let $(3, -2)$ be (x_1, y_1) and let $(5, 4)$ be (x_2, y_2). Then

$$x_1 = 3$$
$$y_1 = -2$$
$$x_2 = 5$$
$$y_2 = 4$$

So

$$m = \frac{y_2 - y_1}{x_2 - x_1} = \frac{4 - (-2)}{5 - (3)} = \frac{6}{2} = 3 \quad \checkmark$$

If instead we let $(5, 4)$ be (x_1, y_1) and $(3, -2)$ be (x_2, y_2), then

$$x_1 = 5$$
$$y_1 = 4$$
$$x_2 = 3$$
$$y_2 = -2$$

So

$$m = \frac{y_2 - y_1}{x_2 - x_1} = \frac{-2 - (4)}{3 - (5)} = \frac{-6}{-2} = 3 \quad \checkmark$$

The slope is 3, and we have shown that it does not matter which point is called (x_1, y_1) and which is (x_2, y_2).

EXAMPLE 13   Compute the slope of the line through $(0, -5)$ and $(-2, -1)$.

Solution   Let $(0, -5)$ be (x_1, y_1) and let $(-2, -1)$ be (x_2, y_2). Then

$$x_1 = 0$$
$$y_1 = -5$$
$$x_2 = -2$$
$$y_2 = -1$$

So

$$m = \frac{y_2 - y_1}{x_2 - x_1} = \frac{-1 - (-5)}{-2 - (0)} = \frac{4}{-2} = -2 \quad \checkmark$$

The slope of the line is -2.

EXAMPLE 14   Graph the line that passes through the point $(1, 4)$ and has a slope of 3.

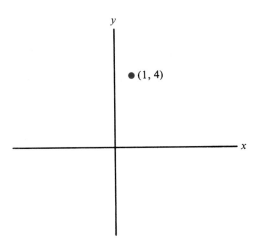

Solution   There are an infinite number of lines through $(1, 4)$, but only one such line also has a slope of 3. If the line has a slope of 3, then

$$\frac{\text{rise}}{\text{run}} = 3 = \frac{3}{1}$$

This means that x must increase by 1 when y increases by 3 in order to produce a point on the line.

We have the point $(1, 4)$ already. Taking this point and

1. Increasing x by 1, making it 2
2. Increasing y by 3, making it 7

gives us another point on the line—namely, $(2, 7)$. Now we can pass a line through the two points.

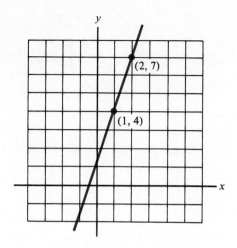

EXAMPLE 15   Sketch a graph of the line through $(0, 3)$ with slope of -4.

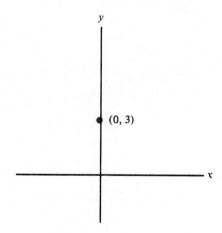

Solution   The slope -4 can be considered as $\dfrac{-4}{+1}$ or $\dfrac{+4}{-1}$.

From $(0, 3)$ we can get another point by changing x by $+1$ and y by -4. Doing so gives

$$(0 + 1, 3 - 4) = (1, -1)$$

Although we don't actually need a third point, the fraction $\dfrac{+4}{-1}$ could be used to find one. If we take the point $(0, 3)$ and change x by -1 and y by $+4$, we obtain

$$(0 - 1, 3 + 4) = (-1, 7)$$

Here is a graph of the line.

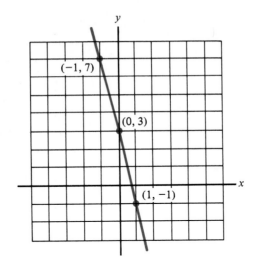

You can see from this example that *negative-sloped lines fall as they go from left to right,* whereas *positive-sloped lines rise as they go from left to right.*

EXERCISES 14.3

1. Compute the slope of the line through each pair of points.
   *(a) (5, 7) and (3, 6)   (b) (10, 8) and (6, 4)
   *(c) (4, 2) and (6, 8)   *(d) (5, 1) and (6, 0)
   *(e) (0, 0) and (1, 5)   (f) (7, 0) and (4, 3)
   *(g) (5, 2) and (4, 6)   *(h) (5, 2) and (1, −3)
   *(i) (1, −1) and (−1, 1)   (j) (5, 0) and (0, −6)
   *(k) (2, −3) and (−7, 4)   (l) (−5, 3) and (−3, −1)

2. Sketch a graph of the line through the given point and having the given slope.
   *(a) (0, 1), $m = 2$ (b) (2, 0), $m = 3$

   *(c) (4, 1), $m = \dfrac{5}{3}$ (d) (1, 2), $m = \dfrac{2}{3}$

   *(e) (−2, 1), $m = 4$ (f) (−3, 4), $m = 1$
   *(g) (1, 3), $m = -2$ (h) (4, 5), $m = -3$

   *(i) (5, 3), $m = -\dfrac{1}{2}$ (j) (4, 6), $m = -\dfrac{3}{2}$

   *(k) (1, 0), $m = -3$ (l) (0, 6), $m = -2$

   *(m) (−1, 6), $m = -\dfrac{5}{2}$ (n) (1, −2), $m = -\dfrac{7}{3}$

3. Determine the slope of the line passing through the points (6, 4) and (8, 4). Then sketch a graph of the line. The graph should help you understand the numerical slope in this instance.

4. What can be said about the slope of a vertical line, such as the line through $(7, 2)$ and $(7, 4)$?

14.4 THE SLOPE-INTERCEPT FORM OF A LINE

The equation of a straight line can be obtained from the slope concept

$$m = \frac{y_2 - y_1}{x_2 - x_1}$$

First, however, you must become familiar with the notion of y-intercept. The point where a line crosses the y axis is called the **y-intercept** of the line.

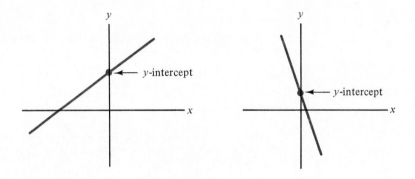

At the y-intercept, x is always 0, since the point is in fact on the y axis. Thus, the y-intercept is a point of the form $(0, b)$, with b being the y-coordinate where the line crosses the y axis.

EXAMPLE 16 Determine the y-intercept of each line.
(a) $y = 9x - 4$ (b) $y = 5x + 1$

Solution (a) The y-intercept is the point of the line where $x = 0$. So let $x = 0$ in the equation $y = 9x - 4$.

$$\begin{aligned} y &= 9(0) - 4 \\ &= 0 - 4 \\ &= -4 \end{aligned}$$

The y-intercept is $(0, -4)$.

(b) Letting $x = 0$ in $y = 5x + 1$ yields

$$\begin{aligned} y &= 5(0) + 1 \\ &= 0 + 1 \\ &= 1 \end{aligned}$$

The y-intercept is $(0, 1)$.

NOTE

We have defined the y-intercept of a line to be $(0, b)$, the point where the line crosses the y axis. *Some mathematicians consider the y-intercept to be just the y value when x is 0, rather than the entire point.* They would say that the y-intercept of the line of Example 16(a) is -4 rather than $(0, -4)$.

Now let us return to

$$m = \frac{y_2 - y_1}{x_2 - x_1}$$

To obtain the equation of a line from this formula, use any point (x, y) rather than the specific point (x_2, y_2), and let (x_1, y_1) be the y-intercept $(0, b)$. This yields

$$m = \frac{y - b}{x - 0}$$

or

$$m = \frac{y - b}{x}$$

Multiply both sides by x and the result is

$$y - b = mx$$

Finally, adding b to both sides yields the result

$$y = mx + b$$

The equation $y = mx + b$ is called the slope-intercept form of a line, since m is the slope and b, or $(0, b)$, is the y-intercept.

> *Slope-Intercept Form of a Line*
>
> $$y = mx + b$$
>
> $m = $ slope, $\quad b = y$-intercept

When the equation of a line is written in the slope-intercept form, the slope and the y-intercept can be read directly from the equation.

EXAMPLE 17 Determine the slope and y-intercept of each line.
(a) $y = 3x - 2$ (b) $y = -x + 4$

Solution (a) The line is in the form $y = mx + b$. Here $m = 3$. The slope is 3. Also, $b = -2$, so the y-intercept is -2 or $(0, -2)$.

(b) The line is $y = -1x + 4$. The slope is -1 and the y-intercept is 4 or $(0, 4)$.

EXAMPLE 18 Determine the equation of the line having slope 3 and y-intercept 7.

Solution The equation of a line with slope m and y-intercept b is $y = mx + b$. In this example, we are given that $m = 3$ and $b = 7$. This means the equation of the line is $y = 3x + 7$.

EXAMPLE 19   Determine the equation of the line having slope 4 and passing through the point (2, 9).

Solution   The slope is given as 4. This means the line, being of $y = mx + b$ form, must be $y = 4x + b$. The value of b can be determined by using the fact that the point (2, 9) is on the line. Since (2, 9) is on the line, it follows that $x = 2$ and $y = 9$ must satisfy the equation of the line. Substituting 2 for x and 9 for y into the equation $y = 4x + b$ yields

$$9 = 4(2) + b$$
$$9 = 8 + b$$
$$b = 1$$

Thus the equation of the line is $y = 4x + 1$.

EXAMPLE 20   Find the equation of the line that passes through the points (1, 6) and (4, 0).

Solution   The two points can be used to obtain the slope of the line. After that, the procedure will be the same as that of Example 19. The slope of the line is

$$m = \frac{0 - 6}{4 - 1} = \frac{-6}{3} = -2$$

We will now use this slope (-2) and the point (4, 0) in order to obtain the equation of the line. Since the slope is -2, we have

$$y = -2x + b$$

Substituting 4 for x and 0 for y yields

$$0 = -2(4) + b$$
$$0 = -8 + b$$
$$b = 8$$

Thus the line is $y = -2x + 8$.

As a final note, we could have used the point (1, 6) rather than (4, 0). The choice was purely arbitrary.

Parallel Lines

We will end this section with some observations about parallel lines. Any two lines that are *parallel* will have the same slope. If the slopes of two lines are different, the two lines will intersect.

parallel lines (same slope)   intersecting lines (different slopes)

The lines $y = 6x + 5$ and $y = 6x - 1$ are parallel, since they both have the same slope, 6. By contrast, the lines $y = 3x + 1$ and $y = 2x - 8$ are not parallel. They are intersecting lines because their slopes are different: the first line has slope 3 and the second line has slope 2.

EXAMPLE 21   Determine the equation of the line parallel to the line $y = 3x - 4$ and passing through the point (5, 11).

Solution   Any line that is parallel to the line $y = 3x - 4$ must have the same slope as it, namely, 3. This means that the line we seek has slope 3 and is of the form

$$y = 3x + b$$

Since the line must pass through the point (5, 11), $x = 5$ and $y = 11$ must satisfy the equation $y = 3x + b$. Thus we have

$$11 = 3(5) + b$$
$$11 = 15 + b$$
$$b = -4$$

The line we seek is $y = 3x - 4$.

EXERCISES 14.4

1. Determine the y-intercept of each line.
   *(a) $y = 6x + 9$ (b) $y = -3x + 2$
   *(c) $y = x - 6$ (d) $y = 2x - 1$
   *(e) $y = -5x + 1$ (f) $y = 12x + 15$
   *(g) $y = -7x$ (h) $y = 4x$

2. Determine the slope and y-intercept of each line.
   *(a) $y = 5x + 3$ (b) $y = 7x + 2$
   *(c) $y = -9x + 4$ (d) $y = -2x + 1$
   *(e) $y = -x + 7$ *(f) $y = x - 8$
   *(g) $y = 12x - 7$ *(h) $y = 6x$

3. Determine the equation of the line having the given slope and y-intercept.
   *(a) $m = 5, b = 2$ *(b) $m = 4, (0, b) = (0, 7)$
   *(c) $m = -2, b = 1$ *(d) $m = 6, (0, b) = (0, -1)$
   *(e) $m = 3, b = 0$ *(f) $m = -3, (0, b) = (0, 0)$
   *(g) $m = -1, b = 4$ (h) $m = -1, b = 0$
   *(i) $m = 1, b = 0$ (j) $m = 1, b = -10$

4. Find the equation of the line having the given slope and passing through the given point.
   *(a) $m = 3, (4, 1)$ (b) $m = 2, (3, 10)$
   *(c) $m = 1, (3, 7)$ (d) $m = 1, (4, 4)$
   *(e) $m = -4, (-3, 14)$ (f) $m = -5, (-2, 4)$
   *(g) $m = -2, (1, -5)$ (h) $m = -1, (-8, 2)$
   *(i) $m = 12, (-1, -3)$ (j) $m = 9, (-3, -4)$

5. Determine the equation of the line that passes through the two given points.
   *(a) $(3, 5), (7, 13)$ (b) $(1, 2), (4, 11)$
   *(c) $(1, 3), (2, 7)$ (d) $(4, 7), (5, 12)$
   *(e) $(2, 11), (1, 14)$ (f) $(4, 8), (2, 14)$
   *(g) $(-1, 6), (2, 12)$ (h) $(-2, -4), (0, 10)$
   *(i) $(3, -4), (-1, 8)$ (j) $(5, -1), (3, -7)$

6. Indicate whether the two lines are parallel or intersecting.
   *(a) $y = 4x + 5$ and $y = 4x - 9$
   (b) $y = 2x + 13$ and $y = 2x + 1$
   *(c) $y = -3x + 1$ and $y = 3x$
   *(d) $y = 5x$ and $y = 5x - 2$
   *(e) $y = 4x - 9$ and $y = -4x + 9$
   *(f) $y = 9x + 1$ and $y = -9x + 1$
   *(g) $y = x$ and $y = -x$
   (h) $y = -4x - 5$ and $y = -4x + 5$

7. Find the equation of the line parallel to the given line and passing through the given point.
   *(a) $y = 2x + 3, (4, 15)$ (b) $y = 5x - 4, (2, 10)$
   *(c) $y = -3x + 1, (-1, 2)$ *(d) $y = -4x - 1, (-3, 0)$
   *(e) $y = -x + 2, (1, -6)$ (f) $y = -x - 6, (4, 1)$
   *(g) $y = -7x - 4, (2, 0)$ (h) $y = x - 11, (0, 0)$

CHAPTER 14 REVIEW EXERCISES

*1. Sketch the graph of each line.
   (a) $3x + 4y = 12$ (b) $y = 2x - 5$ (c) $x = 3y - 2$
   (d) $y = 4$ (e) $x - 2y = 8$ (f) $2x - y = 0$
   (g) $x = -1$ (h) $x + y = 1$ (i) $y = \frac{1}{2}x + 1$
   (j) $x = 3 + y$

*2. Compute the slope of the line through each pair of points.
   (a) $(1, 7)$ and $(3, 19)$ (b) $(-1, 4)$ and $(3, -8)$

*3. Determine the slope and y-intercept of each line.
   (a) $y = -7x + 9$ (b) $y = 6x - 2$

*4. Determine the equation of the line described.
   (a) $m = 4, b = 3$ (b) $m = -2$, through $(3, 4)$
   (c) through $(3, 2)$ and $(5, 10)$ (d) through $(-1, 0)$ and $(2, 15)$

CUMULATIVE REVIEW EXERCISES (CHAPTERS 10–14)

*1. Simplify each expression completely.
   (a) $+7 - 13 + 4 - 8 + 16 - 2$ (b) $(-3)(-1)(-5) - (-8)^2$
   (c) -7^2 (d) $(-8) \div (-1)$

*2. Simplify, if possible.

(a) $(-6x^3)(-x^4y^8)$ (b) $(8m^6n^7)(-3m^2n)$

*3. Multiply; then simplify, if possible.

(a) $(3x - 2)(5x + 4)$ (b) $(t + 4)^2$

*4. Divide.

$$x + 5)\overline{\smash{\big)}\, x^3 - 4x + 20}$$

*5. Solve each linear equation.

(a) $-9x + 4 = -15$ (b) $2x - 19 = 7x + 11$

(c) $6(1 - 2x) = x + 1$ (d) $20 - 3(x + 4) = 0$

*6. Solve each word problem.

(a) The sum of four consecutive numbers is 346. Determine the numbers.

(b) Hank earns a 3% commission on all TV sets he sells. If his commission this week was $171, how many dollars worth of TV sets did he sell?

(c) Determine two complementary angles such that one angle is twice the other angle.

(d) Sylvia will invest $4500, part at 7% annual interest and the rest at 8%. How much should she invest at each rate in order to have an annual return that totals $328?

*7. Sketch the graph of each line.

(a) $3x - 4y = 12$ (b) $y = -2x + 5$

*8. Determine the slope of the line that passes through the points $(1, -4)$ and $(4, 2)$.

*9. Determine the equation of the line that passes through the points $(3, 1)$ and $(5, 9)$.

15

SYSTEMS OF LINEAR EQUATIONS

15.1 INTRODUCTION

We have solved equations that contain one unknown, such as

$$2x - 5 = 3$$

and

$$5 + 12(1 - x) = 7x - 9$$

But suppose that we must solve an equation containing two unknowns, such as

$$x + y = 10$$

There are *many solutions*; for example, $x = 7$ and $y = 3$ satisfy the equation. And here are four more solutions:

$$x = 6 \quad \text{and} \quad y = 4$$
$$x = 0 \quad \text{and} \quad y = 10$$
$$x = 10 \quad \text{and} \quad y = 0$$
$$x = 5 \quad \text{and} \quad y = 5$$

If these xy pairs were taken as points (x, y) and plotted, they would fall on a line, and the x, y coordinates of any point on the line would be a solution of the equation. As you can see then, there are an infinite number of such solutions.

On the other hand, if the problem is to solve a *pair* of equations containing two variables, such as

$$\begin{cases} x + y = 10 \\ x - y = 8 \end{cases}$$

then there is only *one solution*. The **solution** is a pair of numbers (a value for x and a value for y) that satisfy *both* equations. Here the solution is $x = 9$ and $y = 1$. You might want to check that this pair satisfies both equations.

The graph of each linear equation is a straight line. If you were to graph two lines, the point where they meet would be the solution. However, as noted in Chapter 14, graphing usually gives only an *approximate* solution.

If the lines happen to be *parallel* (and don't meet), then there is *no solution.* If the two lines are the *same,* all their points are common to each other so that there are an *infinite number of solutions.* Such unusual cases are presented in Section 15.5.

15.2 THE ADDITION METHOD (ELIMINATION)

One way to get an exact solution is to use the **addition method** of solving two equations in two unknowns. This method is also called **elimination.** It is explained next.

It would be easy to solve

$$\begin{cases} x + y = 10 \\ x - y = 8 \end{cases}$$

if we could eliminate either x or y, because then there would be only one unknown. Suppose that we add the two equations term for term.

$$\begin{array}{r} x + y = 10 \\ x - y = 8 \\ \hline 2x + 0 = 18 \end{array}$$
$\left\{ \begin{array}{l} \text{Since } x - y \text{ } is \text{ } equal \text{ } to \text{ } 8, \text{ we are} \\ \text{adding equal quantities to both sides} \\ \text{of the equation } x + y = 10. \end{array} \right.$

The result is an equation in one unknown, x.

$$2x = 18$$

This equation is easily solved for x.

$$x = 9$$

The first original equation indicates that $x + y = 10$. So if x is 9, then y must be 1. Therefore the solution is

$$x = 9 \quad \text{and} \quad y = 1 \quad \checkmark \quad \text{ or } \quad (9, 1)$$

This pair of equations was easy to solve because the y's added to zero and were thus eliminated. Usually neither of the variables can be immediately eliminated. In such cases, you must multiply either one equation by an appropriate constant or both equations by appropriate constants. This approach is demonstrated in the next two examples.

EXAMPLE 1   Solve by the addition method.

$$\begin{cases} 2x - 3y = 4 \\ 5x + y = 27 \end{cases}$$

Solution   If we add right away, we fail.

$$\begin{cases} 2x - 3y = 4 \\ 5x + y = 27 \end{cases}$$
$$\underline{}$$
$$7x - 2y = 31 \quad ???$$

Instead multiply both sides of the second equation by 3 so that the coefficients of y will be equal in magnitude (both 3 in this case) and opposite in sign (one $-$, one $+$). Then the y's will add to zero.

$$\begin{cases} 2x - 3y = 4 \xrightarrow{\text{same}} 2x - 3y = 4 \\ 5x + y = 27 \xrightarrow{\times 3} \underline{15x + 3y = 81} \end{cases}$$
$$17x = 85$$
$$x = 5$$

We can solve for y now by using original equation $5x + y = 27$. Since x is 5, we get $5(\underline{5}) + y = 27$, or $y = 2$. The solution is $x = 5$, $y = 2$, or $(5, 2)$. Note that it does not matter which of the original equations is used to obtain the value of the second variable.

In the next example you will see a system in which both equations must be multiplied in order to eliminate one of the unknowns.

EXAMPLE 2   Solve by the addition method.

$$\begin{cases} 2x + 7y = 3 \\ 3x + 13y = 7 \end{cases}$$

Solution   We might multiply the top equation by -3 and the bottom one by 2. As a result, the coefficients of x will be equal in magnitude and opposite in sign. Then the x terms will add to zero.

$$\begin{cases} 2x + 7y = 3 \xrightarrow{-3} -6x - 21y = -9 \\ 3x + 13y = 7 \xrightarrow{2} \underline{6x + 26y = 14} \end{cases}$$
$$5y = 5$$
$$y = 1 \quad \checkmark$$

Next, use one of the original equations, say $3x + 13y = 7$, and substitute 1 for y. This step will give us the corresponding value of x.

$$3x + 13y = 7$$
$$3x + 13(\underline{1}) = 7$$
$$3x + 13 = 7$$
$$3x = -6$$
$$x = -2 \quad \checkmark$$

The solution is $x = -2$, $y = 1$, or $(-2, 1)$.

EXAMPLE 3   (**optional**) Solve by the addition method.

$$\begin{cases} 6x + 13y = 5 \\ y = 2x - 7 \end{cases}$$

Solution   As you can see, this system is not actually set up (lined up) to be solved by the addition method. In fact, *in the next section we'll solve it very easily by a method called substitution.* However, if we did decide to solve the system by the addition method, we would proceed as follows.

If $-2x$ is added to both sides of the second equation, it will be changed so that it has the same form as the top one, namely, variables lined up on one side of the equal sign and constants on the other side. Then both equations will line up for the addition method.

$$\begin{cases} 6x + 13y = 5 \\ y = 2x - 7 \end{cases}$$
$$\downarrow$$

$$\begin{cases} 6x + 13y = 5 \xrightarrow{\text{same}} 6x + 13y = 5 \\ -2x + y = -7 \xrightarrow{3} -6x + 3y = -21 \end{cases}$$
$$16y = -16$$
$$y = -1 \ \checkmark$$

Next, take one of the original equations and substitute -1 for y.

$$\underline{y} = 2x - 7$$
$$-1 = 2x - 7$$
$$6 = 2x$$
$$x = 3 \ \checkmark$$

The solution is $x = 3$, $y = -1$, or $(3, -1)$.

Some word problems lead naturally to the use of *two* unknowns, as shown in the next example.

EXAMPLE 4   Suppose that 5 hamburgers and 3 soft drinks cost \$4.35 and that 3 hamburgers and 2 soft drinks cost \$2.70. Determine the cost of one hamburger.

Solution   Let $x =$ cost of one hamburger
$y =$ cost of one soft drink
Then

$$5x + 3y = \$4.35$$
and
$$3x + 2y = \$2.70$$

Since expressing cost in cents will make the arithmetic easier, we will replace \$4.35 by 435 and replace \$2.70 by 270. The resulting system:

$$\begin{cases} 5x + 3y = 435 \\ 3x + 2y = 270 \end{cases}$$

This system can be solved by the addition method to obtain $x = 60$ and $y = 45$. This means that *hamburgers cost 60¢ each* and soft drinks cost 45¢ each.

EXERCISES 15.2

Answers to starred exercises are given in the back of the book.

1. Solve each system of equations by using the addition method (elimination).

*(a) $\begin{cases} 2x + y = 19 \\ 3x - y = 11 \end{cases}$ (b) $\begin{cases} 5x + y = 20 \\ 2x - y = -6 \end{cases}$

*(c) $\begin{cases} 2x + 5y = 13 \\ -2x + 11y = 3 \end{cases}$ *(d) $\begin{cases} -3x + 4y = 2 \\ 3x - 5y = -1 \end{cases}$

*(e) $\begin{cases} 7x - 3y = 11 \\ 4x + y = 9 \end{cases}$ *(f) $\begin{cases} 7x - y = 10 \\ 3x + 4y = -9 \end{cases}$

*(g) $\begin{cases} 5x + 8y = 14 \\ x + 5y = 13 \end{cases}$ (h) $\begin{cases} 7x - 2y = 8 \\ 5x - y = 7 \end{cases}$

*(i) $\begin{cases} -4x + 3y = 10 \\ x + 5y = 9 \end{cases}$ (j) $\begin{cases} 3x - 5y = 45 \\ -x - 2y = 7 \end{cases}$

*(k) $\begin{cases} 6m + 7n = 30 \\ -2m + 9n = -10 \end{cases}$ (l) $\begin{cases} 3s + 2t = -12 \\ 5s + 4t = -22 \end{cases}$

2. Solve each system of equations by using the addition method (elimination).

*(a) $\begin{cases} 7x - 4y = 10 \\ 4x + 3y = 11 \end{cases}$ *(b) $\begin{cases} -3x + 5y = 9 \\ 2x + 3y = 13 \end{cases}$

*(c) $\begin{cases} 4x - 6y = 0 \\ 7x + 5y = 31 \end{cases}$ (d) $\begin{cases} -3p + 8q = 6 \\ 5p - 7q = 9 \end{cases}$

*(e) $\begin{cases} 3x + 2y = 6 \\ 7x + 5y = 13 \end{cases}$ (f) $\begin{cases} 5x + 8y = 3 \\ 4x + 9y = -8 \end{cases}$

*(g) $\begin{cases} 6m - 7n = 9 \\ 9m - 10n = 15 \end{cases}$ (h) $\begin{cases} 3x - 7y = -12 \\ 4x - 11y = -16 \end{cases}$

*(i) $\begin{cases} 2x + 3y = 5 \\ 5x + 4y = 2 \end{cases}$ *(j) $\begin{cases} 7a + 4b = -1 \\ 3a + 5b = 16 \end{cases}$

*(k) $\begin{cases} 4x - 3y = -5 \\ 5x - 2y = 6 \end{cases}$ (l) $\begin{cases} 8x - 5y = 6 \\ 5x - 3y = 5 \end{cases}$

*(m) $\begin{cases} 8x + 3y = 13 \\ 10x - 7y = -16 \end{cases}$ (n) $\begin{cases} 7x + 4y = 24 \\ 5x + 7y = 13 \end{cases}$

3. (optional) Solve each system of equations by using the addition method.

*(a) $\begin{cases} 5x + 6y = 11 \\ x = 3y - 2 \end{cases}$ (b) $\begin{cases} y = 3x + 1 \\ 7x - 6y = 5 \end{cases}$

*(c) $\begin{cases} 5x + 2y = 15 \\ y = 3x + 13 \end{cases}$ (d) $\begin{cases} 4x - 3y = 9 \\ y = 7 - 2x \end{cases}$

4. Solve each of the following word problems by establishing two equations in two unknowns and then solving that system of equations.

*(a) Suppose that 4 hot dogs and 3 milkshakes cost $4.80 and that 3 hot dogs and 2 milkshakes cost $3.45. Determine the cost of 1 hot dog and the cost of 1 milkshake.

(b) A doughnut shop will sell you 5 doughnuts and 2 cups of coffee for $2.65. They will also sell you 7 doughnuts and 3 cups of coffee for $3.80. How much does 1 doughnut cost?

*(c) The sum of two numbers is 41. Their difference is 13. Find the numbers.

(d) The sum of two numbers is 151. Their difference is 65. Find the numbers.

15.3 SUBSTITUTION

A better way to solve the system of Example 3 is by a method called **substitution.** The system is

$$\begin{cases} 6x + 13y = 5 \\ y = 2x - 7 \end{cases}$$

The second equation says that y is the same as $2x - 7$. So let's substitute $2x - 7$ for y *in the other equation.*

$$6x + 13y = 5$$

becomes

$$6x + 13(2x - 7) = 5$$

The equation $6x + 13(2x - 7) = 5$ contains only *one* variable, x, and is easily solved. In steps,

$$6x + 13(2x - 7) = 5$$
$$6x + 26x - 91 = 5$$
$$32x - 91 = 5$$
$$32x = 96$$
$$x = 3 \quad \checkmark$$

It is now a simple matter to determine y, since we have a $y = \cdots$ equation— namely, $y = 2x - 7$.

becomes
$$\begin{aligned} y &= 2x - 7 \\ y &= 2(3) - 7 \qquad \text{when we substitute 3 for } x \\ y &= 6 - 7 \\ y &= -1 \quad \checkmark \end{aligned}$$

The solution is $x = 3$, $y = -1$, or $(3, -1)$.

EXAMPLE 5   Solve by the substitution method.

$$\begin{cases} x = 5y + 8 \\ 2x - 3y = 2 \end{cases}$$

Solution   Use $x = 5y + 8$ to substitute for x in $2x - 3y = 2$.

$$
\begin{aligned}
2x - 3y &= 2 \\
2(\underline{5y + 8}) - 3y &= 2 \qquad \text{by substitution of } 5y + 8 \text{ for } x \\
10y + 16 - 3y &= 2 \\
7y + 16 &= 2 \\
7y &= -14 \\
y &= -2 \quad \checkmark
\end{aligned}
$$

We now find x.

$$
\begin{aligned}
x &= 5y + 8 \\
x &= 5(\underline{-2}) + 8 \qquad \text{when we substitute } -2 \text{ for } y \\
x &= -10 + 8 \\
x &= -2 \quad \checkmark
\end{aligned}
$$

The solution is $x = -2$, $y = -2$, or $(-2, -2)$.

EXAMPLE 6   **(optional)** Solve by substitution.

$$\begin{cases} 5x + 4y = 13 \\ 2x + 3y = \ \ 8 \end{cases}$$

Solution   It would be better (easier) to solve this system by using the addition method, since it is already aligned for addition. Nevertheless, we *can* use substitution. The work that follows shows the procedure to use if you really wanted to solve this system by substitution.

In order to be able to make a substitution, we begin by solving the second equation for x in terms of y.

$$
\begin{aligned}
2x + 3y &= 8 \\
2x &= -3y + 8 \\
x &= \frac{-3y + 8}{2} \\
x &= -\frac{3}{2}y + 4
\end{aligned}
$$

Next, substitute $-\frac{3}{2}y + 4$ for x in $5x + 4y = 13$ to get

$$5\left(-\frac{3}{2}y + 4\right) + 4y = 13$$

or

$$-\frac{15}{2}y + 20 + 4y = 13$$

Multiply both sides of the equation by 2 to eliminate fractions and proceed.

$$-15y + 40 + 8y = 26$$
$$-7y + 40 = 26$$
$$-7y = -14$$
$$y = 2 \ \checkmark$$

To solve for x, substitute 2 for y in the equation $x = -\frac{3}{2}y + 4$.

$$x = -\frac{3}{2}y + 4$$

$$x = -\frac{3}{2}(2) + 4$$

$$x = -3 + 4$$

$$x = 1 \ \checkmark$$

The solution is $x = 1$, $y = 2$, or $(1, 2)$.

In Chapter 14 we saw how two simultaneous linear equations can be solved by graphing to determine *approximately* where the two lines (each an equation) meet. So we have three methods for solving simultaneous linear equations.

1. Addition (or elimination)
2. Substitution
3. Graphing

EXERCISES 15.3

1. Solve each system of equations by using substitution.

*(a) $\begin{cases} 2x + 3y = 13 \\ y = x + 1 \end{cases}$ (b) $\begin{cases} y = 3x + 13 \\ 5x + 2y = 15 \end{cases}$ *(c) $\begin{cases} x = 4y - 4 \\ 2x + 3y = 14 \end{cases}$

(d) $\begin{cases} 5x + 6y = 11 \\ x = 3y - 2 \end{cases}$ *(e) $\begin{cases} 7x - 6y = 5 \\ y = 3x + 1 \end{cases}$ (f) $\begin{cases} x - 7y = 9 \\ y = x - 3 \end{cases}$

*(g) $\begin{cases} x = y \\ 7x + 4y = 44 \end{cases}$ (h) $\begin{cases} 8x - 3y = 20 \\ y = x \end{cases}$ *(i) $\begin{cases} y = 5x + 1 \\ 4x + 3y = 10 \end{cases}$

(j) $\begin{cases} x = y + 2 \\ 6x - 8y = 6 \end{cases}$ *(k) $\begin{cases} x = 3y - 3 \\ y = 2x - 4 \end{cases}$ (l) $\begin{cases} x = 2y - 16 \\ y = 3x + 18 \end{cases}$

*(m) $\begin{cases} a = 1 - b \\ 5a + 4b = 2 \end{cases}$ (n) $\begin{cases} 4x - 3y = 9 \\ y = 7 - 2x \end{cases}$ *(o) $\begin{cases} y = x \\ 5x - 2y = 33 \end{cases}$

2. Solve each system of equations. Use the addition method (elimination) if the system is aligned for addition. Use substitution if one of the equations is already solved for x in terms of y or y in terms of x.

*(a) $\begin{cases} y = 4x - 1 \\ 10x - 2y = 2 \end{cases}$ (b) $\begin{cases} 7x - 5y = 25 \\ 2x + 3y = 16 \end{cases}$

*(c) $\begin{cases} -3x + 2y = 7 \\ 4x + 6y = -5 \end{cases}$ (d) $\begin{cases} 3x - 10y = 16 \\ x = 4y + 5 \end{cases}$

*(e) $\begin{cases} 4x + 5y = 12 \\ y = 1 - x \end{cases}$ (f) $\begin{cases} 10x + 3y = 14 \\ 6x + 5y = 18 \end{cases}$

3. (optional) Solve each system of equations by using substitution.

*(a) $\begin{cases} 4x + y = 9 \\ 7x - 3y = 11 \end{cases}$ (b) $\begin{cases} 7x - 2y = 8 \\ 5x - y = 7 \end{cases}$

*(c) $\begin{cases} 7x + 5y = 13 \\ 3x + 2y = 6 \end{cases}$ (d) $\begin{cases} 2x + 3y = 5 \\ 5x + 4y = 2 \end{cases}$

4. Solve each of the following word problems by establishing two equations in two unknowns and then solving that system of equations.

*(a) The length (y) of a rectangular field is twice the width (x). The perimeter is 690 meters. Find the length and width of the field.

(b) The length of a rectangle is 7 inches more than the width. The perimeter is 114 inches. Find the length and width of the rectangle.

*(c) One number is four times another number. The sum of the numbers is 120. Determine the two numbers.

(d) One number is three times another number. The difference between the numbers is 94. Determine the two numbers.

*(e) There are two complementary angles such that one angle is eight times the other angle. Determine the angles. (Recall that two angles are complementary if their sum is 90°.)

(f) There are two supplementary angles such that one angle is three times the other angle. Determine the angles. (Recall that two angles are supplementary if their sum is 180°.)

15.4 CHECKING A SOLUTION

To check your solution to a pair of simultaneous linear equations, check the two values in both equations. *Both equations must balance.* From a graphing point of view, you are checking to see if the point is in fact on both lines.

EXAMPLE 7 Is $x = 5$, $y = 2$ the solution of the following equations?

$$\begin{cases} 2x - 3y = 4 \\ 5x + y = 27 \end{cases}$$

Solution Check 5 and 2 in the first equation.

$$2x - 3y = 4$$
$$2(5) - 3(2) \overset{?}{=} 4$$
$$10 - 6 \overset{?}{=} 4$$
$$4 = 4 \quad \checkmark \qquad \text{Checks in the first equation.}$$

Now check 5 and 2 in the second equation.

$$5x + y = 27$$
$$5(5) + (2) \stackrel{?}{=} 27$$
$$25 + 2 \stackrel{?}{=} 27$$
$$27 = 27 \quad \checkmark \quad \text{Checks in the second equation, too.}$$

The solution is $x = 5$, $y = 2$; it checks in *both* equations.

EXAMPLE 8 Is $x = 2$, $y = 3$ the solution of the following equations?

$$\begin{cases} 5x + 3y = 19 \\ 2x + 7y = 24 \end{cases}$$

Solution Check 2 and 3 in the first equation.

$$5x + 3y = 19$$
$$5(2) + 3(3) \stackrel{?}{=} 19$$
$$10 + 9 \stackrel{?}{=} 19$$
$$19 = 19 \quad \checkmark \quad \text{Checks in the first equation.}$$

Now check 2 and 3 in the second equation.

$$2x + 7y = 24$$
$$2(2) + 7(3) \stackrel{?}{=} 24$$
$$4 + 21 \stackrel{?}{=} 24$$
$$25 \neq 24 \quad \text{The check fails in the second equation.}$$

Since the check fails in at least one of the equations, we know that $x = 2$ and $y = 3$ is not the correct x, y pair—they are not the solution.

EXERCISES 15.4

1. Solve and check each system of equations. In each instance use the method (addition or substitution) that appears easier.

*(a) $\begin{cases} 7x + 2y = 1 \\ 5x + 3y = 7 \end{cases}$
(b) $\begin{cases} x = 1 - 3y \\ y = 2x - 16 \end{cases}$
*(c) $\begin{cases} 9y - 5x = 2 \\ 4y + 3x = 27 \end{cases}$

(d) $\begin{cases} 10a - 7b = 24 \\ 4a - 3b = 10 \end{cases}$
*(e) $\begin{cases} 5x - 6y = 13 \\ x = 2y - 3 \end{cases}$
(f) $\begin{cases} 7x - 2y = 1 \\ y = 1 + 5x \end{cases}$

15.5 UNUSUAL CASES

You will not always succeed when you attempt to solve a system of two linear equations in two unknowns. There are two situations in which you will fail to obtain a solution.

  If two lines are parallel, then they do not meet. So when you attempt to solve a system of equations that represents parallel lines, you will not succeed. Instead of a solution, you will get such results as $5 = 0$ or $3 = 7$—that is, "equations" consisting of two unequal constants, one on each side of the equal sign.

Although two equations may appear different, they may only be different forms of the same equation. This is the other situation in which you will fail to obtain a unique solution. In terms of lines, the two lines are in reality only one line. So all the points that are on the one line are also on the "other" line. Consequently, there is not one solution in this case but rather an infinite number of solutions. Instead of obtaining a solution when you try to solve the system algebraically, you obtain results like $0 = 0$ or $4 = 4$. That is, you get an equation consisting of two equal constants, one on each side of the equal sign.

EXERCISES 15.5

*1. Solve each system of equations *if possible*. Use whichever method you prefer—addition or substitution. *Beware of unusual cases*.

(a) $\begin{cases} x + 4y = 8 \\ x = 3y + 1 \end{cases}$ (b) $\begin{cases} 3x + 8y = 13 \\ 7x - 6y = 18 \end{cases}$

(c) $\begin{cases} y = 5x - 1 \\ 10x - 2y = 2 \end{cases}$ (d) $\begin{cases} x = 6 \\ 5x - 7y = 16 \end{cases}$

(e) $\begin{cases} 4x + 2y = 17 \\ y = 3 - 2x \end{cases}$ (f) $\begin{cases} 3x - 2y = 10 \\ -6x + 4y = 7 \end{cases}$

(g) $\begin{cases} y = x \\ 2x + 3y = 15 \end{cases}$ (h) $\begin{cases} -2x + 5y = 8 \\ 4x - 10y = -16 \end{cases}$

15.6 SYSTEMS OF THREE EQUATIONS (OPTIONAL)

Systems of three equations in three unknowns can be solved by reducing the system to two equations in two unknowns. This we can do by using the addition method.

EXAMPLE 9   Solve the system below.

$$\begin{cases} 5x + 2y + z = 5 \\ 2x + 3y - z = -6 \\ 7x + 4y + 2z = 4 \end{cases}$$

Solution   A brief study of the coefficients of x, y, and z suggests that z will be easiest to eliminate.

If the first and second equations are added, z will be eliminated.

$$\begin{array}{r} 5x + 2y + z = 5 \\ 2x + 3y - z = -6 \\ \hline 7x + 5y = -1 \end{array}$$

Returning to the original equations, note that if twice the second equation is added to the third equation, z will be eliminated.

$$2x + 3y - z = -6 \xrightarrow{\ 2\ } 4x + 6y - 2z = -12$$
$$7x + 4y + 2z = \ \ 4 \longrightarrow \underline{7x + 4y + 2z = \ \ \ 4}$$
$$11x + 10y \qquad = -8$$

We now have two equations *in the same two unknowns*, x and y: $7x + 5y = -1$ and $11x + 10y = -8$. They can be combined as one system and easily solved for x and y. Specifically, multiply $7x + 5y = -1$ by -2 in order to eliminate y.

$$\begin{cases} 7x + 5y = -1 \xrightarrow{\ -2\ } -14x - 10y = \ \ 2 \\ 11x + 10y = -8 \longrightarrow \ \ \underline{11x + 10y = -8} \end{cases}$$
$$-3x \qquad = -6$$
$$x \qquad = \ \ 2 \ \checkmark$$

Next, y can be determined by using 2 for x in the equation $7x + 5y = -1$.

$$7(2) + 5y = -1$$
$$14 + 5y = -1$$
$$5y = -15$$
$$y = -3 \quad \checkmark$$

Finally, return to the original system, select any equation, and determine z by substituting 2 for x and -3 for y. The first equation, $5x + 2y + z = 5$, appears to be the easiest to use because the coefficient of the unknown z is simply 1.

$$5x + 2y + z = 5$$
$$5(2) + 2(-3) + z = 5$$
$$10 - 6 + z = 5$$
$$4 + z = 5$$
$$z = 1 \ \checkmark$$

The solution, therefore, is

$$x = \ \ 2$$
$$y = -3$$
$$z = \ \ 1$$

Each linear equation in three variables represents a plane in three dimensions. The solution to the system then can be visualized as the point where the three planes meet. This is an (x, y, z) point in three dimensions. For the preceding example, the planes meet at $(2, -3, 1)$.

EXERCISES 15.6

*1. Solve each system of equations.

(a) $\begin{cases} x + y + z = \ \ 6 \\ 2x + 3y - z = \ \ 5 \\ 3x + 2y + z = 10 \end{cases}$ (b) $\begin{cases} 5x - y + 2z = 8 \\ 2x + y + 5z = 6 \\ 3x - y + 3z = 7 \end{cases}$

$$\text{(c)} \begin{cases} 5x + 3y + z = 7 \\ x + 4y - 2z = 8 \\ 3x + 7y - 4z = 18 \end{cases} \quad \text{(d)} \begin{cases} x - 4y + 3z = 1 \\ -2x + 3y + 2z = -7 \\ 3x + 2y + 5z = 17 \end{cases}$$

$$\text{(e)} \begin{cases} 7x + 3y + 2z = 9 \\ 3x - y + 3z = 17 \\ 5x + 4y + 4z = 13 \end{cases} \quad \text{(f)} \begin{cases} 2x - 3y + 4z = -3 \\ 4x + 5y - 5z = 31 \\ 6x + 2y + 3z = 20 \end{cases}$$

$$\text{(g)} \begin{cases} 2r + 3s + 2t = 3 \\ 3r + 5s + 7t = 19 \\ 6r + 4s + 3t = 12 \end{cases} \quad \text{(h)} \begin{cases} 5m + 2n - 3p = -8 \\ 3m - 3n + 5p = 13 \\ -4m + 2n - 2p = -6 \end{cases}$$

CHAPTER 15 REVIEW EXERCISES

*1. Solve each system of equations by the addition method. Check your solution.

$$\text{(a)} \begin{cases} 6x - 5y = 6 \\ 5x - 3y = -2 \end{cases} \quad \text{(b)} \begin{cases} 5a + 7b = 13 \\ 7a + 4b = 24 \end{cases}$$

*2. Solve each system of equations by substitution. Check your solution.

$$\text{(a)} \begin{cases} y = 5 - 2x \\ x = 2 - y \end{cases} \quad \text{(b)} \begin{cases} x = 19 - 3y \\ 7x - 4y = 8 \end{cases}$$

*3. Solve each system of equations. Use whichever method you prefer—addition or substitution.

$$\text{(a)} \begin{cases} 7m + 9n = 2 \\ n = 2m + 3 \end{cases} \quad \text{(b)} \begin{cases} 4x - 3y = 7 \\ 12x + 5y = -7 \end{cases}$$

CUMULATIVE REVIEW EXERCISES (CHAPTERS 10–15)

*1. Simplify each expression completely.
   (a) $10 - 2(-9) + (-3)^3$ (b) $(-2)(+1)(-5) - (-4)(+3)$

*2. Multiply.
   (a) $5x^4(3x^3 + 7x^5)$ (b) $(x + 8)(x - 7)$

*3. Divide.
$$x - 2 \overline{)\, x^3 + 3x^2 - 7x - 6}$$

*4. Combine like terms.
$$-3m + m + 8m - 11m + 4m - m$$

*5. Solve each linear equation.
   (a) $8x - 3 = 15x + 1$ (b) $3(2 + 7x) + x = 13$

*6. Solve each word problem.
   (a) 294 feet of fence will be used to enclose a rectangular garden area. The length of the garden will be 3 feet more than twice the width. Determine the length and width of the garden.
   (b) Two trains leave a station at the same time and go in opposite directions. One train travels 25 mph faster than the other. If they are 870 miles apart after 6 hours, how fast is each train traveling?

*7. Sketch a graph of the lines $y = 3x + 2$ and $y = -2x + 7$ and estimate where they meet.

*8. Determine the equation of the line with slope -2 that passes through the point $(5, 4)$.

*9. Determine the equation of the line that passes through the points $(-2, 7)$ and $(1, 19)$.

*10. Solve the system of linear equations by using the addition method.
$$\begin{cases} 2x + 5y = 19 \\ 3x + 7y = 26 \end{cases}$$

*11. Solve the system of linear equations by using substitution.
$$\begin{cases} x = 2y + 14 \\ 7x - 3y = 54 \end{cases}$$

16

INEQUALITIES

16.1 INTRODUCTION

Much of the algebra presented thus far has dealt with linear equations—solving a linear equation in one unknown, sketching the graph of a linear equation in two variables, and solving systems of linear equations. The concern has been to find numbers that make one expression equal to another.

In this chapter, we digress to study **inequalities,** statements in which one quantity is greater than or less than another. The inequalities will be linear, just as previous equations have been linear.

We use the symbol $>$ to mean **greater than** and $<$ to mean **less than.**

- $5 > 3$ says that five is greater than three.
- $3 < 5$ says that three is less than five.

In general, the statements $a > b$ and $b < a$ are equivalent; they mean the same thing.

The symbol $\geq$ means *greater than or equal to;* $\leq$ means *less than or equal to.* Thus

$$5 \geq 3 \quad \text{and} \quad 3 \leq 5$$
$$5 \geq 5 \quad \text{and} \quad 5 \leq 5$$

The inequality $x > 5$ refers to all numbers x that are greater than 5. It says that x is greater than 5. This can be drawn as

The statement $x \geq 5$ refers to all numbers x that are either greater than or equal to 5. It says that x is greater than or equal to 5. This can be represented as

An open circle is used to indicate that the endpoint is not included and a closed circle indicates that the endpoint is included.

All numbers between 2 and 6, including 2 and 6, would be all numbers x such that

$$x \geq 2 \quad \text{and} \quad x \leq 6$$

These two inequalities can be combined as $2 \leq x \leq 6$. You can see in it the separate statements

$$2 \leq x \quad \text{same as } x \geq 2$$

and

$$x \leq 6$$

The statement $2 \leq x \leq 6$ can be read as either

- All x between (and including) 2 and 6, or
- All x greater than or equal to 2 and less than or equal to 6.

This is graphed as

The **interval notation** [2, 6] can be used to mean $2 \leq x \leq 6$. If we want to represent $2 < x \leq 6$, we can write (2, 6] or draw

Similarly, $2 < x < 6$ is (2, 6) or

Finally, $2 \leq x < 6$ is [2, 6) or

EXAMPLE 1   Draw $-5 < m \le 3$.

Solution

EXAMPLE 2   Draw x satisfying $x > 3$ or $x < -5$.

Solution   We use *or* rather than *and* because a number cannot be both greater than 3 and less than -5. This situation is clarified by the drawing below.

It is tempting but quite *wrong* to write $x > 3$ or $x < -5$ as

$$3 < x < -5 \quad \text{(wrong)}$$

which reads $x > 3$ *and* $x < -5$, which cannot be.

EXERCISES 16.1

Answers to starred exercises are given in the back of the book.

*1. Write each inequality in interval notation.
   (a) $3 < x < 8$ (b) $5 \le x \le 6$ (c) $0 \le t < 5$
   (d) $1 < y \le 19$ (e) $10 \le x \le 100$ (f) $-6 \le x < 0$
   (g) $-4 < w \le -1$ (h) $-10 < x < 10$

*2. Write each interval using an inequality.
   (a) $(2, 5)$ (b) $[2, 5]$ (c) $(1, 7]$ (d) $[3, 11)$
   (e) $[-3, 0]$ (f) $(-2, 4]$ (g) $[-7, -3]$ (h) $(0, 7)$

*3. Make a drawing to represent each interval.
   (a) $4 < x < 7$ (b) $0 \le y < 7$ (c) $0 \le x \le 1$
   (d) $-5 < t \le 3$ (e) $[4, 12]$ (f) $[5, 8)$
   (g) $(9, 10)$ (h) $(-4, 1]$ (i) $x < 4$
   (j) $x \ge -1$ (k) $x > 2$ *or* $x < -2$ (l) $t \le -3$ *or* $t > 2$

16.2 SOLVING INEQUALITIES

Inequalities resemble equations except that they involve $> < \ge \le$ instead of $=$. The following are examples of linear inequalities.

$$2x > 6$$
$$5x - 1 \le 19$$
$$4(x - 1) + 3 \ge x - 3$$

In order to solve such inequalities, we use three basic properties. For the most part, these properties resemble those used to solve equations. The key distinction is given in the box below.

1. If the same number is added to both sides of an inequality, the direction of the inequality ($>$ or $<$) remains the same.

Example. $5 > 2$; so $5 + \underline{8} > 2 + \underline{8}$—that is, $13 > 10$.

Example. $8 > -4$; so $8 + (\underline{-3}) > -4 + (\underline{-3})$—that is, $5 > -7$.

Example. $-2 < 4$; so $-2 + (\underline{-6}) < 4 + (\underline{-6})$—that is, $-8 < -2$.

Example. $-8 < -3$; so $-8 + \underline{17} < -3 + 17$—that is, $9 < 14$.

2. If both sides of an inequality are multiplied or divided by a positive number, the direction of the inequality ($>$ or $<$) remains the same.

Example. $5 > -2$; so $5 \cdot \underline{3} > -2 \cdot \underline{3}$—that is, $15 > -6$.

Example. $4 < 7$; so $4 \cdot \underline{2} < 7 \cdot \underline{2}$—that is, $8 < 14$.

Example. $10 > -4$; so $\dfrac{10}{2} > \dfrac{-4}{2}$—that is, $5 > -2$.

Example. $6 < 9$; so $\dfrac{6}{3} < \dfrac{9}{3}$—that is, $2 < 3$.

3. If both sides of an inequality are multiplied or divided by a negative number, the direction of the inequality is reversed. That is, $>$ becomes $<$ and $<$ becomes $>$.

Example. $-2 > -3$, so $(-2)(\underline{-5}) < (-3)(\underline{-5})$—that is, $10 < 15$.

Example. $3 < 4$, so $(3)(\underline{-5}) > (4)(\underline{-5})$—that is, $-15 > -20$.

Example. $-12 < -4$, so $\dfrac{-12}{-2} > \dfrac{-4}{-2}$—that is, $6 > 2$.

Example. $24 > 16$, so $\dfrac{24}{-2} < \dfrac{16}{-2}$—that is, $-12 < -8$.

Only this last property can be troublesome. The others are essentially the same as those we used when solving *equations*. But note the one difference again.

> If you multiply or divide both sides of an inequality by a negative number, the direction of the inequality is reversed: $>$ becomes $<$ and $<$ becomes $>$.

As you follow the next three examples, notice how the method of solving these inequalities is similar to the method of solving linear equations.

EXAMPLE 3   Solve $2x + 5 < 9$.

Solution   Add -5 to both sides to get the x term alone.

$$\begin{array}{rcr} 2x + 5 < & 9 \\ -5 & & -5 \\ \hline 2x & < & 4 \end{array}$$

Now divide both sides by 2 to get a coefficient of 1 for x.

$$x < 2 \quad \checkmark$$

As a partial check, any number less than 2 should check in $2x + 5 < 9$. For instance,

$$x = 0: \quad 2 \cdot 0 + 5 \overset{?}{<} 9 \qquad\qquad x = -2: \quad 2(-2) + 5 \overset{?}{<} 9$$
$$0 + 5 \overset{?}{<} 9 \qquad\qquad\qquad\qquad -4 + 5 \overset{?}{<} 9$$
$$5 < 9 \quad \checkmark \qquad\qquad\qquad\qquad\quad 1 < 9 \quad \checkmark$$

Also, any number greater than or equal to 2 should not check in $2x + 5 < 9$. For instance,

$$x = 3: \quad 2 \cdot 3 + 5 \overset{?}{<} 9$$
$$6 + 5 \overset{?}{<} 9$$
$$11 \overset{?}{<} 9 \quad \textit{No!}$$

EXAMPLE 4   Solve $5x + 2 - 9x \geq 17$.

Solution   If we combine like terms on the left side, we obtain

$$-4x + 2 \geq 17$$

Add -2 to both sides to get the x term alone.

$$-4x \geq 15$$

Divide both sides by -4 in order to get a coefficient of 1 for x.

$$x \leq -\frac{15}{4} \quad \left\{ \begin{array}{l} \text{Division by } -4, \text{ a } \textit{negative number,} \text{ reverses the} \\ \text{direction of the inequality; } \geq \text{ becomes } \leq. \end{array} \right.$$

EXAMPLE 5   Solve $3(x + 1) - 7 > 8x + 9$.

Solution   Distribute the 3 to eliminate parentheses.

$$3x + 3 - 7 > 8x + 9$$

Combine the like terms on the left side.

$$3x - 4 > 8x + 9$$

To get the x term alone, first add $+4$ to both sides.

$$3x > 8x + 13$$

Then add $-8x$ to both sides.

$$-5x > 13$$

Obtain a coefficient of 1 for x.

$$x < -\frac{13}{5} \qquad \begin{cases} \text{after division of both sides by } -5 \\ \text{and consequent change of } > \text{ to } < \end{cases}$$

EXAMPLE 6 How long must you make the sides of a square if the perimeter of the square is to be at least 108 centimeters?

Solution Let

$$x = \text{the length of a side of the square}$$

The perimeter is then $4x$, and we insist that the perimeter be at least 108 centimeters. The words "at least" mean the same as the symbol $\geq$. Thus we have the inequality

$$4x \geq 108$$

The inequality is readily solved by dividing both sides by 4. The result is

$$x \geq 27$$

Thus, for the perimeter of the square to be at least 108 centimeters, *each side of the square must be at least 27 centimeters.*

EXERCISES 16.2

*1. Solve each inequality.

(a) $x + 1 < 6$
(b) $t + 15 > -9$
(c) $5x + 1 > 7$
(d) $3x - 4 \leq -8$
(e) $9y + 5 < 5y - 2$
(f) $-8n - 9 < 6n$
(g) $5x + 3 + 2x < -4$
(h) $2x - 7 - 9x \leq 6$
(i) $4(x - 2) - 3 > x$
(j) $4x + 2(3 - x) \geq 6x + 7$

2. Solve each inequality.

*(a) $y - 7 > -1$
(b) $3m - 6 \geq 0$
*(c) $-2x + 3 < 6$
(d) $2x - 9 > -3$
*(e) $3x + 1 - x \geq 13$
(f) $2m + 6 > 7m + 4$
*(g) $7t - 4 - 10t < 12$
(h) $3(x + 1) + 5 < 5$
*(i) $2(x + 3) - 6 \geq 9$
(j) $2(3 + 2x) - 4 > 7x$
*(k) $4x + 3 + 2x < 5 - 2x$
(l) $5x + 1 < 2(3x - 4) + 3$

3. Solve each word problem by establishing an inequality and then solving the inequality.

*(a) How long should you make the sides of a square if the perimeter of the square is to be at least 248 inches?

*(b) If the sum of three consecutive numbers is to be at least 195, how large should the smallest of the numbers be?

*(c) A rectangular garden area to be fenced in must be 84 feet long.

What can be the width of the rectangle if the perimeter cannot exceed 300 feet?

(d) If the perimeter of an equilateral triangle (that is, a triangle with all sides equal) must be less than 351 centimeters, what can be the lengths of the sides?

*(e) Three more than twice a number is greater than 51. Find all such numbers.

*(f) Seven less than five times a number is less than 833. Find all such numbers.

16.3 GRAPHS OF INEQUALITIES (OPTIONAL)

EXAMPLE 7   Sketch the graph of the inequality $x + y < 4$.

Solution   The points on the line $x + y = 4$ are points where $x + y$ *equals* 4. So perhaps $x + y$ is less than 4 either above the line or below the line but certainly not on the line. Let's graph the line $x + y = 4$ and consider the possibilities.

$$x + y = 4$$

x	y
0	4
4	0

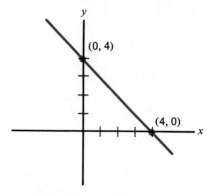

Points *above* the line, such as $(6, 0)$, $(0, 5)$, $(4, 3)$, and $(-1, 6)$, do not satisfy $x + y < 4$, since for each of these points $x + y > 4$. Observe.

x	y	$x + y$	$x + y < 4$
6	0	6	no
0	5	5	no
4	3	7	no
-1	6	5	no

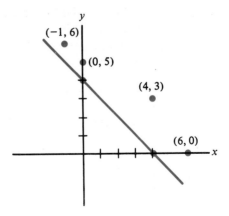

On the other hand, the following points, all of which are below the line $x + y = 4$, satisfy the inequality $x + y < 4$. Accordingly, we shade the area below the line, the desired area. The line itself is drawn dashed because its points do not satisfy $x + y < 4$. If the inequality were $x + y \leq 4$ instead, then the line would be drawn solid in order to suggest that points on the line are to be included.

x	y	$x + y$	$x + y < 4$
0	0	0	yes
1	0	1	yes
0	3	3	yes
-2	2	0	yes
0	-1	-1	yes
1	-2	-1	yes
-3	5	2	yes

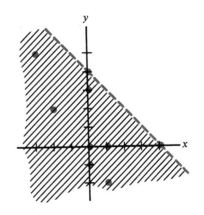

EXAMPLE 8   Find graphically the solution to the system below.

$$\begin{cases} y \geq x + 2 \\ 3x + y \geq 6 \end{cases}$$

Solution   First, obtain two points for each line $y = x + 2$ and $3x + y = 6$. Graph them and shade each inequality. The solution, then, is all points that satisfy both inequalities—that is, all points that are shaded twice.

$$y = x + 2 \qquad\qquad 3x + y = 6$$

x	y
0	2
−2	0

x	y
0	6
2	0

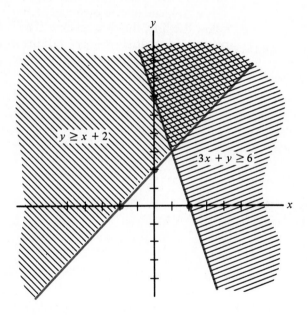

Thus the solution is all points in the V-shaped region above the lines, including the portions of the lines that make up the outline of the V-shaped region.

EXAMPLE 9   Find graphically the solution of the system below.

$$\begin{cases} y - 2x - 5 > 0 \\ 1 + 2x \geq y \end{cases}$$

Solution Points for the lines:

$$y - 2x - 5 = 0 \qquad 1 + 2x = y$$

x	y
0	5
-1	3

x	y
0	1
1	3

The graph is shown next. The graph shows that the two inequalities have no points in common, and thus there is no solution to this system of linear inequalities.

EXERCISES 16.3

1. Graph each inequality in the xy plane.
   *(a) $y > x + 1$ (b) $y > x - 5$
   *(c) $y \le 2x + 1$ (d) $y < 3x + 5$
   *(e) $x + y \le 6$ (f) $x + y > 6$
   *(g) $x - y < 4$ (h) $x - y > 7$
   *(i) $x \ge 2y - 6$ (j) $x > y + 2$
   *(k) $2x + 3y < 12$ (l) $2x - 3y \le 12$
   *(m) $x \ge 5$ (n) $x < -3$
   *(o) $y \ge 0$ (p) $y < 5$

2. Solve each system of linear inequalities by graphing.
   *(a) $\begin{cases} y > x + 1 \\ y < 2x - 2 \end{cases}$ (b) $\begin{cases} x + y \ge 7 \\ y - x \le 5 \end{cases}$

   *(c) $\begin{cases} x - y \ge 0 \\ 2x - y \le 3 \end{cases}$ (d) $\begin{cases} y \le 4x + 1 \\ 4x - y \le -4 \end{cases}$

*(e) $\begin{cases} y \geq 4 \\ x \leq -3 \end{cases}$ (f) $\begin{cases} 2x + y < 6 \\ 3x - 2y \geq 6 \end{cases}$

*(g) $\begin{cases} y \geq 1 - 2x \\ y \leq 2 \end{cases}$ (h) $\begin{cases} x \geq 0 \\ y \geq 0 \end{cases}$

*(i) $\begin{cases} 2x + 3y \leq 6 \\ 3x + 2y \geq 6 \end{cases}$ (j) $\begin{cases} y - x \leq -4 \\ x - y \geq -2 \end{cases}$

3. Represent each situation by one or more inequalities. If only one variable is involved, then call it x. If two variables are needed, use x and y.

*(a) Employees work at least 40 hours a week.
*(b) The temperature never gets as hot as 100°
 (c) The temperature never gets as cold as 0°.
 (d) The temperature never gets as hot as 90° or as cold as 5°.
*(e) The supply is always greater than the demand.
*(f) Employees work at least 40 hours but no more than 60 hours per week.
 (g) The stock number of an auto part cannot be negative.
*(h) John works at least 20 hours and his boss works at least 40 hours. The total number of hours both work cannot exceed 75 hours.
 (i) The cost of 10 pencils and 15 pens is at least $3.50. The cost of 15 pencils and 5 pens is at most $3.

CHAPTER 16 REVIEW EXERCISES

*1. Write each inequality in interval notation.
  (a) $-4 < x < 6$ (b) $-5 \leq t < 0$ (c) $78 \leq y \leq 100$
  (d) $-9 < m \leq 7$

*2. Write each interval using an inequality.
  (a) $[-3, 10]$ (b) $(2, 11]$

*3. Solve each inequality.
  (a) $5x - 8 < 17$ (b) $-4x - 7 > 19$
  (c) $8x - 8 \leq 3x - 5$ (d) $9 - 7x < x + 2$
  (e) $7(1 + 4x) > -6$ (f) $8(x + 1) - 3 \geq 2x - 7$

*4. Graph each inequality in the xy plane.
  (a) $y \leq 3x - 2$ (b) $x < y + 5$ (c) $2x + 5y > 10$
  (d) $x \leq 7$

*5. Solve each system of linear inequalities by graphing.
  (a) $\begin{cases} y > 2x + 1 \\ x + y < 7 \end{cases}$ (b) $\begin{cases} y \geq -x + 2 \\ y \geq -5 \end{cases}$

CUMULATIVE REVIEW EXERCISES (CHAPTERS 10–16)

*1. Simplify each expression completely.
  (a) $-4 + 7 - 3 + 10 - 5 + 1$
  (b) $(-3)(-7) - (+2)(-9) + (-6)(+2)$

*2. Evaluate the expression $3x^2 - 5x$ for $x = -2$.

*3. Combine like terms.
$$7x + 8y - 19x + x - 3y + y$$

*4. Simplify the expression $(-6x^7y)(12x^2y^8)$.

*5. Simplify the expression $5x - 3(2 - 9x) + 15$.

*6. Multiply $(4x - 9)(x + 15)$.

*7. Solve each equation.
(a) $17 = 1 + 5x$ (b) $x - 8 = 5x + 14$
(c) $5(4x + 3) = 1$ (d) $5 + 2(1 - x) + x = 17$

*8. Write an algebraic expression to represent each phrase.
(a) The interest on $3h$ dollars invested at 7% per year for k years.
(b) The total number of cents in $3x$ nickels and $7x$ dimes.
(c) Half of the sum of x and $3y$.

*9. Solve each word problem.
(a) If the length of a rectangle is 25 centimeters and the width is represented by $2x - 3$, determine the actual width if the area is 525 square centimeters.
(b) Jill buys some 25¢ stamps and some 40¢ stamps. If she pays $14.00 and receives a total of 47 stamps, how many of each did she get?

*10. Sketch a graph of the line $x = 2y + 5$.

*11. Determine the equation of the line that passes through the points $(-3, -1)$ and $(-1, 9)$.

*12. Solve each system of equations.
(a) $\begin{cases} y = 2x + 12 \\ 3x + 8y = 1 \end{cases}$ (b) $\begin{cases} 7x - 2y = 15 \\ 8x + 3y = -4 \end{cases}$

*13. Solve each linear inequality.
(a) $6x + 2 > 4x - 17$ (b) $3(1 - 2x) + 6 \le 13$

17

FACTORING

17.1 INTRODUCTION TO FACTORING

In Chapter 11 we began the study of algebra by using the distributive property to combine like terms. Recall the first example,

$$3x + 8x = (3 + 8)x = 11x$$

Both terms, $3x$ and $8x$, contain the same factor x. Using the distributive property to take the common factor x and place it outside parentheses is an example of the process known as *factoring*.

In this chapter we will present some basic factoring suggested by the example involving $3x + 8x$. We will also consider quadratic factoring, which involves a totally different approach. Then you will see how factoring is used to solve certain types of equations. In Chapter 18 you will see how factoring enables us to reduce algebraic fractions and carry out various operations with fractions.

EXAMPLE 1   Factor $3x + mx$.

Solution   Here x is a factor common to both $3x$ and mx. So x can be factored out according to the distributive property.

$$3x + mx = (3 + m)x \quad \checkmark$$

The factoring can be checked by applying the distributive property to multiply out the factored form. Here

$$(3 + m)x = 3 \cdot x + m \cdot x = 3x + mx$$

And $3x + mx$ is indeed the expression we started with.

EXAMPLE 2   Factor $ay + by$.

Solution   Here y is common to both terms, so it can be factored out.

$$ay + by = (a + b)y \quad \checkmark$$

EXAMPLE 3   Factor the expression $5mx + 10my$.

Solution   When 10 is considered as $5 \cdot 2$, you can see that $5 \cdot m$, or $5m$, is common to both terms and can therefore be factored out.

$$5mx + 10my = \underline{5} \cdot \underline{m} \cdot x + \underline{5} \cdot 2 \cdot \underline{m} \cdot y$$
$$= \underline{5m} \cdot x + \underline{5m} \cdot 2y$$
$$= \underline{5m}(x + 2y) \quad \checkmark$$

NOTE
The direction "Factor" means *factor completely*. In Example 3, you should not factor out *only* the 5 or *only* the m. You should factor out *both* 5 and m. In this type of factoring, where a monomial factor common to all terms is factored out, the monomial factor should be the greatest factor common to all terms.

EXAMPLE 4   Factor the expression $15x^2y + 10xy^2$.

Solution   You can see that the factor $5xy$ is present in both $15x^2y$ and $10xy^2$. Thus

$$15x^2y + 10xy^2 = 5 \cdot 3 \cdot x \cdot x \cdot y + 5 \cdot 2 \cdot x \cdot y \cdot y$$
$$= \underline{5xy} \cdot 3x + \underline{5xy} \cdot 2y$$
$$= \underline{5xy}(3x + 2y) \quad \checkmark$$

EXAMPLE 5   Factor the expression $x^2y - xy$.

Solution   Here xy is common to both terms. The factoring is

$$xy(x - 1) \quad \checkmark$$

Check: $xy(x - 1) = xy \cdot x - xy \cdot 1 = x^2y - xy$

EXAMPLE 6   Factor the expression $x^2 + 3x$.

Solution   Notice that x is common to both terms, so it can be factored out. Thus

$$x^2 + 3x = x(x + 3) \quad \checkmark$$

Again, you can check this result by multiplying out the factored form and seeing that it does indeed produce the original expression $x^2 + 3x$.

EXAMPLE 7   Factor the expression $m^6 + m^2$.

Solution   The greatest factor common to both terms is m^2. Thus

$$m^6 + m^2 = \underline{m^2} \cdot m^4 + \underline{m^2} \cdot 1$$
$$= m^2(m^4 + 1) \quad \checkmark$$

EXAMPLE 8   Factor the expression $12mn^2p - 16m^2n$.

Solution   The greatest common factor is $4mn$. A common mistake is to look too quickly at the expression and see only a 2 in the 12 and 16. There is a 4 common to both 12 and 16. Thus

$$12mn^2p - 16m^2n = \underline{4mn} \cdot 3np - \underline{4mn} \cdot 4m$$
$$= 4mn(3np - 4m) \quad \checkmark$$

EXAMPLE 9   Factor the expression $4abc + 2acd + 3ace$.

Solution   This expression consists of three terms, so factor out whatever is common to all three terms, namely, ac.

$$4abc + 2acd + 3ace = \underline{ac} \cdot 4b + \underline{ac} \cdot 2d + \underline{ac} \cdot 3e$$
$$= ac(4b + 2d + 3e) \quad \checkmark$$

Notice that there is a factor of 2 in each of the first two terms. But since there is no factor of 2 in the third term, no 2 can be factored out. A factor must appear in *every term* of an expression in order to be factored out of the expression.

EXAMPLE 10   Factor the expression $7mn + 13uv$.

Solution   There is no factor common to both terms of this expression, so the expression cannot be factored.

EXERCISES 17.1

Answers to starred exercises are given in the back of the book.
*1. Factor each expression.

(a) $7x + nx$ (b) $mx + nx$ (c) $pr + qr$
(d) $3y - ty$ (e) $5wx + 10wy$ (f) $3xy + 12xz$
(g) $12xy - 18yz$ (h) $15mn + 20mnp$ (i) $x^2 + 6x$
(j) $y^2 - 8y$ (k) $a^2 + a$ (l) $y^2 - y$
(m) $5m - m^2$ (n) $3x^2 - x$ (o) $20m^2n + 15mn^2$
(p) $18x^2y^2 - 24xy^2$ (q) $6x + 7x^2$ (r) $17t^2 - 6t$
(s) $3x + 3$ (t) $10 - 10y$

2. Factor each expression, if possible.

*(a) $10x + 5$ (b) $6 - 3w$
*(c) $8 + 40x^2$ (d) $5x^2 + 30$
*(e) $x^5 + x^3$ (f) $y^7 + y^2$
*(g) $a^2b - 2abc^4$ (h) $16w^2x^2 - 24wx^2$
*(i) $ab + cd$ (j) $ab + bc$
*(k) $pc + qc + rc$ (l) $17xy - 33wz + 14wx$
*(m) $m^3 + m^2 + m$ (n) $x^4 + x^3 + x^2$
*(o) $xyz + 3yz + wyz$ (p) $wxy + wxz - xyz$
*(q) $100m^2n + 50m^2n^2 + 20m^2$ *(r) $5ab^2 + 10a^2b + a^2b^2$
*(s) $15uv^2 - 6u^2 + 12uv$ *(t) $mnp - npq + mqr$
*(u) $m^2x + mx^2 + mx$ *(v) $15x^2y^2 + 5x^2 + 3y^2$

17.2 FACTORING QUADRATIC EXPRESSIONS

A **quadratic** expression is one that contains the second power of the variable and no higher power. For example, the following expressions are quadratic in x.

$$x^2 + 5x - 4$$
$$3x^2 - 7x + 11$$
$$x^2 + 3x$$
$$x^2 - 9$$

In Chapter 11 you learned how to multiply binomials. We now need to examine a special case, the one in which each binomial is the sum of a particular variable (say x) and a constant. Here are some examples.

$$(x + 2)(x + 4)$$
$$(x - 3)(x + 5)$$
$$(x + 1)(x - 9)$$
$$(y + 6)(y + 7)$$

Observe the multiplication.

$$(x + 2)(x + 4) = x^2 + 4x + 2x + 8$$
$$= x^2 + 6x + 8 \quad \checkmark$$

In the final result the coefficient of x (the $+6$) is the *sum* of the $+2$ and $+4$ from the original factors $(x + 2)$ and $(x + 4)$. Note also that the constant $+8$ is the *product* of the $+2$ and $+4$ from those original factors.

Let's see another example.

$$(x + 6)(x - 3) = x^2 - 3x + 6x - 18$$
$$= x^2 + 3x - 18 \quad \checkmark$$

The sum of $+6$ and -3 is $+3$; and that is the coefficient of x. The product of $+6$ and -3 is -18; and that is the constant term.

EXAMPLE 11 Determine the product of $(x + 2)$ and $(x + 5)$ by using the shortcut suggested above.

Solution

$$(x + 2)(x + 5) = x^2 + 7x + 10$$

EXAMPLE 12 Find the product: $(x - 9)(x - 4)$.

Solution

$$(x - 9)(x - 4) = x^2 - 13x + 36$$

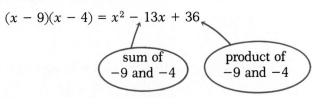

The examples of this section are all of the form

$$(x + \underline{\quad})(x + \underline{\quad})$$

that is,

$$(1x + \underline{\quad})(1x + \underline{\quad})$$

The shortcut that we have used applies only when the coefficient of x is one.

> *If the coefficient of x is not one,*
> *then the shortcut does not apply.*

A very important problem in algebra is to *reverse* this shortcut multiplication process, that is, to *factor* a quadratic expression such as $x^2 + 7x + 12$ into $(x + 3)(x + 4)$ by using what you know about sum and product from the shortcut multiplication process. If you examine both $(x + 3)(x + 4)$ and $x^2 + 7x + 12$, you realize that

- x^2 comes from $x \cdot x$.
- $+7$ comes from the *sum* $+3 + 4$.
- $+12$ comes from the *product* $(+3)(+4)$.

So when given a quadratic expression such as $x^2 + 7x + 12$ to factor, the thinking might go as follows:

1. The factors are of the form $(x \text{ ___})(x \text{ ___})$.
2. The product is $+12$. Using integers, this could be any of the following pairs:

$$(+12)(+1)$$
$$(-12)(-1)$$
$$(+6)(+2)$$
$$(-6)(-2)$$
$$(+4)(+3)$$
$$(-4)(-3)$$

3. *But* the sum must be $+7$, and only $(+4)$ and $(+3)$ add up to $+7$.

$$(+12)(+1)$$
$$(-12)(-1)$$
$$(+6)(+2)$$
$$(-6)(-2)$$
$$(+4)(+3) \quad \longleftarrow$$
$$(-4)(-3)$$

So the factors are $(x + 4)$ and $(x + 3)$. Therefore $x^2 + 7x + 12 = (x + 4)(x + 3)$.

As a check, multiply out the $(x + 4)(x + 3)$ to show that it does in fact equal $x^2 + 7x + 12$.

EXAMPLE 13   Factor the expression $x^2 + 8x + 7$.

Solution   Again the form is $(x \text{ ___})(x \text{ ___})$. From the expression you can see that the product is $+7$. Using integers, this is either $(+7)(+1)$ or $(-7)(-1)$. Since the sum must be $+8$, the choice is $(+7)$ and $(+1)$. Thus

$$x^2 + 8x + 7 = (x + 7)(x + 1) \quad \checkmark$$

The factorization $(x + 7)(x + 1)$ can also be written $(x + 1)(x + 7)$.

EXAMPLE 14   Factor the expression $x^2 - 5x + 4$.

Solution   The form is $(x \text{ ___})(x \text{ ___})$.

The product is $+4$. This could be any of the following.

$$(+4)(+1)$$
$$(-4)(-1)$$
$$(+2)(+2)$$
$$(-2)(-2)$$

The choice is $(-4)(-1)$, since the sum must be -5 in order to produce the $-5x$ term. So

$$x^2 - 5x + 4 = (x - 4)(x - 1) \quad \checkmark$$

EXAMPLE 15   Factor the expression $x^2 - 12x + 20$.

Solution   The form is $(x \underline{\ \ })(x \underline{\ \ })$.

The product, $+20$, could come from any of the following.

$$(+20)(+1)$$
$$(-20)(-1)$$
$$(+10)(+2)$$
$$(-10)(-2)$$
$$(+5)(+4)$$
$$(-5)(-4)$$

The choice is (-10) and (-2), since the sum is -12. Thus

$$x^2 - 12x + 20 = (x - 10)(x - 2) \quad \checkmark$$

EXAMPLE 16   Factor the expression $x^2 + 3x - 10$.

Solution   The form is $(x \underline{\ \ })(x \underline{\ \ })$. The product is -10. This could be any of the following.

$$(-10)(+1)$$
$$(+10)(-1)$$
$$(-5)(+2)$$
$$(+5)(-2)$$

The choice is the last one, $(+5)(-2)$, since the sum must be $+3$ (from $+3x$). Thus

$$x^2 + 3x - 10 = (x + 5)(x - 2) \quad \checkmark$$

NOTE
$(x + 5)(x - 2)$ can be written $(x - 2)(x + 5)$, using the commutative property, but it *cannot* be written $(x + 2)(x - 5)$ or $(x - 5)(x + 2)$. These last two are not commuted forms of $(x + 5)(x - 2)$; they are completely different—and they do not check. Multiply the factors to see that each multiplies to yield $x^2 - 3x - 10$, which is not the original expression.

EXAMPLE 17   Factor the expression $n^2 - n - 12$.

Solution   The form is $(n \underline{\ \ })(n \underline{\ \ })$. The product is -12. Consider

$$(+12)(-1)$$
$$(-12)(+1)$$
$$(-6)(+2)$$
$$(+6)(-2)$$
$$(+4)(-3)$$
$$(-4)(+3)$$

The choice is $(-4)(+3)$, since the sum must be -1 (because of the $-n$ term, where $-n$ means $-1n$). Thus

$$n^2 - n - 12 = (n - 4)(n + 3) \quad \checkmark$$

EXAMPLE 18   Factor the expression $x^2 + 6x + 12$.

Solution   The form is $(x \underline{\quad})(x \underline{\quad})$. The product is 12. The choices:

$$(+12)(+1)$$
$$(-12)(-1)$$
$$(+6)(+2)$$
$$(-6)(-2)$$
$$(+4)(+3)$$
$$(-4)(-3)$$

But *none* of these pairs gives a sum of $+6$. This means the expression $x^2 + 6x + 12$ *cannot be factored* by the methods that we have studied so far.

EXAMPLE 19   Factor the expression $x^2 + 7x$.

Solution   This expression may have caught you off guard, but it is a simple one to factor. It is of the type factored in Section 17.1, before we even began to study quadratic expressions. There are only two terms in $x^2 + 7x$, and there is an x common to both terms. So factor out the x. The result is

$$x^2 + 7x = x(x + 7) \quad \checkmark$$

EXAMPLE 20   Factor the expression $x^2 - 9$.

Solution   The form is $(x \underline{\quad})(x \underline{\quad})$. The product is -9. The choices:

$$(-9)(+1)$$
$$(+9)(-1)$$
$$(+3)(-3)$$

Note that $x^2 - 9$ is the same as $x^2 + 0x - 9$. There are *zero* x's; that is why the x term does not appear in $x^2 - 9$. Thus the sum (which is the coefficient of x) is 0. So $(+3)(-3)$ is the proper choice, since $(+3) + (-3) = 0$. And therefore

$$x^2 - 9 = (x + 3)(x - 3) \quad \checkmark$$

Consider the following similar factorizations.

$$x^2 - 4 \longrightarrow (x + 2)(x - 2)$$
$$x^2 - 25 \longrightarrow (x + 5)(x - 5)$$
$$y^2 - 16 \longrightarrow (y + 4)(y - 4)$$

In each case, the factors involve square roots. Notice the pattern for such factoring. These are examples of factoring the **difference of two squares.** Consider the examples below.

$x^2 - 9$ (x squared, 3 squared) $\longrightarrow (x + 3)(x - 3)$
$m^2 - 4$ (m squared, 2 squared) $\longrightarrow (m + 2)(m - 2)$
$x^2 - 25$ (x squared, 5 squared) $\longrightarrow (x + 5)(x - 5)$
$49 - x^2$ (7 squared, x squared) $\longrightarrow (7 + x)(7 - x)$
$y^2 - 16$ (y squared, 4 squared) $\longrightarrow (y + 4)(y - 4)$
$x^2 - a^2$ (x squared, a squared) $\longrightarrow (x + a)(x - a)$
$x^2 - 9y^2$ (x squared, $3y$ squared) $\longrightarrow (x + 3y)(x - 3y)$
$4m^2 - 25n^2$ ($2m$ squared, $5n$ squared) $\longrightarrow (2m + 5n)(2m - 5n)$

In general,

> Difference of Two Squares
> $a^2 - b^2 = (a + b)(a - b)$

EXAMPLE 21   Factor the expression $16x^2 - y^2$.

Solution   The expression $16x^2 - y^2$ is the difference of two squares, since $16x^2$ is the square of $4x$, y^2 is the square of y, and the connecting sign is a minus. Thus

$$16x^2 - y^2 = (4x + y)(4x - y) \quad \checkmark$$

EXAMPLE 22   Factor the expression $x^2 + y^2$.

Solution   Although this expression contains two squares, it is not the difference of two squares, rather it is the sum of two squares. *The sum of two squares cannot be factored.*

EXERCISES 17.2

1. Determine the product by using the shortcut explained in Examples 11 and 12.

*(a) $(x + 2)(x + 3)$ *(b) $(x + 5)(x + 9)$
*(c) $(m + 6)(m + 4)$ *(d) $(m + 1)(m + 1)$
*(e) $(x - 5)(x - 2)$ (f) $(x - 7)(x - 1)$
*(g) $(m - 6)(m - 4)$ (h) $(m - 2)(m - 3)$
*(i) $(y - 3)(y + 5)$ (j) $(y - 6)(y + 2)$
*(k) $(x - 7)(x + 7)$ *(l) $(x + 5)(x - 5)$
*(m) $(m + 3)(m - 8)$ (n) $(m + 6)(m - 5)$
*(o) $(y + 8)(y - 3)$ (p) $(y + 3)(y - 3)$
*(q) $(x - 9)(x + 2)$ (r) $(x + 5)(x + 5)$
*(s) $(m - 2)(m - 10)$ (t) $(m + 9)(m - 3)$

*2. Factor each expression.

(a) $x^2 + 5x + 6$ (b) $x^2 + 6x + 5$
(c) $x^2 + 4x + 4$ (d) $x^2 - 6x + 5$
(e) $t^2 - 6t - 7$ (f) $x^2 - 12x + 20$
(g) $m^2 - 5m - 6$ (h) $x^2 + 3x$
(i) $x^2 + x - 12$ (j) $a^2 - 9a$
(k) $y^2 - 4y - 12$ (l) $m^2 + m - 30$
(m) $x^2 + x - 2$ (n) $y^2 + 5y - 24$
(o) $x^2 - 8x + 15$ (p) $x^2 - 12x + 35$
(q) $y^2 - 10y - 11$ (r) $x^2 + 12x - 13$
(s) $x^2 - 8x$ (t) $x^2 - 4x + 3$
(u) $n^2 - 10n + 21$ (v) $n^2 - n - 6$
(w) $x^2 + 4x - 21$ (x) $x^2 + 3x - 4$
(y) $m^2 + 10m$ (z) $x^2 - 7x + 10$

3. Factor each expression or indicate that it cannot be factored.

*(a) $x^2 - 9x + 20$ (b) $x^2 - 7x - 18$
*(c) $m^2 - 5m + 16$ (d) $y^2 - 14y + 24$
*(e) $x^2 - 2x + 1$ (f) $x^2 - 15x - 24$
*(g) $x^2 - 36$ (h) $n^2 - 64$
*(i) $x^2 + 36$ (j) $y^2 - 11y + 18$
*(k) $m^2 + 2m - 18$ (l) $m^2 - 5m + 14$
*(m) $x^2 + 10x + 25$ (n) $y^2 + 7y + 10$
*(o) $x^2 - 2x - 15$ (p) $16 - m^2$
*(q) $t^2 + t - 20$ (r) $9 + x^2$
*(s) $x^2 - 9x$ (t) $y^2 - 25y$
*(u) $x^2 - 14x + 45$ (v) $x^2 + 17x + 16$
*(w) $m^2 - 8m$ (x) $m^2 + 13m - 14$
*(y) $x^2 + 9x$ (z) $x^2 - 13x + 36$

4. Factor each expression or indicate that it cannot be factored.

*(a) $x^2 - m^2$ (b) $x^2 - t^2$
*(c) $x^2 - 81y^2$ (d) $y^2 - 16x^2$
*(e) $x^2 + m^2$ (f) $t^2 - u^2$
*(g) $9x^2 - 25y^2$ (h) $16a^2 - 81b^2$
*(i) $100y^2 - 49x^2$ (j) $9a^2 + b^2$
*(k) $4x^2 + 9y^2$ (l) $25a^2 + 4b^2$
*(m) $25 - 4t^2$ (n) $81 - 25y^2$
*(o) $16 - 9x^2$ (p) $9 - 49x^2$
*(q) $1 - n^2$ (r) $1 - c^2$
*(s) $9 - x^2$ (t) $16 - x^2$
*(u) $64 - y^2$ (v) $25 - t^2$

17.3 MORE QUADRATIC FACTORING

Sometimes the methods of Sections 17.1 and 17.2 must be combined to completely factor a quadratic expression. When applicable,

> *First:*   Factor out the greatest common monomial factor
> (as in Section 17.1).
> *Second:* Factor the remaining quadratic expression
> (as in Section 17.2).

EXAMPLE 23   Factor the expression $2t^2 - 4t - 16$.

Solution   Before attempting to apply the quadratic factoring techniques of the preceding section, you should examine the expression for common monomial factors. Notice the factor of 2 in each term of $2t^2 - 4t - 16$. As it happens, 2 is the greatest common monomial factor. Thus

$$2t^2 - 4t - 16 = 2(t^2 - 2t - 8)$$

Now proceed to factor the quadratic expression $t^2 - 2t - 8$. The result is

$$= 2(t - 4)(t + 2) \quad \checkmark$$

EXAMPLE 24   Factor the expression $3x^2 + 9x - 12$.

Solution
$$\begin{aligned} 3x^2 + 9x - 12 &= 3(x^2 + 3x - 4) \\ &= 3(x + 4)(x - 1) \quad \checkmark \end{aligned}$$

EXAMPLE 25   Factor the expression $5y^2 + 35y + 10$.

Solution
$$5y^2 + 35y + 10 = 5(y^2 + 7y + 2)$$

The quadratic expression $y^2 + 7y + 2$ cannot be factored, so the factored form $5(y^2 + 7y + 2)$ is the final result.

A more involved problem arises when the coefficient of the x^2 term is not 1 *and* it cannot be factored out. Here are a few examples of quadratic expressions of that type.

$$2x^2 + 5x + 3$$
$$3x^2 - 11x + 6$$
$$4x^2 - 13x + 9$$

Consider the expression $2x^2 + 5x + 3$. The factors take the form

$$(2x \quad)(x \quad)$$

since you need $2x \cdot x$ to get the $2x^2$ term of $2x^2 + 5x + 3$. The product is 3, and it can only come from $1 \cdot 3$. The sum is +5 (or +5x), and this seems impossible to obtain using 3 and 1 until you examine the form

$$(2x \underline{\quad})(x \underline{\quad})$$

and realize that part of your sum is *twice* (from the $2x$) the number placed after x. There are two possibilities:

$$(2x + 1)(x + 3) \quad \text{and} \quad (2x + 3)(x + 1)$$

The sum from $\quad (2x + 1)(x + 3) \quad$ is $7x$.

$$x + 6x$$

The sum from $\quad (2x + 3)(x + 1) \quad$ is $5x$, as needed.

$$3x + 2x$$

So $(2x + 3)(x + 1)$ is the correct factorization, since it yields the sum $5x$.

EXAMPLE 26 Factor the expression $3y^2 - 11y + 6$.

Solution The $3y^2$ comes from $3y \cdot y$, so we will proceed with the form $(3y\_\_)(y\_\_)$. The product $+6$ suggests $(+6)(+1)$, $(-6)(-1)$, $(+3)(+2)$, and $(-3)(-2)$. Considering that the coefficient of y is a *negative* number, there is no way that $(+6)(+1)$ or $(+3)(+2)$ could work. So the only possible factors of $3y^2 - 11y + 6$ are:

$$(3y - 6)(y - 1)$$
$$(3y - 1)(y - 6)$$
$$(3y - 3)(y - 2)$$
$$(3y - 2)(y - 3)$$

Only the last one multiplies out to yield $3y^2 - 11y + 6$. Thus

$$3y^2 - 11y + 6 = (3y - 2)(y - 3) \ \checkmark$$

EXAMPLE 27 Factor the expression $6x^2 - x - 12$.

Solution The $6x^2$ could have come from either $6x \cdot x$ or $3x \cdot 2x$. This means the form of the factoring could be either $(6x\_\_)(x\_\_)$ or $(3x\_\_)(2x\_\_)$. Also, the product -12 could come from any of these: $(-12)(+1)$, $(+12)(-1)$, $(-6)(+2)$, $(+6)(-2)$, $(-3)(+4)$, or $(+3)(-4)$. So there are a lot of cases to try, unless you get lucky and find the right one soon. (Actually, there are 24 cases here.) After several tries, the correct factoring is produced, namely,

$$6x^2 - x - 12 = (2x - 3)(3x + 4) \ \checkmark$$

EXERCISES 17.3

1. Factor each quadratic expression completely. Begin by factoring out the greatest common monomial factor.

   *(a) $3m^2 + 15m + 18$ (b) $2x^2 + 6x + 4$
   *(c) $2x^2 - 18x - 20$ (d) $2x^2 - 14x + 24$
   *(e) $3x^2 - 3$ (f) $4x^2 - 144$
   *(g) $5x^2 - 30x + 45$ (h) $3x^2 - 24x + 45$
   *(i) $7x^2 - 21x$ (j) $5x^2 + 140x$
   *(k) $2t^2 - 10t + 8$ (l) $3x^2 + 12x - 36$
   *(m) $4u^2 + 52u - 56$ (n) $5x^2 - 5x - 60$
   *(o) $20 - 5x^2$ (p) $18 - 2x^2$
   *(q) $3y^2 + 3y$ (r) $2m^2 + 6m$

2. Factor each quadratic expression.

   *(a) $2x^2 + 13x + 15$ *(b) $3x^2 + 5x + 2$
   *(c) $3x^2 - 5x - 2$ *(d) $5y^2 - 17y + 6$
   *(e) $5n^2 + 13n - 6$ (f) $7x^2 + 3x - 4$
   *(g) $6x^2 - 5x - 6$ (h) $8x^2 - 2x - 3$
   *(i) $6x^2 + 61x + 10$ (j) $3x^2 - 61x + 20$
   *(k) $4x^2 + 4x - 15$ (l) $4x^2 - 5x - 6$
   *(m) $5t^2 + 22t + 8$ (n) $6x^2 + 11x - 10$
   *(o) $3y^2 - y - 14$ (p) $5x^2 - 29x + 20$
   *(q) $2x^2 + 7x - 15$ (r) $3y^2 + 7y - 6$

3. Factor each quadratic expression, if possible.

   *(a) $5y^2 + 22y + 8$ *(b) $2m^2 + 5m - 12$
   *(c) $6x^2 + 61x + 10$ *(d) $5x^2 + 2x + 1$
   *(e) $2y^2 - 12y + 18$ (f) $12x^2 - 10x - 8$
   *(g) $3m^2 - 75$ (h) $4x^2 + 8x + 3$
   *(i) $6x^2 - 16x + 10$ (j) $3x^2 - 15x - 42$
   *(k) $4x^2 - 13x + 9$ (l) $6x^2 + 9x - 6$
   *(m) $2x^2 + 2$ (n) $5x^2 - 2x - 4$
   *(o) $2x^2 - 2$ (p) $3x^2 + 8x - 2$
   *(q) $5x^2 + 2x + 7$ (r) $5x^2 - 17x + 6$
   *(s) $3x^2 - x - 10$ (t) $3y^2 - y - 24$
   *(u) $5x^2 - 16x + 12$ (v) $2x^2 - 7x - 8$
   *(w) $4t^2 + t - 3$ (x) $4y^2 + 13y - 12$

17.4 SOLVING QUADRATIC EQUATIONS

Let m and n represent numbers. If $m \cdot n = 0$, then at least one of the numbers (m or n) must be zero. Also, if either of the numbers (m or n) is zero, then the product $m \cdot n$ is zero.

Suppose we know that $(x + 3)(x + 2) = 0$. The product is zero if either $(x + 3)$ or $(x + 2)$ is zero. Furthermore, if $(x + 3) = 0$, then $x = -3$, as shown next.

$$x + 3 = 0$$
$$\underline{-3 = -3}$$
$$x \quad = -3 \ \checkmark$$

Similarly, if $(x + 2) = 0$, then $x = -2$. Therefore -3 and -2 are the solutions (or **roots**) of $(x + 3)(x + 2) = 0$, since the substitution of either -3 or -2 for x makes $(x + 3)(x + 2)$ equal to zero. You can check this statement by actually making the substitution. First, substitute -3 for x in $(x + 3)(x + 2)$.

$$(\underline{x} + 3)(\underline{x} + 2) = 0$$
$$(\underline{-3} + 3)(\underline{-3} + 2) \overset{?}{=} 0$$
$$(0)(-1) \overset{?}{=} 0$$
$$0 = 0 \ \checkmark$$

Next, substitute -2 for x in $(x + 3)(x + 2) = 0$.

$$(\underline{x} + 3)(\underline{x} + 2) = 0$$
$$(\underline{-2} + 3)(\underline{-2} + 2) \overset{?}{=} 0$$
$$(1)(0) \overset{?}{=} 0$$
$$0 = 0 \ \checkmark$$

Both solutions check.

EXAMPLE 28   Solve the equation $(n - 5)(n + 1) = 0$.

Solution   Setting $(n - 5)$ equal to zero gives one solution. Setting $(n + 1)$ equal to zero gives the other.

$$n - 5 = 0 \quad \Big| \quad n + 1 = 0$$
$$\text{yields } n = 5 \ \checkmark \Big| \ \text{yields } n = -1 \ \checkmark$$

EXAMPLE 29   Solve the equation $x^2 - 7x + 12 = 0$.

Solution $x^2 - 7x + 12$ is easily factored. So $x^2 - 7x + 12 = 0$ can be written as $(x - 3)(x - 4) = 0$. Thus

$$x - 3 = 0 \quad \text{or} \quad x = 3 \ \checkmark$$

and

$$x - 4 = 0 \quad \text{or} \quad x = 4 \ \checkmark$$

The solutions are 3 and 4.

   The solutions should be checked in the original, $x^2 - 7x + 12 = 0$, rather than in the factored form, since an error could have been made in the factoring. To check 3 as a solution:

$$x^2 - 7x + 12 = 0$$
$$(\underline{3})^2 - 7(\underline{3}) + 12 \overset{?}{=} 0$$
$$9 - 21 + 12 \overset{?}{=} 0$$
$$0 = 0 \ \checkmark$$

It is left for you to check 4 as a solution.

EXAMPLE 30   Solve the equation $x^2 - 5x = 0$.

Solution   Because there is no constant term, we can factor out an x.

$$x^2 - 5x = 0$$

becomes

$$(x)(x - 5) = 0$$

Setting each factor equal to 0 yields

$$x = 0 \ \checkmark \quad \bigg| \quad x - 5 = 0$$
$$x = 5 \ \checkmark$$

The solutions are 0 and 5.

The examples we have seen suggest that *quadratic equations always have two solutions.* You should realize that there will always be two factors for the quadratic expression and hence two solutions. It is possible, however, for both solutions to be the same. This situation will occur whenever both factors are the same. Here is an example.

EXAMPLE 31   Solve the equation $m^2 - 6m + 9 = 0$.

Solution
$$m^2 - 6m + 9 = 0$$
$$(m - 3)(m - 3) = 0$$

$$m - 3 = 0 \quad \bigg| \quad m - 3 = 0$$
$$m = 3 \ \checkmark \quad \bigg| \quad m = 3 \ \checkmark$$

The solutions are $m = 3$ and $m = 3$. We can say that 3 is a "double root" of the equation $m^2 - 6m + 9 = 0$.

EXAMPLE 32   Solve the equation $y^2 + 5y + 1 = 0$.

Solution   We cannot factor $y^2 + 5y + 1$. This means we cannot solve the equation $y^2 + 5y + 1 = 0$ by the methods that we have studied. More advanced methods are examined in Chapter 22.

EXAMPLE 33   Solve the equation $6y^2 + 17y + 5 = 0$.

Solution   When the expression is factored, the equation $6y^2 + 17y + 5 = 0$ becomes

$$(3y + 1)(2y + 5) = 0$$

The solutions follow by setting the factors equal to 0 and solving the resulting equations.

$$3y + 1 = 0 \quad \bigg| \quad 2y + 5 = 0$$
$$3y = -1 \quad \bigg| \quad 2y = -5$$
$$y = -\frac{1}{3} \ \checkmark \quad \bigg| \quad y = -\frac{5}{2} \ \checkmark$$

EXAMPLE 34   Solve the equation $3x^2 - 12 = 0$.

Solution   A 3 can be factored out of the expression $3x^2 - 12$. So

$$3x^2 - 12 = 0$$

becomes

$$3(x^2 - 4) = 0$$

The 3 can be eliminated from the equation by dividing both sides by 3. The result is

$$x^2 - 4 = 0$$

The expression on the left side is the difference of two squares and is readily factored. The equation is then easily solved.

$$(x + 2)(x - 2) = 0$$

$$
\begin{array}{c|c}
x + 2 = 0 & x - 2 = 0 \\
x = -2 \ \checkmark & x = 2 \ \checkmark
\end{array}
$$

EXAMPLE 35   Solve the equation $x^2 + 3x = 28$.

Solution   In order to factor the quadratic expression and consider what makes it equal to zero, the equation must be in an $= 0$ form. So add -28 to both sides of the equation $x^2 + 3x = 28$ in order to obtain the desired form.

$$
\begin{array}{rr}
x^2 + 3x = & 28 \\
-28 = & -28 \\
\hline
x^2 + 3x - 28 = & 0
\end{array}
$$

Now we can proceed as in previous examples.

$$x^2 + 3x - 28 = 0$$
$$(x - 4)(x + 7) = 0$$

$$
\begin{array}{c|c}
x - 4 = 0 & x + 7 = \ \ 0 \\
x = 4 \ \checkmark & x = -7 \ \checkmark
\end{array}
$$

EXAMPLE 36   Solve the equation $x^2 = 36 - 9x$.

Solution   In order to solve a quadratic equation by factoring, the expression must be equal to 0. So add $9x$ and -36 to both sides of this equation in order to obtain the desired form. First, add $9x$ to both sides.

$$
\begin{array}{rr}
x^2 = 36 - 9x \\
+9x \qquad + 9x \\
\hline
x^2 + 9x = 36
\end{array}
$$

Then add -36 to both sides.

$$x^2 + 9x = 36$$
$$\underline{ - 36 -36}$$
$$x^2 + 9x - 36 = 0$$

Now we can proceed to factor and solve.

$$x^2 + 9x - 36 = 0$$
$$(x + 12)(x - 3) = 0$$

$x + 12 = 0$	$x - 3 = 0$
$x = -12$ ✓	$x = 3$ ✓

EXAMPLE 37   Solve the equation $-x^2 + x + 12 = 0$.

Solution   The equation $-x^2 + x + 12 = 0$ appears strange because we are accustomed to working with quadratic expressions that begin x^2, *not* $-x^2$. If the equation is multiplied by -1, then the $-x^2$ term will become x^2.

$$\underline{-1}(-x^2 + x + 12) = \underline{-1}(0)$$

becomes

$$x^2 - x - 12 = 0$$

which is easily factored and solved.

$$(x - 4)(x + 3) = 0$$

$x - 4 = 0$	$x + 3 = 0$
$x = 4$ ✓	$x = -3$ ✓

EXAMPLE 38   A small rocket is shot upward from the edge of a cliff and travels according to $s = -16t^2 + 352t + 768$, where s is the distance of the rocket from the ground and t is time in seconds. When does the rocket strike the ground?

Solution   The rocket strikes the ground when its distance from the ground is 0, that is, when $s = 0$. Since s is $-16t^2 + 352t + 768$, we want to know when this expression is zero. So we shall solve

$$-16t^2 + 352t + 768 = 0$$

To begin, multiply the equation by -1 in order to obtain a positive coefficient for t^2.

$$16t^2 - 352t - 768 = 0$$

Next, factor 16 out of each term of the expression. That will make the quadratic expression easier to work with.

$$16(t^2 - 22t - 48) = 0$$

The quadratic expression is now readily factored.

$$16(t - 24)(t + 2) = 0$$

The solutions of this equation are, then, 24 and -2. But since these t values are units of time (seconds), we must reject -2 as nonsense in this setting. Thus the solution is 24. *The rocket will strike the ground 24 seconds after launch.*

EXERCISES 17.4

*1. Solve each quadratic equation by factoring.
   (a) $x^2 + 7x + 6 = 0$ (b) $y^2 + 4y + 3 = 0$
   (c) $x^2 - 6x + 5 = 0$ (d) $y^2 - 7y + 10 = 0$
   (e) $m^2 - m - 6 = 0$ (f) $x^2 + 7x - 8 = 0$
   (g) $n^2 + 2n + 1 = 0$ (h) $t^2 - 2t - 3 = 0$
   (i) $x^2 - 16 = 0$ (j) $b^2 - 49 = 0$
   (k) $x^2 + 7x = 0$ (l) $y^2 - 8y = 0$
   (m) $x^2 - 11x + 24 = 0$ (n) $m^2 - m - 30 = 0$
   (o) $x^2 - 4x - 12 = 0$ (p) $x^2 + 6x + 9 = 0$

2. Solve each quadratic equation by factoring, if the expression can be factored.
   *(a) $y^2 - y - 20 = 0$ *(b) $x^2 - x = 0$
   *(c) $x^2 - 1 = 0$ *(d) $x^2 - 15x - 14 = 0$
   *(e) $x^2 + 25 = 0$ (f) $x^2 + 8x + 16 = 0$
   *(g) $2x^2 - 18x + 16 = 0$ (h) $3x^2 + 3x - 60 = 0$
   *(i) $2x^2 + 11x + 5 = 0$ (j) $3x^2 + 13x + 12 = 0$
   *(k) $3x^2 + 10x - 8 = 0$ (l) $2x^2 - 13x + 20 = 0$

3. Solve each quadratic equation by factoring, if possible.
   *(a) $5x^2 + 2x - 3 = 0$ *(b) $3w^2 + 8w + 4 = 0$
   *(c) $4x^2 + 8x - 5 = 0$ (d) $x^2 + 25x = 0$
   *(e) $6y^2 - 54 = 0$ (f) $7p^2 - 175 = 0$
   *(g) $-x^2 + 10x - 24 = 0$ (h) $-x^2 - 5x + 36 = 0$
   *(i) $6n^2 + 17n + 12 = 0$ (j) $2y^2 + 9y + 7 = 0$
   *(k) $x^2 + 16x + 1 = 0$ (l) $m^2 - 4m - 6 = 0$
   *(m) $4x^2 + 8x - 60 = 0$ *(n) $5x^2 - 40x = 0$
   *(o) $-x^2 + 5x + 14 = 0$ (p) $-2x^2 + 13x + 24 = 0$
   *(q) $-2x^2 + x + 15 = 0$ *(r) $-x^2 + 5x - 8 = 0$

4. Solve each quadratic equation by factoring, if possible.
    *(a) $x^2 + 8x = 16$ (b) $y^2 + 3y = 4$
    *(c) $n^2 - 8n = -7$ (d) $x^2 + 9x = 10$
    *(e) $3x^2 + x = 10$ *(f) $3x^2 + 36x = -96$
    *(g) $2x^2 = 13x + 7$ (h) $m^2 = 2m + 8$
    *(i) $4x^2 + 6 = 11x$ (j) $2m^2 = 11m - 12$
    *(k) $10x + 4 = -6x^2$ (l) $x^2 = 7x$
    *(m) $b^2 = 9$ (n) $a^2 = 49$
    *(o) $x^2 - 7x + 2 = 20$ (p) $x^2 - 6x + 1 = -8$

5. Solve each problem. Assume that t represents time in seconds and s represents distance in feet above the ground.
    *(a) A small rocket is shot upward from the edge of a cliff and travels according to $s = -16t^2 + 208t + 768$. When does the rocket strike the ground?
    (b) A small rocket is shot upward from the top of a building and travels according to $s = -16t^2 + 80t + 384$. When does the rocket strike the ground?
    *(c) A rocket is shot upward from the ground and travels according to $s = -16t^2 + 480t$. When will the rocket hit the ground?
    (d) A ball is thrown directly downward from the top of a building and travels according to $s = -16t^2 + 40t + 416$. When does the ball reach the ground?
    *(e) At what time in its flight will the rocket of part (c) be at a height of 3200 feet?

6. Solve each problem by establishing a quadratic equation and then solving the equation.
    *(a) The length of a rectangle is three times the width. The area is 147 square meters. What are the length and width of the rectangle?
    (b) The length of a rectangle is five times the width. The area is 320 square yards. What are the length and width of the rectangle?
    *(c) The width of a rectangle is 4 feet less than the length. The area is 96 square feet. What are the length and width of the rectangle?
    (d) The length of a rectangle is 3 centimeters more than the width. The area is 88 square centimeters. What are the length and width of the rectangle?
    *(e) If you add the square of a certain number and nine times the number, the sum is 36. What is the number(s)?
    *(f) If you square a number and then subtract 57, the result is 24. Determine the number(s).
    *(g) If you add the square of a number and 20 times the number, the result is zero. What is the number(s)?
    (h) The square of a number is added to seven times the number. That sum is 18. What is the number(s)?
    (i) Seven times a number is subtracted from the square of the number. That difference is -10. What is the number(s)?

(j) If you add the square of a number and 10 times the number, the result is zero. What is the number(s)?

CHAPTER 17 REVIEW EXERCISES

*1. Factor each expression, if possible.
  (a) $mx + nx + px$ (b) $a^2b^2c + ab^2c^2$
  (c) $26ab + 39bc$ (d) $x^2 - 10xy$
  (e) $10 - 30y$ (f) $x^2 + y^2$
  (g) $x^2 + 3x - 10$ (h) $x^2 - 14x + 13$
  (i) $x^2 - 21x$ (j) $m^2 + 30m$
  (k) $n^2 - 100$ (l) $y^2 - 1$
  (m) $4x^2 + 8x + 3$ (n) $2x^2 - 10x - 28$
  (o) $4x^2 - 5x - 9$ (p) $3x^2 - 48$

*2. Solve each quadratic equation, if the quadratic expression can be factored.
  (a) $x^2 + 13x + 12 = 0$ (b) $x^2 - 2x + 1 = 0$
  (c) $t^2 - 10t + 25 = 0$ (d) $y^2 - 5y - 14 = 0$
  (e) $x^2 - 9x + 12 = 0$ (f) $t^2 + 10t + 16 = 0$
  (g) $m^2 + 14m + 24 = 0$ (h) $x^2 + 9x + 14 = 0$
  (i) $3x^2 - 3x - 18 = 0$ (j) $2n^2 - 8n - 32 = 0$
  (k) $2m^2 - 16m + 30 = 0$ (l) $2x^2 - 3x - 14 = 0$
  (m) $5x^2 + 14x - 3 = 0$ (n) $m^2 - 4m = 5$
  (o) $m^2 = 49$ (p) $x^2 = 12x$

CUMULATIVE REVIEW EXERCISES (CHAPTERS 10–17)

*1. Simplify each expression.
  (a) $(-8)(+2)(-5) + (-3)(+2)$
  (b) $-9 + 2 - 6 + (-3)^2 - 1 + 9(-5)$
  (c) $(-3)(+4 - 2 - 7) - 8(-9 + 15)$
  (d) $-8^2 + 6^2 + 9(-1)^{15} - 1^{11}$
  (e) $\dfrac{(-6)(+4)}{(-2)^3} + \dfrac{(+1)}{(-1)} - 1$

*2. Simplify each expression.
  (a) $4x + 9x - 3x + x - 12x$ (b) $5x + 4(1 - 3x) + 2$

*3. Solve each equation.
  (a) $3x + 9 = -2$ (b) $\frac{2}{3}x - 18 = 7$
  (c) $5x - 9 = 7x + 18$ (d) $8(1 - 3x) = 15x + 1$
  (e) $6 - (x - 7) = 3x + 4$

*4. Simplify each expression.
  (a) x^5x^9 (b) y^4y^{34} (c) $m^6m^7m^8$
  (d) $\dfrac{x^{10}}{x^6}$ (e) $\dfrac{y^{15}}{y^7}$ (f) $\dfrac{m^9}{m^3}$

*5. Perform the long division.

(a) $x - 3 \overline{) x^3 - 2x^2 + 4x - 18}$ (b) $x + 2 \overline{) x^3 - 4x^2 - 9x + 6}$

*6. Write an algebraic expression to represent each phrase.

(a) Ten more than z.

(b) Nine less than five times x.

(c) The number of feet in $7x$ inches.

(d) The interest on q dollars invested at 7% per year for r years.

*7. Sketch the graph of each straight line.

(a) $y = 3x - 2$ (b) $x = 2y + 3$

(c) $x - y = 1$ (d) $3x + y = 9$

*8. Determine the slope of the line that passes through the points $(5, 2)$ and $(7, -1)$.

*9. Solve each system of linear equations.

(a) $\begin{cases} 2x + 3y = 3 \\ 5x - 2y = 17 \end{cases}$ (b) $\begin{cases} 5x + 2y = -3 \\ y = x + 2 \end{cases}$

*10. Solve each inequality.

(a) $-x + 6 > 19$ (b) $5x - 3 \le -9$

*11. Factor each expression, if possible.

(a) $x^2 + x$ (b) $7y + ny$

(c) $x^2 - 9x - 18$ (d) $x^2 + 10x - 24$

(e) $x^2 - y^2$ (f) $5x^2 + 10x$

(g) $3x^2 + 17x - 28$ (h) $2x^2 + 3x - 20$

*12. Solve each equation.

(a) $x^2 - 12x + 36 = 0$ (b) $y^2 + 16y - 36 = 0$

(c) $n^2 - 2n = 15$ (d) $t^2 + 8t + 15 = 0$

(e) $m^2 - 11m + 18 = 0$ (f) $x^2 = x$

(g) $t^2 = 49$ (h) $x^2 = 7x - 10$

18

FRACTIONS CONTAINING VARIABLES

18.1 INTRODUCTION AND MULTIPLICATION

The factoring you have studied in Chapter 17 will be applied throughout this chapter as you add, subtract, multiply, and divide fractions.

Recall from Chapter 6 that a fraction consists of a *numerator* and a *denominator*.

$$\frac{3}{5} \begin{array}{l} \longleftarrow \text{numerator} \\ \longleftarrow \text{denominator} \end{array}$$

Two fractions are multiplied as follows: Multiply numerator times numerator to create the new numerator *and* multiply denominator times denominator to create the new denominator.

Several examples are given on the next page, where we begin with an arithmetic (review) example and then continue with algebra examples.

EXAMPLE 1 Multiply $\frac{3}{5} \cdot \frac{4}{7}$.

Solution

$$\frac{3}{5} \cdot \frac{4}{7} = \frac{3 \cdot 4}{5 \cdot 7}$$

$$= \frac{12}{35} \quad \checkmark$$

Multiplication of fractions in algebra is similar to multiplication of fractions in arithmetic, as demonstrated by the examples that follow.

EXAMPLE 2 Multiply $\frac{a}{b} \cdot \frac{x}{y}$.

Solution

$$\frac{a}{b} \cdot \frac{x}{y} = \frac{a \cdot x}{b \cdot y}$$

$$= \frac{ax}{by} \quad \checkmark$$

EXAMPLE 3 Multiply $\frac{x}{2} \cdot \frac{x + 3}{y}$.

Solution

$$\frac{x}{2} \cdot \frac{x + 3}{y} = \frac{x(x + 3)}{2y} \quad \checkmark \quad \text{or} \quad \frac{x^2 + 3x}{2y} \quad \checkmark$$

EXAMPLE 4 Multiply $\frac{x + 4}{x} \cdot \frac{x + 3}{5x}$.

Solution

$$\frac{x + 4}{x} \cdot \frac{x + 3}{5x} = \frac{(x + 4)(x + 3)}{x \cdot 5x}$$

$$= \frac{(x + 4)(x + 3)}{5x^2} \quad \checkmark \quad \text{or} \quad \frac{x^2 + 7x + 12}{5x^2} \quad \checkmark$$

EXERCISES 18.1

Answers to starred exercises are given in the back of the book.

*1. Multiply.

(a) $\frac{2}{3} \cdot \frac{5}{7}$

(b) $\frac{4}{9} \cdot \frac{2}{5}$

(c) $\frac{x}{3} \cdot \frac{y}{5}$

(d) $\frac{7}{t} \cdot \frac{b}{6}$

(e) $\frac{5}{7} \cdot \frac{x + 1}{2}$

(f) $\frac{2}{x} \cdot \frac{x - 4}{5}$

(g) $\frac{x + 5}{x} \cdot \frac{4}{x}$

(h) $\frac{x + 4}{x} \cdot \frac{x + 1}{x}$

(i) $\frac{mn}{5} \cdot \frac{3mn}{t}$

(j) $\frac{xy}{4w} \cdot \frac{xy}{4w}$

(k) $\frac{x}{3} \cdot \frac{x - 2}{7}$

(l) $\frac{3w}{2} \cdot \frac{w}{x}$

2. Multiply.

*(a) $\frac{3xy}{7} \cdot \frac{2x}{5}$

(b) $\frac{7wx}{3} \cdot \frac{4w}{5}$

*(c) $\frac{5r^2w}{3x} \cdot \frac{2rw}{x}$

(d) $\dfrac{8a^3b}{7c} \cdot \dfrac{5a^2b^6}{3c}$ *(e) $\dfrac{7}{y^2} \cdot \dfrac{5x}{y^3}$ (f) $\dfrac{4x}{y^6} \cdot \dfrac{3x}{y^2}$

*(g) $\dfrac{5xy^2}{7z} \cdot \dfrac{-3x}{2}$ (h) $\dfrac{-6m^2n^2}{5k} \cdot \dfrac{7mn^3}{13}$ *(i) $\dfrac{-3tu^2}{v} \cdot \dfrac{4u^3}{7v^2}$

3. Multiply.

*(a) $\dfrac{x^4 + 1}{3x} \cdot \dfrac{17}{4x}$ (b) $\dfrac{x^2 + 3}{5x} \cdot \dfrac{6}{x}$ *(c) $\dfrac{x + 1}{x - 4} \cdot \dfrac{x + 3}{x - 4}$

(d) $\dfrac{x + 9}{x + 1} \cdot \dfrac{x - 9}{x - 1}$ *(e) $\dfrac{5x - 3}{3x} \cdot \dfrac{x + 7}{x}$ (f) $\dfrac{7x + 2}{x + 1} \cdot \dfrac{6}{x}$

18.2 REDUCTION OF FRACTIONS

Reduction of fractions in algebra is similar to reduction of fractions in arithmetic (presented in Chapter 6). *Factors* that appear in both numerator and denominator can be separated and divided out. We use the idea that *a number, variable, or expression divided by itself equals 1.*

$$\boxed{\dfrac{\text{number}}{\text{itself}} = 1}$$

For example,

$$\dfrac{3}{3} = 1 \qquad \dfrac{x}{x} = 1 \qquad \dfrac{y + 2}{y + 2} = 1$$

EXAMPLE 5 Reduce $\dfrac{20}{25}$.

Solution Reduce $\frac{20}{25}$ by finding a common factor, 5, in both numerator and denominator; then divide out the common factor.

$$\dfrac{20}{25} = \dfrac{5 \cdot 4}{5 \cdot 5} = \dfrac{5}{5} \cdot \dfrac{4}{5} = 1 \cdot \dfrac{4}{5} = \dfrac{4}{5} \; \checkmark$$

We are allowed to reduce because we can write the fraction as *one*, $\frac{5}{5}$, times $\frac{4}{5}$. Many people prefer to show fewer steps and do the reduction as follows.

$$\dfrac{20}{25} = \dfrac{\cancel{5} \cdot 4}{\cancel{5} \cdot 5} = \dfrac{4}{5}$$

EXAMPLE 6 Reduce $\dfrac{5x}{7x}$.

Solution Here the factor x is common to both numerator and denominator.

$$\dfrac{5x}{7x} = \dfrac{5 \cdot x}{7 \cdot x} = \dfrac{5}{7} \cdot \dfrac{x}{x} = \dfrac{5}{7} \cdot 1 = \dfrac{5}{7} \; \checkmark$$

EXAMPLE 7 Reduce $\dfrac{12x^2y}{30y^2}$.

Solution Here are the steps. Additional explanation follows.

$$\frac{12x^2y}{30y^2} = \frac{2 \cdot 6 \cdot x^2 \cdot y}{5 \cdot 6 \cdot y \cdot y}$$

$$= \frac{2 \cdot \cancel{6} \cdot x^2 \cdot \cancel{y}}{5 \cdot \cancel{6} \cdot y \cdot \cancel{y}}$$

$$= \frac{2x^2}{5y} \quad \checkmark$$

We factored 12 into $2 \cdot 6$, 30 into $5 \cdot 6$, and y^2 into $y \cdot y$ because we saw what would divide out. If you don't "see it," then factor everything into prime factors and you should see it then.

$$\frac{12x^2y}{30y^2} = \frac{\cancel{2} \cdot 2 \cdot \cancel{3} \cdot x \cdot x \cdot \cancel{y}}{\cancel{2} \cdot \cancel{3} \cdot 5 \cdot y \cdot \cancel{y}} = \frac{2 \cdot x \cdot x}{5 \cdot y} = \frac{2x^2}{5y}$$

As we proceed, you will see the role of factoring in the reduction of fractions. In fact,

> Fractions are reduced
> *only* by factoring.

EXAMPLE 8 Reduce $\dfrac{c}{ac - c^2}$.

Solution Clearly there are c's in both the numerator and denominator.

$$\frac{c}{ac - c^2} = \frac{c \cdot 1}{a \cdot \underline{c} - c \cdot \underline{c}}$$

The c in the numerator is a *factor*. There is also a *factor* of c in the denominator. Factor it out so that we can then divide it out with the c in the numerator.

$$= \frac{c \cdot 1}{c(a - c)}$$

$$= \frac{\cancel{c} \cdot 1}{\cancel{c}(a - c)}$$

$$= \frac{1}{a - c} \quad \checkmark$$

EXAMPLE 9 Reduce $\dfrac{x^2 - x - 12}{x^2 + x - 20}$.

Solution Both numerator and denominator are quadratic expressions. They should be factored if possible.

$$\frac{x^2 - x - 12}{x^2 + x - 20} = \frac{(x + 3)(x - 4)}{(x + 5)(x - 4)}$$

Now divide out any factors common to both numerator and denominator.

$$= \frac{(x + 3)(x - 4)}{(x + 5)(x - 4)}$$

$$= \frac{x + 3}{x + 5} \quad \checkmark$$

And you cannot reduce $\dfrac{x + 3}{x + 5}$, as explained in the next example.

EXAMPLE 10 Reduce $\dfrac{x + 3}{x + 5}$.

Solution It is tempting to reduce $\dfrac{x + 3}{x + 5}$ as $\dfrac{\cancel{x} + 3}{\cancel{x} + 5} = \dfrac{3}{5}$. But doing so is *incorrect*. Remember, *you can only reduce fractions in which you can separate a fraction that is equal to one.* So, if $\dfrac{x + 3}{x + 5}$ can be written $\dfrac{x}{x} \cdot \dfrac{\quad}{\quad}$, then $\dfrac{x}{x}$ can be removed without changing the value of the fraction. Otherwise not.

It can also be reasoned numerically that such a reduction is wrong. After all, suppose that x was equal to 2. Then

$$\frac{x + 3}{x + 5} = \frac{2 + 3}{2 + 5} = \frac{5}{7}$$

And $\dfrac{5}{7} \neq \dfrac{3}{5}$. So reducing $\dfrac{x + 3}{x + 5}$ to $\dfrac{3}{5}$ is incorrect.

If the fraction had been $\dfrac{3x}{5x}$ it would have reduced to $\dfrac{3}{5}$, as

$$\frac{3x}{5x} = \frac{3}{5} \cdot \frac{x}{x} = \frac{3}{5} \cdot 1 = \frac{3}{5}$$

EXAMPLE 11 Reduce $\dfrac{2x^2 + 12x + 10}{4x + 20}$.

Solution Note first the common factor 2 in each term of the numerator and in each term of the denominator.

$$\frac{2x^2 + 12x + 10}{4x + 20} = \frac{2(x^2 + 6x + 5)}{2(2x + 10)}$$

$$= \frac{\cancel{2}(x^2 + 6x + 5)}{\cancel{2}(2x + 10)}$$

$$= \frac{x^2 + 6x + 5}{2x + 10}$$

Next, note that the quadratic expression factors. Also, 2 can be factored out of $2x + 10$.

$$= \frac{(x + 5)(x + 1)}{2(x + 5)}$$

$$= \frac{(\cancel{x + 5})(x + 1)}{2(\cancel{x + 5})}$$

$$= \frac{x + 1}{2} \quad \checkmark$$

EXAMPLE 12   Reduce $\dfrac{x + 3}{3 + x}$.

Solution   The numerator and denominator are the same, since $x + 3 = 3 + x$. So the fraction reduces to 1.

$$\frac{x + 3}{3 + x} = 1 \quad \checkmark$$

EXAMPLE 13   Reduce $\dfrac{x - 5}{5 - x}$.

Solution   Clearly $(x - 5)$ and $(5 - x)$ are different and yet they do resemble one another. In fact, with a little algebraic manipulation we can show that $(x - 5) = -(5 - x)$. In other words, $(x - 5)$ and $(5 - x)$ are opposites. Here's the manipulation.

$$-(5 - x) = -5 + x$$
$$= +x - 5$$
$$= x - 5$$

If $-(5 - x)$ is substituted for $(x - 5)$ in the original fraction, the fraction reduces to -1. Here is the work.

$$\frac{x - 5}{5 - x} = \frac{-(5 - x)}{(5 - x)}$$
$$= -1 \quad \checkmark$$

This example points out a useful general result. *A number divided by its opposite yields* -1.

$$\boxed{\dfrac{\text{number}}{\text{opposite}} = -1}$$

EXAMPLE 14   Reduce.

(a) $\dfrac{-y}{y}$ (b) $\dfrac{x - y}{y - x}$

Solution   (a) Since $-y$ and y are opposites, the fraction reduces to -1.

$$\frac{-y}{y} = -1$$

(b) Do you see that $(x - y)$ and $(y - x)$ are opposites?

$$x - y = -y + x = -(y - x).$$

Consequently,

$$\frac{x - y}{y - x} = -1 \quad \checkmark$$

The last two examples were based on a "trick" that you should look for in the future. If you doubt the correctness of the procedure despite the justification presented, consider the arithmetic example given below.

$$\frac{4 - 9}{9 - 4} = \frac{-5}{+5} = -1$$

EXAMPLE 15 Reduce $\dfrac{21 - 3x}{2x^2 - 98}$.

Solution

$$\frac{21 - 3x}{2x^2 - 98} = \frac{3(7 - x)}{2(x^2 - 49)} \qquad \begin{cases} \text{after factoring out constant} \\ \text{factors} \end{cases}$$

$$= \frac{3(7 - x)}{2(x + 7)(x - 7)} \qquad \begin{cases} \text{factoring the difference} \\ \text{of two squares} \end{cases}$$

$$= (-1) \cdot \frac{3(\cancel{7 - x})}{2(x + 7)(\cancel{x - 7})} \quad \text{since } \frac{7 - x}{x - 7} = -1 \quad \text{(opposites)}$$

$$= -\frac{3}{2(x + 7)} \quad \checkmark$$

EXERCISES 18.2

*1. Reduce each fraction, if possible. (*Note:* "Reduce" means to reduce completely.)

(a) $\dfrac{16}{80}$ (b) $\dfrac{12}{600}$ (c) $\dfrac{3x}{4x}$

(d) $\dfrac{8y}{10y}$ (e) $\dfrac{5x^2}{20x}$ (f) $\dfrac{5x}{2x^3}$

(g) $\dfrac{3x}{x^2 - x}$ (h) $\dfrac{m}{2m - 2m^2}$ (i) $\dfrac{m - n}{n - m}$

(j) $\dfrac{m - n}{m + n}$ (k) $\dfrac{x - 3}{3 - x}$ (l) $\dfrac{1 - x}{x - 1}$

(m) $\dfrac{x + 3}{2x + 6}$ (n) $\dfrac{x + 1}{x^2 + 2x + 1}$ (o) $\dfrac{x - 5}{x^2 - 7x + 10}$

(p) $\dfrac{x + 3}{x^2 - 9}$ (q) $\dfrac{x^2 - 7x + 12}{x - 3}$ (r) $\dfrac{x^2 - 16}{x - 4}$

2. Reduce each fraction, if possible.

*(a) $\dfrac{x^2 + 7x + 12}{x^2 + 5x + 4}$ (b) $\dfrac{x^2 - 9x + 20}{x^2 - 4x - 5}$ *(c) $\dfrac{x^2 + 9}{x^2 + 81}$

(d) $\dfrac{2x^2 - 8}{4x + 8}$ *(e) $\dfrac{x^2 + xy}{4x}$ (f) $\dfrac{3c}{bc + c^2}$

*(g) $\dfrac{8 - 2x}{x - 4}$ (h) $\dfrac{(y - x)^2}{x^2 - y^2}$ *(i) $\dfrac{4 + a}{4 + b}$

(j) $\dfrac{1 + b}{1 - b}$ *(k) $\dfrac{10x - 20}{3x - 6}$ (l) $\dfrac{8 - 4w}{6 - 3w}$

*(m) $\dfrac{6x - 6y}{2(y - x)}$ (n) $\dfrac{12x^2 - 48}{2 - x}$ *(o) $\dfrac{5x - 10}{7x - 14}$

(p) $\dfrac{u^2 - 4}{u^2 - 2u}$ *(q) $\dfrac{x^2 - 25}{10 - 2x}$ *(r) $\dfrac{3x^2 + 11x + 6}{3x^2 + 4x - 15}$

(s) $\dfrac{2x^2 - 7x - 15}{2x^2 - 9x - 5}$ *(t) $\dfrac{2x^2 + 12x + 10}{4x + 20}$ (u) $\dfrac{2x + 8}{2x^2 - 14x + 24}$

18.3 MORE MULTIPLICATION

Always look for reduction before you actually multiply numerator times numerator and denominator times denominator.

EXAMPLE 16   Multiply $\dfrac{3}{5} \cdot \dfrac{5}{7}$.

Solution   The 5's divide out; that is, $\dfrac{5}{5} = 1$.

$$\dfrac{3}{\cancel{5}} \cdot \dfrac{\cancel{5}}{7} = \dfrac{3}{7} \quad \checkmark$$

EXAMPLE 17   Multiply $\dfrac{3x^2y}{7x} \cdot \dfrac{5w}{6y}$

Solution   Examination of the fractions shows that

$$\dfrac{3x^2y}{7x} \cdot \dfrac{5w}{6y} = \dfrac{3 \cdot x \cdot x \cdot y}{7 \cdot x} \cdot \dfrac{5 \cdot w}{3 \cdot 2 \cdot y}$$

$$= \dfrac{\cancel{3} \cdot x \cdot \cancel{x} \cdot \cancel{y}}{7 \cdot \cancel{x}} \cdot \dfrac{5 \cdot w}{\cancel{3} \cdot 2 \cdot \cancel{y}} \quad (x, y, \text{ and } 3 \text{ divide out})$$

$$= \dfrac{x}{7} \cdot \dfrac{5w}{2}$$

$$= \dfrac{5wx}{14} \quad \checkmark$$

> When the directions say "reduce the fraction"
> or "multiply," the first thing you should
> think or do is *factor,* if possible.

EXAMPLE 18 Multiply $\dfrac{x^2 + 2x + 1}{x^2 - 2x - 3} \cdot \dfrac{x^2 + 2x - 15}{7x + 35}$.

Solution Factor the expressions and look for factors common to both numerator and denominator.

$$\frac{x^2 + 2x + 1}{x^2 - 2x - 3} \cdot \frac{x^2 + 2x - 15}{7x + 35} = \frac{(x+1)(x + 1)}{(x+1)(x-3)} \cdot \frac{(x+5)(x-3)}{7(x+5)}$$

$$= \frac{x + 1}{7} \quad \checkmark$$

EXAMPLE 19 Multiply $\dfrac{x^2 + 6x + 8}{x - 5} \cdot \dfrac{x}{x^2 - 4}$.

Solution Begin by factoring the expressions.

$$\frac{x^2 + 6x + 8}{x - 5} \cdot \frac{x}{x^2 - 4} = \frac{(x + 4)(x+2)}{(x - 5)} \cdot \frac{x}{(x+2)(x - 2)}$$

$$= \frac{(x + 4)x}{(x - 5)(x - 2)} \quad \checkmark$$

$$\text{or} \quad \frac{x^2 + 4x}{x^2 - 7x + 10} \quad \checkmark$$

EXERCISES 18.3

*1. Multiply. Be sure to reduce when possible.

(a) $\dfrac{3x}{5} \cdot \dfrac{10}{x}$

(b) $\dfrac{15x^2}{y} \cdot \dfrac{y^2}{60x}$

(c) $\dfrac{y + 3}{5} \cdot \dfrac{15}{2y + 6}$

(d) $\dfrac{20a}{3c} \cdot \dfrac{6x}{4a^3b^2c}$

(e) $\dfrac{x^2 - 25}{x + 1} \cdot \dfrac{2}{2x + 10}$

(f) $\dfrac{5}{x^2 - 9} \cdot \dfrac{4x - 12}{7}$

(g) $\dfrac{3x + 3}{x^2 + 7x + 6} \cdot \dfrac{x + 6}{6x^2}$

(h) $\dfrac{n^2 - n - 12}{n^2 - 16} \cdot \dfrac{2n - 8}{n + 3}$

(i) $\dfrac{x^2 + x - 12}{x^2 - 6x + 9} \cdot \dfrac{5x - 20}{x^2 - 16}$

(j) $\dfrac{1 - x^2}{x^2 + 4x + 3} \cdot \dfrac{2x^2 - 18}{2x}$

2. Multiply. Reduce when possible.

*(a) $\dfrac{3x^2 + 15x + 18}{x^2 + 2x} \cdot \dfrac{5x^3}{15x^2}$

(b) $\dfrac{2x^2 - 14x + 20}{x^2 - 25} \cdot \dfrac{x^4}{2x}$

*(c) $\dfrac{2x - 2y}{4x} \cdot \dfrac{3y}{3y - 3x}$

(d) $\dfrac{5m - 5n}{25m} \cdot \dfrac{7mn}{3n - 3m}$

*(e) $\dfrac{x^2 - 3x - 10}{x^2 - 25} \cdot \dfrac{x^2 + 9x + 20}{x^2 + 2x}$

(f) $\dfrac{a^2 - 16}{a + 2} \cdot \dfrac{2a}{a^2 - 4a}$

*(g) $\dfrac{x^2 - 9}{x + 3} \cdot \dfrac{4x}{2x - 6}$

(h) $\dfrac{b - 2a}{b + 2a} \cdot \dfrac{7a}{7b}$

*(i) $\dfrac{4x - 4}{9x} \cdot \dfrac{3x^2}{1 - x}$

(j) $\dfrac{a - b}{x - y} \cdot \dfrac{y - x}{b - a}$

*(k) $\dfrac{x^2 - 3x}{7x} \cdot \dfrac{2x + 6}{9 - x^2}$

(l) $\dfrac{x^2 - 8x + 12}{x^2 + 2x - 8} \cdot \dfrac{x^2 + x - 12}{x^2 - 36}$

*(m) $\dfrac{x^2 - 2x + 1}{x^2 + 5x} \cdot \dfrac{25 - x^2}{x^2 - 1}$

(n) $\dfrac{x^2 - 3x - 4}{x^2 - x - 12} \cdot \dfrac{x^2 + 5x + 6}{x^2 + 8x + 7}$

18.4 DIVISION OF FRACTIONS

Division of fractions in algebra is similar to division of fractions in arithmetic (which was presented in Chapter 6). Invert the second fraction (the divisor) and multiply by it.

EXAMPLE 20 Divide $\dfrac{3}{5} \div \dfrac{7}{4}$.

Solution

$$\dfrac{3}{5} \div \dfrac{7}{4} = \dfrac{3}{5} \cdot \dfrac{4}{7}$$
$$= \dfrac{3 \cdot 4}{5 \cdot 7}$$
$$= \dfrac{12}{35} \quad \checkmark$$

EXAMPLE 21 Divide $\dfrac{4}{9} \div \dfrac{5}{3}$.

Solution

$$\dfrac{4}{9} \div \dfrac{5}{3} = \dfrac{4}{9} \cdot \dfrac{3}{5}$$
$$= \dfrac{4}{\cancel{9} \cdot 3} \cdot \dfrac{\cancel{3}}{5}$$

$$= \frac{4 \cdot 1}{3 \cdot 5}$$

$$= \frac{4}{15} \checkmark$$

As we proceed to algebra examples, note the steps to be followed, *in order,* as division of fractions is carried out.

> *Division of Fractions*
> 1. Change to multiplication.
> 2. Factor wherever possible.
> 3. Reduce whenever possible.
> 4. Multiply what remains.

EXAMPLE 22 Divide $\dfrac{2x + 4}{6x} \div \dfrac{x^2 - 4}{7x^3}$.

Solution You cannot try to reduce until after you invert the second fraction and change to multiplication.

$$\frac{2x + 4}{6x} \div \frac{x^2 - 4}{7x^3} = \frac{2x + 4}{6x} \cdot \frac{7x^3}{x^2 - 4}$$

Now that we have a *multiplication* problem, each expression should be factored, if possible.

$$= \frac{2(x + 2)}{2 \cdot 3 \cdot x} \cdot \frac{7 \cdot x \cdot x^2}{(x + 2)(x - 2)}$$

Identical factors can now be divided out.

$$= \frac{\cancel{2}\cancel{(x + 2)}}{\cancel{2} \cdot 3 \cdot \cancel{x}} \cdot \frac{7 \cdot \cancel{x} \cdot x^2}{\cancel{(x + 2)}(x - 2)}$$

When what remains is multiplied, the result is

$$\frac{7x^2}{3(x - 2)} \checkmark$$

If you prefer, multiply it out completely to get

$$\frac{7x^2}{3x - 6} \checkmark$$

EXAMPLE 23 Divide $\dfrac{x^2 + 7x + 12}{x^2 - 1} \div \dfrac{x^2 + 6x + 9}{2x + 2}$.

Solution

$$\frac{x^2 + 7x + 12}{x^2 - 1} \div \frac{x^2 + 6x + 9}{2x + 2} = \frac{x^2 + 7x + 12}{x^2 - 1} \cdot \frac{2x + 2}{x^2 + 6x + 9}$$

$$= \frac{(x + 3)(x + 4)}{(x + 1)(x - 1)} \cdot \frac{2(x + 1)}{(x + 3)(x + 3)}$$

$$= \frac{(\cancel{x+3})(x+4)}{(\cancel{x+1})(x-1)} \cdot \frac{2(\cancel{x+1})}{(\cancel{x+3})(x+3)}$$

$$= \frac{2(x+4)}{(x-1)(x+3)} \quad \checkmark$$

which you may prefer to multiply out completely and leave as

$$\frac{2x+8}{x^2+2x-3} \quad \checkmark$$

EXAMPLE 24   Divide $\dfrac{2-x}{x+5} \div \dfrac{x^2-4}{2x+10}$.

Solution

$$\frac{2-x}{x+5} \div \frac{x^2-4}{2x+10} = \frac{2-x}{x+5} \cdot \frac{2x+10}{x^2-4}$$

$$= \frac{2-x}{x+5} \cdot \frac{2(x+5)}{(x+2)(x-2)}$$

The $(x+5)$ factors divide out.

$$= \frac{2-x}{\cancel{x+5}} \cdot \frac{2(\cancel{x+5})}{(x+2)(x-2)}$$

There is $2-x$ in the numerator and its opposite, $x-2$, in the denominator. They reduce to -1. To illustrate, we'll cross them out and put a minus in front of the entire product. Compare with Example 15.

$$= -\frac{\cancel{2-x}}{\cancel{x+5}} \cdot \frac{2(\cancel{x+5})}{(x+2)(\cancel{x-2})}$$

$$= -\frac{2}{x+2} \quad \checkmark$$

EXERCISES 18.4

*1. Divide. Be sure to reduce when possible.

(a) $\dfrac{3x}{7} \div \dfrac{x}{x+5}$

(b) $\dfrac{x-6}{8} \div \dfrac{2x-12}{4}$

(c) $\dfrac{5x^2}{3} \div \dfrac{x^2+3x}{9y}$

(d) $\dfrac{x^2+2x+1}{9x} \div \dfrac{x^2-1}{3}$

(e) $\dfrac{x-3}{x+3} \div \dfrac{6-2x}{3+x}$

(f) $\dfrac{x^2-2x-15}{x^2-3x-10} \div \dfrac{9x}{6+3x}$

(g) $\dfrac{m^2-9m+20}{15m} \div \dfrac{m^2-3m-10}{5m+10}$

2. Divide. Reduce when possible.

*(a) $\dfrac{9xy}{y} \div \dfrac{3x^2}{5}$

(b) $\dfrac{15abc}{2c} \div \dfrac{3a}{8}$

*(c) $\dfrac{x^2-6x+8}{3x-6} \div \dfrac{x-4}{x}$

(d) $\dfrac{m^2-9m}{6m} \div \dfrac{m-9}{3}$

$$*(e) \quad \frac{x^2 + x + 12}{5x + 15} \div \frac{x^2 - 16}{2x - 8} \qquad (f) \quad \frac{x^2 + 4x}{2x + 8} \div \frac{x^2 + x - 12}{2x - 6}$$

$$*(g) \quad \frac{x^2 + 9x}{x - 3} \div \frac{x^2 + 13x + 36}{6 - 2x} \qquad (h) \quad \frac{6t - t^2}{9t} \div \frac{t^2 - 5t - 6}{6t^2}$$

$$*(i) \quad \frac{3x + 3y}{x^2 - 64} \div \frac{3xy}{2x^2 + 16x + 32} \qquad (j) \quad \frac{2y - 2x}{x^2 - y^2} \div \frac{4xy^2}{2x}$$

18.5 ADDITION AND SUBTRACTION OF FRACTIONS

Addition of fractions in algebra is similar to addition of fractions in arithmetic. A common denominator must be determined, and all fractions being added must either have this common denominator or be changed so as to have the common denominator. Then the numerators can be added and placed over the common denominator. The presentation here assumes that you are familiar with the concept of least common multiple (LCM) and its use in arithmetic to determine the lowest common denominator. Unless your instructor reviews this concept in class or offers another approach, you should *go back to Chapter 6 and read about LCM and addition of fractions.* Once again, to add fractions:

> 1. Determine common denominator (LCM).
> 2. Change all fractions to fractions having the common denominator.
> 3. Add the numerators.
> 4. Place that sum over the common denominator.

EXAMPLE 25 Add.

(a) $\dfrac{3}{7} + \dfrac{2}{7}$ (b) $\dfrac{a}{x} + \dfrac{b}{x}$

Solution (a) Since the denominators of the two fractions are the same, we can simply add the numerators and place the sum over the (same) denominator.

$$\frac{3}{7} + \frac{2}{7} = \frac{3 + 2}{7} = \frac{5}{7}$$

(b) Again, the denominators are the same.

$$\frac{a}{x} + \frac{b}{x} = \frac{a + b}{x}$$

EXAMPLE 26 Add.

(a) $\dfrac{2}{3} + \dfrac{1}{5}$ (b) $\dfrac{a}{u} + \dfrac{b}{v}$

Solution   (a) The denominators are 3 and 5. The common denominator is $3 \cdot 5$, or 15. Each denominator should be changed to 15, as shown next, and then the fractions can be added.

$$\frac{2}{3} + \frac{1}{5} = \frac{2}{3} \cdot \frac{5}{5} + \frac{1}{5} \cdot \frac{3}{3}$$

$$= \frac{10}{15} + \frac{3}{15}$$

$$= \frac{13}{15}$$

(b) Since the denominators are u and v, the common denominator is uv. Change each fraction to a fraction with denominator uv and then add the fractions.

$$\frac{a}{u} = \frac{a}{u} \cdot \frac{v}{v} = \frac{av}{uv}$$

$$\frac{b}{v} = \frac{b}{v} \cdot \frac{u}{u} = \frac{bu}{uv}$$

Thus

$$\frac{a}{u} + \frac{b}{v} = \frac{av}{uv} + \frac{bu}{uv} = \frac{av + bu}{uv} \quad \checkmark$$

EXAMPLE 27   Add $\dfrac{3}{5x} + \dfrac{7}{10y}$.

Solution   Since

$$5x = 5 \cdot x$$

and

$$10y = 2 \cdot 5 \cdot y$$

$$\text{LCM } (5x, 10y) = 2 \cdot 5 \cdot x \cdot y = 10xy$$

$$\frac{3}{5x} + \frac{7}{10y} = \frac{3}{5x} \cdot \frac{2y}{2y} + \frac{7}{10y} \cdot \frac{x}{x}$$

$$= \frac{6y}{10xy} + \frac{7x}{10xy}$$

$$= \frac{6y + 7x}{10xy} \quad \checkmark$$

The denominator $5x$ must be multiplied by $2y$ to produce $10xy$. The denominator $10y$ must be multiplied by x to produce $10xy$.

As in arithmetic, subtraction of fractions resembles addition of fractions.

EXAMPLE 28   Combine $\dfrac{3a}{bcd} - \dfrac{7}{5c}$.

Solution   Since LCM $(bcd, 5c)$ is $5bcd$, the common denominator is $5bcd$. Change each fraction to one with a denominator of $5bcd$ and then combine them.

$$\frac{3a}{bcd} - \frac{7}{5c} = \frac{3a}{bcd} \cdot \frac{5}{5} - \frac{7}{5c} \cdot \frac{bd}{bd}$$

$$= \frac{15a}{5bcd} - \frac{7bd}{5bcd}$$

$$= \frac{15a - 7bd}{5bcd} \quad \checkmark$$

EXAMPLE 29   Combine $\dfrac{3w}{10x^2y} + \dfrac{25z}{12wxy^3}$.

Solution $\left.\begin{array}{l} 10x^2y = 2 \cdot 5 \cdot x^2 \cdot y \\ 12wxy^3 = 2^2 \cdot 3 \cdot w \cdot x \cdot y^3 \end{array}\right\}$ LCM is $2^2 \cdot 3 \cdot 5 \cdot w \cdot x^2 \cdot y^3 = 60wx^2y^3$.

$$\frac{3w}{10x^2y} + \frac{25z}{12wxy^3} = \frac{3w}{10x^2y} \cdot \frac{6wy^2}{6wy^2} + \frac{25z}{12wxy^3} \cdot \frac{5x}{5x}$$

$$= \frac{18w^2y^2}{60wx^2y^3} + \frac{125xz}{60wx^2y^3}$$

$$= \frac{18w^2y^2 + 125xz}{60wx^2y^3} \quad \checkmark$$

EXAMPLE 30   Combine $\dfrac{3}{x + 1} + \dfrac{5}{x}$.

Solution   The denominators $x + 1$ and x have no *factors* in common. (Note that x is *not a factor* of $x + 1$.) This means that the common denominator for the two fractions is $(x + 1)x$. Thus

$$\frac{3}{x + 1} + \frac{5}{x} = \frac{3}{x + 1} \cdot \frac{x}{x} + \frac{5}{x} \cdot \frac{x + 1}{x + 1}$$

$$= \frac{3x}{(x + 1)x} + \frac{5(x + 1)}{(x + 1)x}$$

$$= \frac{3x + 5(x + 1)}{(x + 1)x}$$

$$= \frac{3x + 5x + 5}{(x + 1)x}$$

$$= \frac{8x + 5}{(x + 1)x} \quad \checkmark$$

Notice that the numerator simplified. This does not always happen, but you should watch for the possibility.

EXAMPLE 31   Combine $\dfrac{7}{2x + 2} + \dfrac{x - 3}{4x}$.

Solution $\left.\begin{array}{l} 2x + 2 = 2(x + 1) \\ 4x = 2^2 \cdot x \end{array}\right\}$ LCM is $2^2 \cdot x \cdot (x + 1) = 4x(x + 1)$.

$$\frac{7}{2x + 2} + \frac{x - 3}{4x} = \frac{7}{2(x + 1)} \cdot \frac{2x}{2x} + \frac{x - 3}{4x} \cdot \frac{x + 1}{x + 1}$$

$$= \frac{14x}{4x(x + 1)} + \frac{(x - 3)(x + 1)}{4x(x + 1)}$$

$$= \frac{14x + (x - 3)(x + 1)}{4x(x + 1)}$$

$$= \frac{14x + x^2 - 2x - 3}{4x(x + 1)}$$

$$= \frac{x^2 + 12x - 3}{4x(x + 1)} \quad \checkmark$$

EXAMPLE 32 Combine $\dfrac{x - 1}{x^2 + 5x + 6} + \dfrac{6}{x^2 + 7x + 10}$.

Solution $\begin{aligned} x^2 + 5x + 6 &= (x + 3)(x + 2) \\ x^2 + 7x + 10 &= (x + 5)(x + 2) \end{aligned} \Bigg\}$ LCM is $(x + 3)(x + 2)(x + 5)$.

$$\frac{x - 1}{x^2 + 5x + 6} + \frac{6}{x^2 + 7x + 10}$$

$$= \frac{x - 1}{(x + 3)(x + 2)} + \frac{6}{(x + 5)(x + 2)}$$

$$= \frac{x - 1}{(x + 3)(x + 2)} \cdot \frac{(x + 5)}{(x + 5)} + \frac{6}{(x + 5)(x + 2)} \cdot \frac{(x + 3)}{(x + 3)}$$

$$= \frac{(x - 1)(x + 5) + 6(x + 3)}{(x + 3)(x + 2)(x + 5)}$$

$$= \frac{x^2 + 4x - 5 + 6x + 18}{(x + 3)(x + 2)(x + 5)}$$

$$= \frac{x^2 + 10x + 13}{(x + 3)(x + 2)(x + 5)} \quad \checkmark$$

EXERCISES 18.5

1. Combine the fractions.

*(a) $\dfrac{t}{x} + \dfrac{3}{x}$

*(b) $\dfrac{5}{y} + \dfrac{x}{y}$

*(c) $\dfrac{7}{x} + \dfrac{5}{y}$

*(d) $4 + \dfrac{1}{x}$ $\left(Hint:\ 4 = \dfrac{4}{1} \right)$

*(e) $3 + \dfrac{5}{m}$

*(f) $\dfrac{2}{c} - 9$

*(g) $\dfrac{5}{2x} + \dfrac{3}{x}$

(h) $\dfrac{3}{10x} - \dfrac{7}{20y}$

*(i) $\dfrac{3}{wx^2} + \dfrac{2x}{5w^2}$

(j) $\dfrac{5}{3x^2} - \dfrac{7c}{12x}$

*(k) $\dfrac{8x}{15ab} + \dfrac{3y}{20abc}$

*(l) $\dfrac{3}{a} + \dfrac{4}{b} + \dfrac{5}{c}$

(m) $\dfrac{a}{b} - \dfrac{c}{d} + \dfrac{e}{f}$

*(n) $\dfrac{a}{2b} + \dfrac{c}{3d} - \dfrac{e}{4f}$

*(o) $\dfrac{1}{3x^2y} + \dfrac{7}{10xy^2}$

(p) $\dfrac{3}{2m^2n} - \dfrac{13}{6mn^2}$

*(q) $\dfrac{x}{18a^3b} + \dfrac{y}{12a^2b^2}$

*(r) $\dfrac{5}{xyz^2} + \dfrac{4}{xy^2z} + \dfrac{3}{x^2yz}$

*(s) $\dfrac{3}{a^2bc^2} + \dfrac{4}{ac} - \dfrac{5}{abc}$

(t) $\dfrac{3}{4x^2y} + \dfrac{1}{12xy^2} - \dfrac{5}{9}$

2. Combine the fractions.

*(a) $\dfrac{5}{3x + 2} + \dfrac{7}{x}$

(b) $\dfrac{3}{x + 5} - \dfrac{8}{x}$

*(c) $2x + \dfrac{7}{x + 1}$

(d) $\dfrac{5}{x - 2} + 7x$

*(e) $\dfrac{3}{x - 2} + 2$

(f) $3 + \dfrac{5}{x + 4}$

*(g) $\dfrac{5}{x + 1} + \dfrac{3}{x + 2}$

(h) $\dfrac{1}{x - 5} - \dfrac{4}{x + 1}$

*(i) $\dfrac{1}{x - 5} - \dfrac{7}{x + 3}$

(j) $\dfrac{2}{x + 5} - \dfrac{4}{x - 1}$

*(k) $\dfrac{4}{3x - 3} + \dfrac{1}{6x}$

(l) $\dfrac{7}{5x + 10} + \dfrac{9}{20x}$

*(m) $\dfrac{5}{2x} + \dfrac{3}{2x + 6}$

(n) $\dfrac{7}{3x} + \dfrac{1}{3x - 6}$

*(o) $\dfrac{3}{2x + 2} + \dfrac{x - 4}{8x}$

(p) $\dfrac{1}{6x - 12} + \dfrac{1 + x}{9x}$

*(q) $\dfrac{x}{3x + 3} + \dfrac{7}{6x}$

(r) $\dfrac{x}{2x - 8} + \dfrac{1}{2x}$

3. Combine the fractions.

*(a) $\dfrac{8}{x^2 - 4} + \dfrac{2}{x - 2}$

(b) $\dfrac{x}{x^2 - 9} + \dfrac{5}{x + 3}$

*(c) $\dfrac{1}{x^2 + 3x + 2} + \dfrac{10}{x + 1}$

*(d) $\dfrac{5x}{x^2 - 9x + 14} - \dfrac{3}{x^2 - 7x}$

*(e) $\dfrac{7}{x^2 + 6x + 8} - \dfrac{5}{x^2 + 4x + 4}$

(f) $\dfrac{2}{x^2 - 5x + 6} + \dfrac{3}{x^2 - x - 2}$

*(g) $\dfrac{9}{x^2 + 4x} - \dfrac{21}{2x + 8} + \dfrac{8}{x^2}$

(h) $\dfrac{4}{x^2 + 5x} + \dfrac{3}{2x + 10} + \dfrac{5}{4x}$

18.6 COMPLEX FRACTIONS

Sometimes the numerator and denominator of fractions are themselves fractions. Such complicated fractions are called **complex fractions.** Although they look very complex, they are not all that difficult to simplify. You merely consider the complex fraction as the division-of-fractions problem that it represents.

EXAMPLE 33 Simplify $\dfrac{\frac{2}{3}}{\frac{5}{7}}$.

Solution
$$\frac{\frac{2}{3}}{\frac{5}{7}} = \frac{2}{3} \div \frac{5}{7} \qquad \begin{cases} \text{Write the complex fraction} \\ \text{as division of fractions.} \end{cases}$$

$$= \frac{2}{3} \cdot \frac{7}{5} \qquad \begin{cases} \text{Change to multiplication} \\ \text{by inverting divisor.} \end{cases}$$

$$= \frac{14}{15} \quad \checkmark$$

EXAMPLE 34 Simplify $\dfrac{\frac{m+5}{m+4}}{\frac{m-6}{m-2}}$.

Solution
$$\frac{\frac{m+5}{m+4}}{\frac{m-6}{m-2}} = \frac{m+5}{m+4} \div \frac{m-6}{m-2}$$

$$= \frac{m+5}{m+4} \cdot \frac{m-2}{m-6}$$

$$= \frac{(m+5)(m-2)}{(m+4)(m-6)} \quad \checkmark \qquad \text{or} \qquad \frac{m^2+3m-10}{m^2-2m-24} \quad \checkmark$$

EXAMPLE 35 Simplify $\dfrac{\frac{x+3}{9}}{x}$.

Solution   First, write the denominator x as the fraction $\frac{x}{1}$ so that you can easily see the division of one fraction by another. Then proceed by changing the complex fraction to a division of fractions.

$$\frac{\dfrac{x+3}{9}}{x} = \frac{\dfrac{x+3}{9}}{\dfrac{x}{1}} = \frac{x+3}{9} \div \frac{x}{1}$$

$$= \frac{x+3}{9} \cdot \frac{1}{x}$$

$$= \frac{x+3}{9x} \quad \checkmark$$

EXAMPLE 36   Simplify $\dfrac{\dfrac{x+1}{x}}{\dfrac{1-x}{x}}.$

Solution

$$\frac{\dfrac{x+1}{x}}{\dfrac{1-x}{x}} = \frac{x+1}{x} \div \frac{1-x}{x}$$

$$= \frac{x+1}{x} \cdot \frac{x}{1-x}$$

$$= \frac{x+1}{\cancel{x}} \cdot \frac{\cancel{x}}{1-x}$$

$$= \frac{x+1}{1-x} \quad \checkmark$$

When the numerator or denominator of a complex fraction contains *more than one term*, as in the next example, a special procedure should be used, namely,

1. Consider all the fractions appearing in the numerator and denominator, and determine the least common denominator of all their denominators.
2. Multiply both the numerator and denominator (all terms) by that common denominator.

EXAMPLE 37   Simplify $\dfrac{2 - \dfrac{x}{y}}{y - \dfrac{1}{x}}.$

Solution　Notice that the numerator contains two terms and so does the denominator. Since the numerator and/or denominator contain more than one term, consider all the fractions appearing, namely $\dfrac{x}{y}$ and $\dfrac{1}{x}$. The common denominator of these two fractions is xy. So multiply both numerator and denominator (all terms) by xy.

$$\frac{2 - \dfrac{x}{y}}{y - \dfrac{1}{x}} = \frac{xy\left(2 - \dfrac{x}{y}\right)}{xy\left(y - \dfrac{1}{x}\right)}$$

$$= \frac{xy \cdot 2 - xy \cdot \dfrac{x}{y}}{xy \cdot y - xy \cdot \dfrac{1}{x}} \qquad \text{after distributing the } xy$$

At this point you can see why this special procedure will indeed simplify the complex fraction. The denominator y will divide out with the y of xy. The denominator x will divide out with the x of the xy.

$$= \frac{2xy - x^2}{xy^2 - y} \quad \checkmark \qquad \text{after simplification}$$

EXERCISES 18.6

1. Simplify each complex fraction.

*(a) $\dfrac{\dfrac{3}{4}}{\dfrac{5}{11}}$ 　　 (b) $\dfrac{\dfrac{7}{4}}{\dfrac{5}{3}}$ 　　 *(c) $\dfrac{\dfrac{a}{b}}{\dfrac{c}{d}}$

(d) $\dfrac{\dfrac{3}{x}}{\dfrac{5}{y}}$ 　　 *(e) $\dfrac{\dfrac{3}{x}}{7}$ 　　 (f) $\dfrac{\dfrac{x-2}{3}}{x}$

*(g) $\dfrac{\dfrac{x+3}{x}}{\dfrac{y+2}{y}}$ 　　 (h) $\dfrac{\dfrac{n-2}{n}}{\dfrac{n+2}{n}}$ 　　 *(i) $\dfrac{\dfrac{x^2-9}{x+1}}{\dfrac{x+3}{x^2-1}}$

(j) $\dfrac{\dfrac{x^2+6x+9}{x^2+7x+12}}{\dfrac{x^2+15x+36}{2x+8}}$

2. Simplify each complex fraction.

*(a) $\dfrac{1 + \dfrac{1}{x}}{1 - \dfrac{1}{x}}$

(b) $\dfrac{x + \dfrac{1}{2}}{y + \dfrac{1}{3}}$

*(c) $\dfrac{1 + \dfrac{a}{b}}{1 - \dfrac{a}{b}}$

(d) $\dfrac{\dfrac{x}{y} + 2}{3 + \dfrac{1}{y}}$

*(e) $\dfrac{\dfrac{3}{xy} + 1}{\dfrac{2}{x} + \dfrac{3}{y}}$

(f) $\dfrac{\dfrac{5}{x} - \dfrac{2}{y}}{\dfrac{4}{y} + \dfrac{15}{x}}$

*(g) $\dfrac{\dfrac{1}{2} - \dfrac{7}{x}}{\dfrac{8}{y} + \dfrac{1}{3}}$

(h) $\dfrac{\dfrac{x}{2y} + \dfrac{y}{x}}{\dfrac{x}{y} - \dfrac{y}{x}}$

*(i) $\dfrac{m + \dfrac{x}{y}}{\dfrac{y}{x} - m}$

(j) $\dfrac{p - \dfrac{q}{r}}{p + \dfrac{r}{q}}$

*(k) $\dfrac{9}{\dfrac{a}{b} + 4}$

(l) $\dfrac{7}{1 - \dfrac{x}{y}}$

18.7 EQUATIONS WITH FRACTIONS

If an equation contains fractions, begin solving it by first eliminating all fractions. To accomplish this, *multiply both sides by the lowest common denominator (LCM) of all fractions involved.*

EXAMPLE 38   Solve for x: $7 = \dfrac{2}{3x}$.

Solution   The common denominator is $3x$; so multiply both sides of the equation by $3x$ in order to eliminate fractions.

$$7 \cdot \mathbf{3x} = \dfrac{2}{3x} \cdot \mathbf{3x}$$

or $21x = 2$

Thus

$$x = \dfrac{2}{21} \quad \checkmark$$

EXAMPLE 39   Solve for x: $x + \dfrac{x + 1}{5} = \dfrac{1}{2}$.

Solution   The common denominator is LCM $(5, 2)$, which is 10. So multiply each term by 10 to eliminate all denominators.

$$\mathbf{10} \cdot x + \mathbf{10} \cdot \dfrac{x + 1}{5} = \mathbf{10} \cdot \dfrac{1}{2} \quad \begin{cases} 10 \cdot x = 10x \\[2mm] 10 \cdot \dfrac{x + 1}{5} = \dfrac{2 \cdot \cancel{5} \cdot (x + 1)}{\cancel{5}} = 2(x + 1) \\[2mm] 10 \cdot \dfrac{1}{2} = 5 \end{cases}$$

or
$$10x + 2(x + 1) = 5$$
$$10x + 2x + 2 = 5$$
$$12x + 2 = 5$$
$$12x = 3$$
$$x = \frac{3}{12}$$

or
$$x = \frac{1}{4} \;\checkmark$$

EXAMPLE 40   Solve for n: $\dfrac{5}{n + 1} - \dfrac{3}{20} = \dfrac{9}{5}$.

Solution   Since LCM $(n + 1, 20, 5) = (20)(n + 1)$, multiply all terms by $(20)(n + 1)$.

$$\frac{5}{n + 1} \cdot (20)(n + 1) - \frac{3}{20} \cdot (20)(n + 1) = \frac{9}{5} \cdot (20)(n + 1)$$

or

$$\frac{(5)}{(n+1)} \cdot (20)(n+1) - \frac{(3)}{(20)} \cdot (20)(n + 1) = \frac{(9)}{(5)} \cdot \overset{4}{(20)}(n + 1)$$

which simplifies to
$$(5)(20) - (3)(n + 1) = (9)4(n + 1)$$
$$100 - 3(n + 1) = 36(n + 1)$$
$$100 - 3n - 3 = 36n + 36$$
$$97 - 3n = 36n + 36$$
$$97 = 39n + 36$$
$$61 = 39n$$
$$39n = 61$$
$$n = \frac{61}{39} \;\checkmark$$

EXAMPLE 41   Solve for x: $\dfrac{7}{3x + 1} + 5 = \dfrac{5x}{x - 1}$.

Solution   LCM $(3x + 1, x - 1) = (3x + 1)(x - 1)$. So multiply all terms by the LCM $(3x + 1)(x - 1)$.

$$\frac{7}{3x + 1} \cdot (3x + 1)(x - 1) + 5 \cdot (3x + 1)(x - 1) = \frac{5x}{x - 1} \cdot (3x + 1)(x - 1)$$

Notice now how the denominators divide out and that no fractions are left.

$$\frac{7}{3x + 1} \cdot (3x + 1)(x - 1) + 5 \cdot (3x + 1)(x - 1) = \frac{5x}{x - 1} \cdot (3x + 1)(x - 1)$$

The simplified equation is

$$7(x - 1) + 5(3x + 1)(x - 1) = 5x(3x + 1)$$

Then

$$7x - 7 + 15x^2 - 10x - 5 = 15x^2 + 5x$$

Although this may seem to be a quadratic equation, when $-15x^2$ is added to each side in order to combine the x^2 terms on the same side, the x^2 terms add up to zero. If they did not, we would indeed have a quadratic equation to solve. As it is, however, the equation is linear. With the $15x^2$ terms removed, the equation is

$$7x - 7 - 10x - 5 = 5x$$

or

$$-3x - 12 = 5x$$

or

$$-8x - 12 = 0$$

$$-8x = 12$$

$$x = \frac{12}{-8} = -\frac{3}{2} \quad \checkmark$$

Two notes are in order here.

NOTE 1
Since division by zero is not defined, *no fraction can have a denominator equal to zero* in any equation with fractions. In Example 38, x cannot be 0 or else the denominator $3x$ would be 0. In Example 41, x cannot be either 1 or $-\frac{1}{3}$ for the same reason: If $x = 1$, then the denominator $x - 1$ is 0; if $x = -\frac{1}{3}$, then the denominator $3x + 1$ is 0.

NOTE 2
Whenever both sides of an equation are multiplied by a variable (as in Examples 38, 40, and 41), the solution(s) obtained must be checked in the original equation. *A number can be a solution of the altered equation even though it may not be a solution of the original.* Such numbers are called **extraneous roots.**

EXAMPLE 42 A student has test grades of 72 and 81. What must be her grade on the next test in order to have an average of 80?

Solution The average of three test grades is computed as the sum of the three grades divided by 3. If we let

$$x = \text{her grade on the next test}$$

then we want the average of 72, 81, and x to be 80. That is,

$$\frac{72 + 81 + x}{3} = 80$$

To solve this equation with fractions, begin by multiplying both sides by 3 in order to eliminate the fraction.

$$3 \cdot \frac{72 + 81 + x}{3} = 3 \cdot 80$$

The result is

$$72 + 81 + x = 240$$

or

$$153 + x = 240$$

Finally,

$$x = 87$$

Thus we find that the student needs a grade of 87 on the next test in order to have an average of 80.

EXAMPLE 43   Darren can paint the fence in 3 hours and Teresa can paint the fence in 4 hours. Assuming those rates, how long will it take them both, working together, to paint the fence?

Solution   This problem can be solved by considering the *hourly rate* of each person and the hourly rate of the two of them combined. That is,

- Since Darren can paint the fence in 3 hours, he can paint $\frac{1}{3}$ of the fence in one hour.
- Since Teresa can paint the fence in 4 hours, she can paint $\frac{1}{4}$ of the fence in one hour.

It follows that if we let x be the number of hours it takes both of them working together to paint the fence, then

$$\frac{1}{x} = \text{portion of the fence they can paint together in one hour}$$

To determine x, we need to form an equation. This can be done based on the following:

$$\frac{\text{portion Darren}}{\text{paints in 1 hr}} + \frac{\text{portion Teresa}}{\text{paints in 1 hr}} = \frac{\text{portion Darren and Teresa}}{\text{paint in 1 hr}}$$

The equation:

$$\frac{1}{3} + \frac{1}{4} = \frac{1}{x}$$

To solve this equation, begin by multiplying both sides (each term) by the lowest common denominator, $12x$.

$$12x \cdot \frac{1}{3} + 12x \cdot \frac{1}{4} = 12x \cdot \frac{1}{x}$$

The equation simplifies to

$$4x + 3x = 12$$
$$7x = 12$$

Finally,

$$x = \frac{12}{7} = 1\frac{5}{7}$$

We conclude that, working together, Darren and Teresa can paint the fence in $1\frac{5}{7}$ hours.

EXERCISES 18.7

1. Solve each equation.

*(a) $\dfrac{4}{9x} = 5$ (b) $9 = \dfrac{2}{7x}$

*(c) $x + \dfrac{x+1}{3} = 7$ *(d) $x + \dfrac{x+3}{5} = \dfrac{1}{3}$

*(e) $\dfrac{m}{4} + \dfrac{m+1}{3} = 5$ *(f) $\dfrac{5}{x} + \dfrac{6}{5x} = 2$

*(g) $\dfrac{x+7}{3} + \dfrac{2}{15} = \dfrac{4}{5}$ (h) $\dfrac{16}{x} + \dfrac{5}{7} = \dfrac{2}{x}$

*(i) $\dfrac{x+1}{2x} - \dfrac{1}{x} = \dfrac{3}{5}$ (j) $\dfrac{x+5}{x} - \dfrac{2}{3} = \dfrac{7}{x}$

2. Solve each equation.

*(a) $\dfrac{8}{x-1} = 4$ (b) $\dfrac{3}{x+5} = 7$

*(c) $\dfrac{3}{x+10} - \dfrac{8}{11} = \dfrac{1}{2}$ (d) $\dfrac{5}{12} + \dfrac{2}{x+7} = \dfrac{1}{4}$

*(e) $\dfrac{5}{x-1} + \dfrac{3x}{x-1} + \dfrac{7}{x-1} = 0$ (f) $\dfrac{x}{x+2} + \dfrac{1}{5} + \dfrac{5}{x+2} = 0$

*3. Solve each equation.

(a) $\dfrac{x}{x+2} + \dfrac{3}{x-5} = 1$ (b) $\dfrac{9}{x+1} + \dfrac{3}{x-2} = 0$

(c) $\dfrac{x-9}{x+2} = \dfrac{x-6}{x-3}$ (d) $\dfrac{x+5}{x-4} = \dfrac{x+1}{x+7}$

4. Solve each word problem by establishing an equation and then solving the equation.

   *(a) A student has test grades of 91 and 76. What must be his grade on the next test in order to have an average of 85?

   (b) A student has test grades of 95, 88, and 83. What must be her next test grade in order to have a 90 average?

   *(c) Jane has test grades of 82, 91, and 85. She would like to have a 90 average. What must she get on the fourth test in order to have the average she wants? If each test is worth a maximum of 100 points, would you guess that she has a reasonable chance of getting what she wants?

5. Solve each word problem by establishing an equation and then solving the equation.

   *(a) If Arnie can mow the lawn in 5 hours and Tina can mow it in 4 hours, how long will it take them to mow the lawn working together?

   (b) Kim can wash and polish her car in 3 hours. Terry can do it in 2 hours. How long will it take them working together?

　　　　*(c) Dave can put a new roof on the house in 8 hours. Tony can do it in 6 hours. Wade can also do it in 6 hours. How long will it take the roofers to complete the job if they all work together?

　6. Solve each word problem.
　　　*(a) One number is 4 more than another. The fraction created by dividing the larger number by the smaller is $\frac{5}{3}$. Find the numbers.
　　　(b) One number is 4 more than another. If the larger number is divided by the smaller number, the result is $\frac{3}{2}$. Find the numbers.
　　　*(c) If 1 is added to the reciprocal of a number, the result is $1\frac{2}{5}$. Find the number.
　　　(d) If 3 is subtracted from the reciprocal of a number, the result is $\frac{1}{4}$. Determine the number.

18.8 RATIO AND PROPORTION

A fraction can be regarded as a comparison of two numbers—that is, a **ratio.** For instance, $\frac{5}{6}$ compares 5 with 6. It can be called the ratio of 5 to 6.

An equation which states that two fractions (or ratios) are equal is called a **proportion.** When you change a fraction to one with a different denominator, you are working with a proportion.

EXAMPLE 44　Change $\dfrac{2}{5}$ to tenths.

Solution
$$\frac{2}{5} = \frac{?}{10}$$

Using elementary arithmetic,
$$\frac{2}{5} = \frac{4}{10}$$

This simple equation is an example of a proportion.
　　In the proportion
$$\frac{a}{b} = \frac{c}{d}$$

a and d are called the **extremes,** while b and c are called the **means.**

$$\frac{\text{extreme}}{\text{mean}} = \frac{\text{mean}}{\text{extreme}}$$

We will show that the product of the means is equal to the product of the extremes. In symbols,

$$\boxed{\begin{array}{l} \text{If } \ \dfrac{a}{b} = \dfrac{c}{d} \\[2mm] \text{then } ad = bc \end{array}}$$

To prove this statement, multiply both sides of

$$\frac{a}{b} = \frac{c}{d}$$

by bd. The result is

$$ad = bc$$

The procedure of going from the proportion to the result $ad = bc$ is sometimes called *cross multiplication*.

EXAMPLE 45   Determine n: $\dfrac{3}{7} = \dfrac{18}{n}$.

Solution   Use the fact that the product of the extremes equals the product of the means to obtain

$$3 \cdot n = 7 \cdot 18$$

or
$$3n = 126$$

Finally,

$$n = \frac{126}{3} = 42 \quad \checkmark$$

EXAMPLE 46   Determine y: $\dfrac{12}{y} = \dfrac{15}{17}$.

Solution   Since the product of the extremes equals the product of the means, we have

$$12 \cdot 17 = y \cdot 15$$

Next, multiply $12 \cdot 17$ to get 204 and write $y \cdot 15$ as $15y$.

$$204 = 15y$$

Interchange 204 and $15y$.

$$15y = 204$$

Finally, divide both sides by 15 to get $y =$

$$y = \frac{204}{15} = \frac{68}{5} \quad \checkmark$$

EXAMPLE 47   In order to cut a 20-foot board into two parts in the ratio of 2 to 3 $(2:3)$, how long must each piece be?

Solution   *One approach:* If x represents the size of the larger piece, then $\frac{2}{3}x$ represents the size of the smaller, and

$$x + \frac{2}{3}x = 20$$

Then, after multiplying both sides by the denominator 3,

$$3x + 2x = 60$$

Next,

$$5x = 60$$
$$x = 12 \qquad \text{The larger piece is 12 feet.} \quad \checkmark$$
$$\frac{2}{3}x = 8 \qquad \text{The smaller piece is 8 feet.} \quad \checkmark$$

Another approach: To avoid fractions, let $3x$ be the length of the larger piece and $2x$ the length of the smaller. As you can see, the ratio of the smaller to the larger is still $\frac{2}{3}$, since

$$\frac{2x}{3x} = \frac{2}{3} \cdot \frac{x}{x} = \frac{2}{3} \cdot 1 = \frac{2}{3}$$

Now

$$2x + 3x = 20$$
$$5x = 20$$
$$x = 4$$
$$3x = 12 \quad \text{larger} \quad \checkmark$$
$$2x = 8 \quad \text{smaller} \quad \checkmark$$

EXAMPLE 48   If 12 pounds of a fertilizer will cover a lawn of 1500 square feet, how much fertilizer is needed for a 2000-square-foot lawn?

Solution   Begin by letting x represent the amount of fertilizer needed for a 2000-square-foot lawn. Next, establish a proportion, such as

$$\frac{12}{1500} = \frac{x}{2000}$$

This proportion states that 12 pounds of fertilizer compares with a 1500-square-foot lawn as x pounds of fertilizer compares with a 2000-square-foot lawn.

Since the product of the means equals the product of the extremes, we have

$$(1500)(x) = (12)(2000)$$

or $$1500x = 24{,}000$$

This means that

$$x = \frac{24{,}000}{1500} = 16 \quad \checkmark$$

Thus, 16 pounds of fertilizer is needed to cover a 2000-square-foot lawn.

EXAMPLE 49   Two brothers decide to go into business and share expenses in the ratio of 3 to 5 with the older brother paying the larger share. If the older brother pays $4500, how much does the younger brother pay?

Solution   The *only* thing we don't know is how much the younger brother must pay. So let x represent that amount. Since they share expenses in a $\frac{3}{5}$ ratio, and they share expenses in the ratio of $\dfrac{x}{4500}$, we get the proportion

$$\frac{3}{5} = \frac{x}{4500}$$

We solve by first multiplying the means and then the extremes. We set these products equal to each other to obtain

$$5 \cdot x = 3 \cdot 4500\text{·}$$

or

$$5x = 13{,}500$$

Thus

$$x = 2700 \quad \checkmark$$

The younger brother pays $2700.

EXERCISES 18.8

1. Solve each proportion for the unknown.

   *(a) $\dfrac{4}{5} = \dfrac{x}{20}$ (b) $\dfrac{5}{7} = \dfrac{x}{21}$

   *(c) $\dfrac{x}{15} = \dfrac{2}{3}$ (d) $\dfrac{x}{25} = \dfrac{80}{40}$

   *(e) $\dfrac{12}{30} = \dfrac{5}{m}$ (f) $\dfrac{15}{20} = \dfrac{4}{n}$

   *(g) $\dfrac{3}{x} = \dfrac{7}{4}$ (h) $\dfrac{4}{y} = \dfrac{17}{12}$

   *(i) $\dfrac{14}{15} = \dfrac{5}{x}$ (j) $\dfrac{m}{6} = \dfrac{12}{5}$

*2. John, Bill, Tom, and Sam all travel. John walks at 4 miles per hour, Bill runs at 10 miles per hour, Tom bicycles at 20 miles per hour, and Sam drives his car at 40 miles per hour.

   (a) How many miles can Sam drive in the time that Bill runs 20 miles?

   (b) How many miles can Sam drive in the time that John walks 20 miles?

   (c) How many miles can Sam drive in the time that Tom bicycles 60 miles?

   (d) How many miles can Tom bicycle in the time that John walks 12 miles?

   (e) How many miles can Bill run in the time that Tom bicycles 35 miles?

3. Solve each problem.

   *(a) A 100-foot wire is cut into two parts in the ratio of 7 to 3. How long is each piece?

   (b) Two numbers are in the ratio of $3 : 5$ and their sum is 872. What are the two numbers?

*(c) A factory turns out defective products in the ratio of $1:1000$ (defective : total). If 25 defective products were made in one week, how many total products were probably made?

(d) A factory turns out defective products in the ratio of $13:500$ (defective : total). If 9500 products were produced, how many would you expect to be defective?

*(e) If 3 pounds of bananas cost 87¢, how much will 5 pounds of bananas cost?

*(f) If 3 gallons of paint will cover 450 square feet, how much will 7 gallons cover?

*(g) If 3 pounds of bananas cost 87¢, how many pounds of bananas will you get for $1.16?

*(h) If 3 gallons of paint will cover 420 square feet, how many gallons are needed to cover 1120 square feet?

*(i) A 9- by 12-inch picture is enlarged so that the shorter side is 15 inches. What is the length of the other side of the enlarged photo?

(j) A 16- by 10-inch photo is reduced so that the larger side is 12 inches. How long is the shorter side of the reduced photo?

*(k) A recipe calls for 5 teaspoons of soy sauce and 2 teaspoons of lemon juice. To make a larger portion, Ron uses 12.5 teaspoons of soy sauce. How much lemon juice should he use?

(l) Doris has invested $3000 in stocks and $7000 in bonds. She then receives a check for $1400 and decides to invest in stocks and bonds, using the same ratio as before. How much should she invest in each?

18.9 DIRECT AND INVERSE VARIATION (OPTIONAL)

A quantity y **varies directly** with respect to another quantity x if the ratio of y to x is constant. In other words, y varies directly with respect to x if $\frac{y}{x} = c$ for some constant c. This can also be seen as $y = cx$.

$$\boxed{y = cx \qquad \text{(direct variation)}}$$

When the variation $\frac{y}{x} = 2$ is considered as $y = 2x$, you readily see that y is always twice x. If x becomes larger, then y becomes correspondingly larger. If x becomes smaller, so does y. Also, if y becomes larger or smaller, x changes accordingly. Thus you can visualize that y and x vary *directly*.

EXAMPLE 50   Suppose that y varies directly as x. If y is 20 when x is 5, what is the value of y when x is 12?

Solution   Because y varies directly as x, we know that

$$y = cx$$

Since $y = 20$ when $x = 5$, c can be determined by substituting 20 for y and 5 for x. Thus

$$20 = c \cdot 5$$

or

$$5c = 20$$

or

$$c = 4 \quad \checkmark$$

Thus $y = cx$ becomes

$$y = 4x$$

Now if x is 12, y is determined from $y = 4x$.

$$y = 4(12)$$
$$y = 48 \quad \checkmark$$

EXAMPLE 51   A worker's wages (y) vary directly as the number of hours worked (x). If he works 10 hours he earns \$45. Determine the constant (in $y = cx$) and determine how much he earns for working 17 hours.

Solution   Since y varies directly as x, we know that

$$y = cx$$

Furthermore, we know that y is 45 when x is 10, which means

$$45 = c \cdot 10$$

or

$$10c = 45$$

or

$$c = \frac{45}{10} = 4.5 \quad \checkmark$$

Thus

$$y = 4.5x$$

To determine the y value when x is 17, substitute 17 for x in $y = 4.5x$.

$$y = 4.5(17) = 76.5 \quad \checkmark$$

He earns \$76.50 for working 17 hours.

A quantity (y) **varies inversely** with respect to another quantity (x) if their product (xy) is constant. In other words, x varies inversely with respect to y if $xy = c$.

$$\boxed{xy = c \quad \text{(inverse variation)}}$$

Suppose that $xy = 12$:

- If $x = 6$, y must be 2, since $6 \cdot 2 = 12$.
- If $x = 1$, y must be 12, since $1 \cdot 12 = 12$.
- If $x = \frac{1}{2}$, y must be 24, since $\frac{1}{2} \cdot 24 = 12$.

You can see that if y increases, x must decrease; if y decreases, x must increase. This is what you should understand as an *inverse* variation.

EXAMPLE 52   Suppose that y varies inversely as x. If y is 5 when x is 4, what is x when y is 2?

Solution   Because y varies inversely as x, we know that

$$xy = c$$

Since $y = 5$ when $x = 4$, c can be determined by substituting 5 for y and 4 for x. Thus

$$4 \cdot 5 = c$$

or $c = 20$ $\checkmark$

Thus $xy = c$ becomes $xy = 20$. Now if y is 2, x can be determined from

$$xy = 20$$
$$x(2) = 20$$
$$x = \frac{20}{2} = 10 \ \checkmark$$

EXAMPLE 53   When a gas is kept at constant temperature, pressure (P) varies inversely with respect to volume (V). If P is 50 when V is 12, determine the constant (in $PV = c$) and determine the pressure when the volume is 30.

Solution   Since P varies inversely with respect to V, we know that

$$PV = c$$

Since $P = 50$ when $V = 12$, we have

$$(50)(12) = c$$

or $c = 600$ $\checkmark$

Thus, $PV = c$ becomes $PV = 600$. Now if V is 30, P can be determined from

$$PV = 600$$
$$P(30) = 600$$
$$P = \frac{600}{30} = 20 \ \checkmark$$

The pressure is 20 when the volume is 30.

EXERCISES 18.9

*1. Suppose that y varies directly as x. If y is 21 when x is 3, what is:
   (a) The value of y when x is 2?
   (b) The value of y when x is 1?

    (c) The value of y when x is 19?
    (d) The value of x when y is 14?
    (e) The value of x when y is 10?

*2. Suppose that y varies inversely as x. If y is 6 when x is 5, what is:
    (a) The value of y when x is 3?   (b) The value of y when x is 10?
    (c) The value of y when x is 1?   (d) The value of x when y is 60?
    (e) The value of x when y is $\frac{1}{2}$?

3. Determine the constant in each *direct* variation between x and y.
    *(a) y is 88 when x is 4.   (b) y is 13 when x is 156.
    *(c) y is 63 when x is $\frac{1}{2}$.   (d) y is 42 when x is 42.

4. Determine the constant in each *inverse* variation between x and y.
    *(a) x is 16 when y is 7.   (b) x is 23 when y is 40.
    *(c) y is $\frac{2}{3}$ when x is 12.   (d) y is 9 when x is 9.

5. Solve each problem.
    *(a) A worker's wages (y) vary directly as the number of hours worked (x). If he works 12 hours he earns \$72. Determine the constant (in $y = cx$) and determine how much he earns for working 32 hours.
    *(b) Same as part (a), except determine how many hours he must work in order to earn \$102.
    (c) Assume the worker in part (a) earns \$69 for working 15 hours. Determine c and the number of hours he must work in order to earn \$124.20.
    *(d) Assume $PV = c$, as in Example 53. If P is 40 when V is 70, determine c and determine the pressure when the volume is 16. Also determine the volume when the pressure is 8.
    (e) Assume $PV = c$, as in Example 53. If the constant is known to be 1200, determine the volume when the pressure is 30. Determine the pressure when the volume is 15.

CHAPTER 18 REVIEW EXERCISES

*1. Perform the indicated operation. Reduce if possible.

(a) $\dfrac{5x}{8 + x} \cdot \dfrac{x + 8}{10y}$ (b) $\dfrac{x^2 - 9}{x + 3} \cdot \dfrac{4x}{3 - x}$

(c) $\dfrac{9xy}{10} \div \dfrac{3y}{5x}$ (d) $\dfrac{4}{2x + 10} \div \dfrac{x + 3}{x^2 + 3x - 10}$

(e) $\dfrac{w}{x} + \dfrac{y}{z}$ (f) $\dfrac{3}{7x} + \dfrac{2}{5x}$

(g) $\dfrac{9}{x + 1} + \dfrac{y}{x}$ (h) $\dfrac{5}{3xy} + \dfrac{3}{xz} - \dfrac{7}{5yz}$

(i) $\dfrac{x^2 + 6x + 5}{x^2 - 1} \cdot \dfrac{3x - 3}{18x}$ (j) $\dfrac{16 - x^2}{x - 4} \div \dfrac{2x + 8}{x + 1}$

(k) $\dfrac{16x^3y^2}{7abx} \cdot \dfrac{21ab^2}{4x}$

(l) $\dfrac{b}{a} + \dfrac{d}{c} + \dfrac{f}{e}$

(m) $\dfrac{t^2 - u^2}{2t + 2u} \div \dfrac{5u}{4t}$

(n) $\dfrac{4x^2 + 8}{4x} \cdot \dfrac{10x}{5x^2 + 10}$

(o) $\dfrac{5}{3xy} + \dfrac{19}{x^2} + \dfrac{1}{2y}$

(p) $\dfrac{5}{18x} + \dfrac{4}{15y}$

(q) $\dfrac{1 - x}{x + y} \cdot \dfrac{y + x}{x - 1}$

(r) $\dfrac{3}{x^2 - x} + \dfrac{4}{5x - 5}$

*2. Simplify.

(a) $\dfrac{\dfrac{x + 1}{y}}{\dfrac{x - 1}{x}}$

(b) $\dfrac{3 - \dfrac{1}{m}}{\dfrac{2}{m} + 1}$

*3. Solve each equation.

(a) $x + \dfrac{x + 5}{2} = 28$

(b) $\dfrac{x}{x + 3} - \dfrac{5}{8} = 0$

4. Reduce each fraction, if possible.

*(a) $\dfrac{2x^2 - 2}{x + 1}$

(b) $\dfrac{2x^2 + 7x + 3}{x^2 + 4x + 3}$

*(c) $\dfrac{2x^2 - 18}{2x^2 + 10x + 12}$

(d) $\dfrac{3x^2 + 7x - 6}{3x^2 - 4x - 4}$

*(e) $\dfrac{2x^2 + 8x + 8}{4x^2 - 16}$

(f) $\dfrac{2x^2 - 13x - 7}{14 - 2x}$

CUMULATIVE REVIEW EXERCISES (CHAPTERS 10–18)

*1. Simplify each expression.
(a) $3 - (-7) + (-3)(+2)$
(b) $5x - 2(3 - 4x) + 1$
(c) $(5x^2z^5)(-2xz^6)$
(d) $y^{30} \div y^{14}$

*2. Multiply.
(a) $(3x - 4)5$
(b) $(r + 4)^2$

*3. Solve each linear equation.
(a) $3x = 5x + 2(3 + x)$
(b) $8x - 3 = x - 14$

*4. Solve each word problem.
(a) 24% of what number is 78?
(b) The sum of two numbers is 280. The larger number is 74 less than twice the smaller. What are the two numbers?
(c) Tammy is three times as old as Kim. Ten years from now she will be 8 years older than Kim. How old are they now?
(d) Two trains leave the station at the same time and travel in opposite directions. Both trains travel at the same speed. If the trains are 504 miles apart after 3 hours, how fast are they traveling?

*5. Determine the equation of the line with slope -3 that passes through the point $(5, 4)$.

*6. Solve the system.
$$\begin{cases} 5x + 7y = 20 \\ 4x + 2y = 16 \end{cases}$$

*7. Solve the inequality.
$$3(2 + 5x) + 4 < 1$$

*8. Factor each expression, if possible.
(a) $x^2 + 12x + 36$ (b) $3y^2 + 7y - 20$

*9. Solve each quadratic equation.
(a) $y^2 - 3y - 4 = 0$ (b) $2x^2 - 128 = 0$
(c) $2m^2 - 7m - 9 = 0$ (d) $r^2 = 17r$

*10. Perform the indicated operation; reduce when possible.

(a) $\dfrac{x}{x - 3} + \dfrac{1}{x}$ (b) $\dfrac{2x^2 - 2}{4y} \cdot \dfrac{5x}{x^2 + x}$

19

FORMULA AND EQUATION MANIPULATION

19.1 INTRODUCTION

In algebra it is often desirable to manipulate an equation into a different form. We might, for example, prefer to have the perimeter formula

$$P = 2l + 2w$$

available in a form in which it is solved for w, as

$$w = \frac{P - 2l}{2}$$

The methods needed to carry out the desired manipulations are already familiar to you. This section concentrates on the techniques introduced in Chapter 12 to solve linear equations. Section 19.2 uses the same methods and, in addition, uses the procedure for solving equations containing fractions that was presented in Section 18.7.

EXAMPLE 1   Solve $y = 3x - 2$ for x; that is, express x in terms of y.

Solution   The problem is to eliminate the -2 from the right-hand side and then change the coefficient of x to 1. First, add $+2$ to both sides to get the x term by itself.

$$\begin{array}{r} y = 3x - 2 \\ \underline{+2 \qquad + 2} \\ y + 2 = 3x \end{array}$$

or $3x = y + 2$ if you interchange $y + 2$ and $3x$. Next, divide both sides by 3 to change $3x = \cdots$ to $x = \cdots$.

$$\frac{3x}{3} = \frac{y + 2}{3}$$

or

$$x = \frac{y + 2}{3} \quad \checkmark$$

We say that $y = 3x - 2$ has been "solved for x," since it now gives x in terms of other numbers.

EXAMPLE 2   Solve $y = mx + b$ for x.

Solution   This problem is identical in form to the problem of the previous example. First, add $-b$ to both sides in order to get the x term by itself.

$$\begin{array}{r} y = mx + b \\ \underline{-b = \qquad - b} \\ y - b = mx \end{array}$$

or $mx = y - b$ after interchanging $y - b$ and mx. Then divide both sides by m to change $mx = \cdots$ to $x = \cdots$.

$$\frac{mx}{m} = \frac{y - b}{m}$$

$$x = \frac{y - b}{m} \quad \checkmark$$

EXAMPLE 3   Solve $C = \pi d$ for π.

Solution   The π term is alone on the right-hand side of the equation. We need only change its coefficient to one in order to solve for π. So divide both sides by d, the coefficient of π. When this is done

$$C = \pi d$$

becomes

$$\frac{C}{d} = \frac{\pi d}{d}$$

or
$$\pi = \frac{C}{d} \quad \checkmark$$

In this formula C is the circumference of a circle and d is its diameter. π is the number that you know to be approximately equal to 3.14. Here you can see an interesting definition of π; it is the ratio of the circumference of a circle to its diameter.

EXAMPLE 4   Solve for y: $2x + 3y - 7 = 0$.

Solution   First, add $+7$ to both sides.

$$\begin{array}{rcl} 2x + 3y - 7 &=& 0 \\ +7 && +7 \\ \hline 2x + 3y &=& 7 \end{array}$$

Then add $-2x$ to both sides to get the y term alone.

$$\begin{array}{rcl} 2x + 3y &=& 7 \\ -2x && -2x \\ \hline 3y &=& -2x + 7 \end{array}$$

Finally, divide both sides by 3 to get $y = \cdots$.

$$\frac{3y}{3} = \frac{-2x + 7}{3}$$

or
$$y = \frac{-2x + 7}{3} \quad \checkmark$$

EXAMPLE 5   Solve for b: $x = a(b + c)$.

Solution   Free the b term from within parentheses by distributing the a. Thus

$$x = a(b + c)$$
becomes
$$x = ab + ac$$

Next, get the b term, ab, by itself by adding $-ac$ to both sides.

$$\begin{array}{rcl} x &=& ab + ac \\ -ac &=& \quad - ac \\ \hline x - ac &=& ab \end{array}$$

Finally, get a coefficient of 1 for b by dividing both sides by a.

$$\frac{x - ac}{a} = \frac{ab}{a}$$

$$\frac{x - ac}{a} = b \quad \checkmark$$

EXERCISES 19.1

Answers to starred exercises are given in the back of the book.

*1. Solve each formula for the letter indicated.

(a) $d = 2r$ for r (b) $d = rt$ for r

(c) $C = \pi d$ for d (d) $C = 2\pi r$ for r

(e) $A = lw$ for l (f) $V = lwh$ for h

(g) $i = prt$ for r (h) $V = \pi r^2 h$ for h

2. Solve each formula for the letter indicated.

*(a) $y = 6x$ for x *(b) $y = -5x$ for x

*(c) $y = ax + b$ for x (d) $y = ax + b$ for b

*(e) $y = ax + b$ for a (f) $P = 2l + 2w$ for l

*(g) $P = 2l + 2w$ for w (h) $R = p + prt$ for t

*(i) $x = 7y - 2$ for y (j) $3x = 2y + 5$ for y

*(k) $5y = 4x - 2$ for x (l) $2x + 6y = 9$ for x

*(m) $5x + 7y - 8 = 0$ for x (n) $5x + 7y - 8 = 0$ for y

*(o) $ax + b = 0$ for b *(p) $ax + b = 0$ for x

*(q) $ax + by + c = 0$ for x (r) $ax + by + c = 0$ for y

*(s) $A = m(1 + p)$ for p (t) $a = p(1 + rt)$ for t

*(u) $l = a + (n - 1)d$ for d *(v) $R = p + prt$ for p

19.2 EQUATIONS CONTAINING FRACTIONS

This section continues the study of formula and equation manipulation by extending the coverage to include equations that include fractions.

EXAMPLE 6   Solve for b: $a = \dfrac{b + c}{d}$.

Solution   If the denominator d were not present, it would be much easier to solve this equation for b; so multiply both sides by d, the denominator, to eliminate the fraction. Recall that to solve equations with fractions it is standard procedure to eliminate the fractions.

$$d \cdot a = \cancel{d} \cdot \frac{(b + c)}{\cancel{d}}$$

or $ad = b + c$

Next, eliminate c from the right-hand side, since we want b alone.

$$\begin{array}{r} ad = b + c \\ \underline{-c \qquad\; - c} \\ ad - c = b \end{array}$$

or $b = ad - c$ $\checkmark$

EXAMPLE 7   Solve for C: $F = \frac{9}{5}C + 32$.

Solution   Begin solving for C by eliminating the fraction $\frac{9}{5}$ from $F = \frac{9}{5}C + 32$ by multiplying both sides of the equation by 5. The result is

$$5F = 9C + 160$$

Next, add -160 to both sides of this equation to obtain

$$5F - 160 = 9C$$

or $$9C = 5F - 160$$

Finally, divide both sides of the equation by 9 to get

$$C = \frac{5F - 160}{9} \quad \checkmark$$

EXAMPLE 8   Solve for c: $a = \dfrac{b}{c} + \dfrac{d}{e}$.

Solution   The lowest common denominator of the fractions is $c \cdot e$ or ce. Multiplying all terms by ce will eliminate the fractions. It will then be easier to solve for c.

$$\boldsymbol{ce \cdot a = \cancel{ce} \cdot \frac{b}{\cancel{c}} + c\cancel{e} \cdot \frac{d}{\cancel{e}}}$$

or $$cea = eb + cd$$

We can get $c = \cdots$ on the left by dividing both sides by ea. *But* there will also be a c term on the right; so we will not have solved for c. So instead of doing that, get the c terms together on the same side by adding $-cd$ to both sides.

$$\begin{array}{r} cea = eb + cd \\ \underline{-cd \qquad\; - cd} \\ cea - cd = eb \end{array}$$

Place the coefficients of the c terms in front of c so that they are easier to work with.

$$ae\underline{c} - d\underline{c} = eb$$

Next, get *one* coefficient for c by factoring out c.

$$(ae - d)c = eb$$

Finally, divide both sides by the coefficient of c, which is $(ae - d)$.

$$\frac{(ae - d)c}{(ae - d)} = \frac{eb}{(ae - d)}$$

So

$$c = \frac{eb}{ae - d} \quad \checkmark$$

The next example includes the use of *subscripts*. The numbers x_1 and x_2 are two numbers we wish to call by similar names, both "x," so we use the subscripts 1 and 2 to distinguish them. The numbers 1 and 2 are *not* exponents. Subscripts are used frequently in statistics and calculus. You will see two examples of subscripts in the exercises—the two bases of a trapezoid will be called b_1 and b_2 and two masses will be called m_1 and m_2.

EXAMPLE 9   Solve for x_1: $y = \dfrac{x_1 x_2}{x_1 + x_2}$.

Solution   This problem is easier to handle without fractions; so multiply both sides by the denominator $(x_1 + x_2)$. The result is

$$y(x_1 + x_2) = x_1 x_2$$

We must get the x_1 terms (arrows below) together on the same side in order to combine them.

$$y(\overset{\uparrow}{x_1} + x_2) = \overset{\uparrow}{x_1} x_2$$

But x_1 cannot be removed or combined until we multiply out the $y(x_1 + x_2)$. Thus

$$y(x_1 + x_2) = x_1 x_2$$

becomes

$$yx_1 + yx_2 = x_1 x_2$$

Now we can get the x_1 terms together by adding $-yx_1$ to both sides.

$$
\begin{array}{r}
yx_1 + yx_2 = x_1 x_2 \\
-yx_1 \qquad\qquad -yx_1 \\
\hline
yx_2 = x_1 x_2 - yx_1
\end{array}
$$

or $x_1 x_2 - yx_1 = yx_2$ $\begin{cases} \text{if we interchange the left} \\ \text{and right sides} \end{cases}$

or $x_2 x_1 - yx_1 = yx_2$ $\begin{cases} \text{if we change the order of} \\ x_1 \text{ and } x_2 \text{ in the first term} \end{cases}$

Next, factor x_1 out of the left-hand expression.

$$(x_2 - y)x_1 = yx_2$$

Finally, divide both sides by $x_2 - y$, the coefficient of x_1.

$$x_1 = \frac{yx_2}{x_2 - y} \quad \checkmark$$

If you are unsure how to proceed in manipulating equations or formulas into the form requested, consider the guidelines given next. Be sure to *carry out the steps in the order listed here.*

Follow these steps, *in order*.

1. If there are fractions, multiply both sides (each term) by the lowest common denominator. This will eliminate fractions from the equation.
2. If the variable (or unknown) being solved for is within parentheses, multiply out to free it.
3. Use opposites to collect on one side all terms containing the desired variable and on the other side all other terms.
4. If more than one term contains the variable, factor out the variable in order to determine its coefficient.
5. Divide both sides of the equation by the coefficient of the variable.

EXERCISES 19.2

1. Solve each formula for the letter indicated.

*(a) $A = \dfrac{1}{2}bh$ for h

(b) $A = \dfrac{1}{2}bh$ for b

*(c) $E = \dfrac{1}{2}mv^2$ for m

(d) $s = \dfrac{1}{2}at^2$ for a

*(e) $I = \dfrac{E}{R}$ for E

*(f) $I = \dfrac{E}{R}$ for R

*(g) $v = \dfrac{s}{t}$ for s

(h) $v = \dfrac{s}{t}$ for t

*(i) $F = \dfrac{m_1 m_2}{d^2}$ for m_1

*(j) $A = \dfrac{1}{2}(b_1 + b_2)h$ for h

*(k) $R = \dfrac{1}{2\pi fc}$ for c

*(l) $S = \dfrac{n}{2}(a + l)$ for l

(m) $B = \dfrac{m}{2}(a + b - c)$ for b

(n) $A = \dfrac{h}{6}(a + 2b + c)$ for c

2. Solve each formula for the letter indicated.

*(a) $C = \dfrac{nE}{R + nr}$ for E

*(b) $C = \dfrac{nE}{R + nr}$ for R

*(c) $C = \dfrac{nE}{R + nr}$ for r

*(d) $C = \dfrac{nE}{R + nr}$ for n

*(e) $A = \dfrac{1}{2}h(b_1 + b_2)$ for b_2

(f) $S = \dfrac{n}{2}(a + l)$ for n

*(g) $\dfrac{1}{f} = \dfrac{1}{p} + \dfrac{1}{q}$ for p

(h) $\dfrac{1}{f} = \dfrac{1}{p} + \dfrac{1}{q}$ for q

*(i) $S = \dfrac{rl - a}{r - 1}$ for l

*(j) $S = \dfrac{rl - a}{r - 1}$ for r

*(k) $R = \dfrac{gs}{g+s}$ for s 　　　　(l) $T = \dfrac{ab}{a-b}$ for a

*(m) $\dfrac{a}{y} + \dfrac{b}{x} = 1$ for x 　　　　(n) $\dfrac{a}{y} + \dfrac{b}{x} = 1$ for y

CHAPTER 19 REVIEW EXERCISES

*1. Solve each formula for the letter indicated.
　(a) $p = 3d + 4e$ for e 　　(b) $t = u(v + w)$ for w

　(c) $m = \dfrac{a+b}{c}$ for a 　　(d) $a = \dfrac{1}{2}(b + c)$ for b

*2. Solve each formula for the letter indicated.
　(a) $T = n(4 + rv)$ for v 　　(b) $x = \dfrac{3a - 2b}{c + d}$ for a

　(c) $\dfrac{1}{w} = \dfrac{1}{x} + \dfrac{1}{y}$ for x 　　(d) $x = \dfrac{a+b}{c+d}$ for c

CUMULATIVE REVIEW EXERCISES (CHAPTERS 10–19)

*1. Simplify each expression.
　(a) $(-5)(+2) - 3^2 - (-1)$ 　　(b) $2(3 - 2x) + 3x - 9$
*2. Multiply. $(3x - 2)(7x + 1)$
*3. Solve each linear equation.
　(a) $1 - 6x = 2(x + 3)$ 　　(b) $17 - (3 + 5x) - x = 0$
*4. Write an algebraic expression to represent each phrase.
　(a) The number of yards in $17mn$ feet.
　(b) The length of a rectangle whose width is x feet, if the length is 3 more than five times the width.
　(c) The perimeter of the rectangle in part (b).
　(d) The area of the rectangle in part (b).
　(e) The even number that immediately precedes (in sequence) the even number $2n$.
*5. Sketch a graph of the lines.
　(a) $y = -x + 4$ 　　(b) $6x - 5y = 30$
*6. Determine the equation of the line that passes through the points $(4, 2)$ and $(5, 1)$.
*7. Solve the system.
$$\begin{cases} 3x + 8y = 12 \\ y = 2x - 27 \end{cases}$$
*8. Solve the inequality.
$$8 - 19x < 11$$
*9. Solve each quadratic equation by factoring.
　(a) $x^2 + 5x - 36 = 0$ 　　(b) $2x^2 - 11x + 15 = 0$

*10. Perform the indicated operation; reduce when possible.

(a) $\dfrac{1}{2x} + \dfrac{5x}{x+1}$ (b) $\dfrac{x}{3x+6} \div \dfrac{3}{x^2+2x}$

*11. Solve each formula for x.

(a) $y = a(1 + bx)$ (b) $m = \dfrac{a+x}{b+x}$

PROPERTIES OF EXPONENTS

20.1 SOME BASIC CONCEPTS

The concept of exponent was introduced in Chapter 1 and was used during the arithmetic presentation. Then in Sections 11.2 and 11.4 two basic properties of exponents were explained and applied. For review, the two properties will be restated here along with some examples.

$$x^a \cdot x^b = x^{a+b}$$

Examples:

$$x^3 \cdot x^4 = x^{3+4} = x^7$$
$$y^2 \cdot y^9 = y^{2+9} = y^{11}$$

$$5x^2 \cdot x^{10} \cdot x^8 = 5 \cdot x^{2+10+8} = 5x^{20}$$
$$(-5y^3)(-7y^7) = (-5)(-7)y^3y^7 = 35y^{10}$$

$$\frac{x^a}{x^b} = x^{a-b}$$

Examples:

$$\frac{x^{15}}{x^4} = x^{15-4} = x^{11}$$

$$\frac{m^7}{m^2} = m^{7-2} = m^5$$

Next, consider $(x^3)^5$, in which x^3 is raised to the fifth power.

$$(x^3)^5 = x^3 \cdot x^3 \cdot x^3 \cdot x^3 \cdot x^3$$
$$= \underline{xxx} \cdot \underline{xxx} \cdot \underline{xxx} \cdot \underline{xxx} \cdot \underline{xxx}$$
$$= x^{15}$$

You could have saved steps by reasoning that there would be 5 products of the form x^3, and so the resulting exponent would be five 3's or 15. In general,

$$\boxed{(x^a)^n = x^{a \cdot n}}$$

EXAMPLE 1   Simplify $(x^{10})^5$.

Solution $$(x^{10})^5 = x^{10 \cdot 5} = x^{50} \quad \checkmark$$

EXAMPLE 2   Simplify $(m^{40})^{20}$.

Solution $$(m^{40})^{20} = m^{40 \cdot 20} = m^{800} \quad \checkmark$$

EXAMPLE 3   Simplify $(x^5)^3(3x^7)$.

Solution $$(x^5)^3(3x^7) = x^{5 \cdot 3} \cdot 3x^7$$
$$= x^{15} \cdot 3 \cdot x^7$$
$$= 3x^{22} \quad \checkmark$$

Now consider $(xy)^5$, a product raised to a power.

$$(xy)^5 = \underline{xy} \cdot \underline{xy} \cdot \underline{xy} \cdot \underline{xy} \cdot \underline{xy}$$

which can be regrouped as

$$x \cdot x \cdot x \cdot x \cdot x \cdot y \cdot y \cdot y \cdot y \cdot y$$

and simplified to

$$x^5 y^5$$

Looking at this example for a mechanical shortcut, you see that the exponent 5 of $(xy)^5$ has been applied to each factor—to x and to y.

$$(xy)^5 = x^5 y^5$$

In general,

$$\boxed{(xy)^n = x^n y^n}$$

EXAMPLE 4   Simplify $(ab)^7$.

Solution $$(ab)^7 = a^7 b^7 \quad \checkmark$$

EXAMPLE 5 Simplify $(2mn)^4$.

Solution
$$(2mn)^4 = 2^4 m^4 n^4$$
$$= 16m^4 n^4 \quad \checkmark$$

EXAMPLE 6 Simplify $(x^2 y^3)^4$.

Solution
$$(x^2 y^3)^4 = (x^2)^4 (y^3)^4$$
$$= x^8 y^{12} \quad \checkmark$$

Example 6 suggests the following general result.

$$\boxed{(x^a y^b)^n = x^{an} y^{bn}}$$

EXAMPLE 7 Simplify $(m^9 n^3)^4$.

Solution
$$(m^9 n^3)^4 = m^{9 \cdot 4} n^{3 \cdot 4}$$
$$= m^{36} n^{12} \quad \checkmark$$

EXAMPLE 8 Simplify $(2x^2 y^5 z)^8$.

Solution
$$(2x^2 y^5 z)^8 = 2^8 x^{2 \cdot 8} y^{5 \cdot 8} z^8$$
$$= 2^8 x^{16} y^{40} z^8$$
$$= 256 x^{16} y^{40} z^8 \quad \checkmark$$

Powers of constants will be worked out unless the result requires too great an effort to obtain. In Example 5 we evaluated 2^4. In Example 8 we evaluated 2^8. But we would not evaluate such expressions as 2^{20}, 5^8, or 3^{10}. This decision is purely arbitrary.

EXAMPLE 9 Simplify $(x^2 + z^5)^9$.

Solution
$$(x^2 + z^5)^9 \neq x^{18} + z^{45}$$

If the *sum* $(x^2 + z^5)$ is raised to the ninth power, it cannot be simplified by the preceding methods. On the other hand, $(x^2 \cdot z^5)^9 = x^{18} z^{45}$.

EXAMPLE 10 Simplify $(-3mn^3)(-4mn^4)^2$.

Solution
$$(-3mn^3)(-4mn^4)^2 = -3mn^3 \cdot (-4)^2 (m)^2 (n^4)^2$$
$$= -3 \cdot m \cdot n^3 \cdot 16 \cdot m^2 \cdot n^8$$
$$= -3 \cdot 16 \cdot m \cdot m^2 \cdot n^3 \cdot n^8 \qquad \text{after rearranging}$$
$$= -48 m^3 n^{11} \quad \checkmark$$

EXAMPLE 11 Simplify $\left(\dfrac{x}{y}\right)^6$.

Solution

$$\left(\frac{x}{y}\right)^6 = \frac{x}{y} \cdot \frac{x}{y} \cdot \frac{x}{y} \cdot \frac{x}{y} \cdot \frac{x}{y} \cdot \frac{x}{y}$$

$$= \frac{x \cdot x \cdot x \cdot x \cdot x \cdot x}{y \cdot y \cdot y \cdot y \cdot y \cdot y}$$

$$= \frac{x^6}{y^6} \quad \checkmark$$

From Example 11 we can generalize about a quotient raised to a power.

$$\boxed{\left(\frac{x}{y}\right)^n = \frac{x^n}{y^n}}$$

EXAMPLE 12 Simplify $\left(\frac{2a}{b}\right)^4$.

Solution

$$\left(\frac{2a}{b}\right)^4 = \frac{(2a)^4}{b^4}$$

$$= \frac{2^4 a^4}{b^4}$$

$$= \frac{16a^4}{b^4} \quad \checkmark$$

EXAMPLE 13 Simplify $\left(\frac{x^2 y^3}{z^4}\right)^5$.

Solution

$$\left(\frac{x^2 y^3}{z^4}\right)^5 = \frac{(x^2 y^3)^5}{(z^4)^5}$$

$$= \frac{x^{10} y^{15}}{z^{20}} \quad \checkmark$$

EXERCISES 20.1

Answers to starred exercises are given in the back of the book.

1. Simplify if possible.
   *(a) $x^3 \cdot x^2$
   (b) $x^5 \cdot x^9$
   *(c) $d \cdot d \cdot d \cdot d \cdot d$
   *(d) $2^{15} 2^{10}$
   *(e) $x^3 \cdot x^2 \cdot x \cdot x \cdot x \cdot x$
   (f) $y \cdot y^{10} \cdot y \cdot y \cdot y$
   *(g) $m^5 m^{12} m^8$
   *(h) $x^{435} x^{154}$
   *(i) $5^{236} 5^{379}$
   *(j) $x^{100} + x^{187}$
   *(k) $a^{198} a^{35} a^{17}$
   *(l) $x + x^6$

*(m) $m^5 + m^7 + m^3$ (n) $m^{15}m^6m^{76}$

*(o) $5x \cdot 8x^{10}$ *(p) $(3x^4y^2)(4x^5y^9)$

*(q) $(-4x^2y^3)(14xy)(x^{15}y^9)$ *(r) $3x^7 \cdot -2x \cdot x^5$

*(s) $3a^4b^67a^23ab$ (t) $2a^53b^45a^24a^3b^2$

*(u) $(-6m^4n^7)(15m^5n^4)$ (v) $(a^7bc^8)(-b^5c^9)(-a^4c)$

2. Simplify if possible.

*(a) $(x^4)^3$ (b) $(x^{15})^2$ *(c) $(r^7)^4$

(d) $(y^5)^{12}$ *(e) $(rs)^{14}$ (f) $(mn)^9$

*(g) $(xy)^{50}$ *(h) $(pqr)^{17}$ *(i) $(x^4y^6)^5$

(j) $(m^5n^7)^3$ *(k) $(a^4c^7)^2$ *(l) $(5x^3y^8)^3$

*(m) $(x^4 + y^5)^3$ (n) $(2m^4n^2)^5$ *(o) $(8xy^6)^2$

*(p) $(-3x^5y^{10})^3$ *(q) $(-2m^3np^2)^4$ (r) $(-4x^2y^5z^3)^4$

*(s) $(-5a^3b^7)^3$ (t) $(-w^6x^4y^5z^3)^{10}$

3. Simplify if possible.

*(a) $(3mn^2)(5m^4n^7)^2$ *(b) $(2x^5y^2)^3(3x^4y^9)^4$

*(c) $(-2p^4q)^3(p^8q^2)^5$ (d) $(-fg^5h^2)^2(3f^4)^3$

*(e) $(3x^4 - y^6)^5$ (f) $(5xy^6)^4(x^5y^{19})$

*(g) $(6b^4c^3d)^2(-2bc^9d^2)^3$ *(h) $3m^2n^8 + 7m^3n^5$

*(i) $\left(\dfrac{x}{y}\right)^9$ (j) $\left(\dfrac{r}{s}\right)^{12}$

*(k) $\left(\dfrac{2m}{n}\right)^5$ (l) $\left(\dfrac{ab^3}{3}\right)^4$

*(m) $\left(\dfrac{x^4y^5}{z^3}\right)^2$ *(n) $\left(\dfrac{b^2c^7}{a^4d^5}\right)^3$

*(o) $\left(\dfrac{m^7n^8p^2}{5t^2}\right)^4$ (p) $\left(\dfrac{3x^4y^6}{2z^5}\right)^3$

*4. Evaluate. Write each as one number.

(a) 2^3 (b) 2^6 (c) 5^4 (d) 3^5

(e) $2 \cdot 3^2$ (f) $3 \cdot 2^4$ (g) $(2^4)^2$ (h) $(3^2)^3$

(i) $\left(\dfrac{2}{3}\right)^4$ (j) $\left(\dfrac{3}{4}\right)^3$ (k) $(2 + 3)^4$ (l) $(4 + 1)^3$

5. *(a) Write 4^5 as a power of 2 by first substituting for 4 a power of 2.

(b) Write 8^3 as a power of 2.

(c) Write 9^4 as a power of 3.

20.2 NEGATIVE EXPONENTS

Consider this example, which involves a quotient of two powers of x.

$$\frac{x^7}{x^3} = x^4$$

It can also be seen as

$$x^7 \cdot \frac{1}{x^3} = x^4$$

Suppose that we want to represent the $\frac{1}{x^3}$ as x to a power. In other words,

replace $\frac{1}{x^3}$ by a power of x, as

$$x^7 \cdot x^? = x^4$$

So that it will agree with the properties of exponents, the exponent that we seek must be -3, since $7 + (-3) = 4$—that is, $x^7 \cdot x^{-3} = x^4$. This example suggests the following notation.

$$\frac{1}{x^3} = x^{-3}$$

Similarly,

$$\frac{1}{m^9} = m^{-9}$$

$$\frac{1}{y^4} = y^{-4}$$

Also,

$$x^{-5} = \frac{1}{x^5}$$

In general,

$$\boxed{x^{-n} = \frac{1}{x^n}}$$

EXAMPLE 14 Evaluate 3^{-2}.

Solution

$$3^{-2} = \frac{1}{3^2} = \frac{1}{9} \quad \checkmark$$

Notice that this expression containing a negative exponent was changed to an expression with a positive exponent so that it could be evaluated.

EXAMPLE 15 Simplify.
(a) $2x^{-5}$ (b) $(2x)^{-5}$

Solution (a) $2x^{-5} = 2 \cdot x^{-5}$

$$= 2 \cdot \frac{1}{x^5}$$

$$= \frac{2}{x^5} \quad \checkmark$$

(b) $(2x)^{-5} = \frac{1}{(2x)^5}$

$$= \frac{1}{2^5 x^5}$$

$$= \frac{1}{32x^5} \quad \checkmark$$

EXAMPLE 16 Write $7x^{-5}y^4z^{-8}$ without using negative exponents.

Solution
$$7x^{-5}y^4z^{-8} = 7 \cdot x^{-5} \cdot y^4 \cdot z^{-8}$$

$$= \frac{7}{1} \cdot \frac{x^{-5}}{1} \cdot \frac{y^4}{1} \cdot \frac{z^{-8}}{1} \quad \left\{ \begin{array}{l} \text{making each a fraction} \\ \text{with denominator 1} \end{array} \right.$$

$$= \frac{7}{1} \cdot \frac{1}{x^5} \cdot \frac{y^4}{1} \cdot \frac{1}{z^8} \quad \left\{ x^{-5} = \frac{1}{x^5} \text{ and } z^{-8} = \frac{1}{z^8} \right.$$

$$= \frac{7 \cdot 1 \cdot y^4 \cdot 1}{1 \cdot x^5 \cdot 1 \cdot z^8}$$

$$= \frac{7y^4}{x^5 z^8} \quad \checkmark$$

Because a negative exponent can appear in the denominator of a fraction, we need to know the meaning of such exponents. Observe.

$$\frac{1}{x^{-n}} = \frac{1}{\frac{1}{x^n}} = x^n$$

Thus

$$\boxed{\frac{1}{x^{-n}} = x^n}$$

For instance,

$$\frac{1}{x^{-3}} = x^3$$

$$\frac{1}{m^{-5}} = m^5$$

$$\frac{1}{2^{-4}} = 2^4$$

$$n^4 = \frac{1}{n^{-4}}$$

EXAMPLE 17 Write $\dfrac{5x^{-2}y^3}{m^2n^{-6}}$ without using negative exponents.

Solution

$$\frac{5x^{-2}y^3}{m^2n^{-6}} = \frac{5x^{-2}y^3}{1} \cdot \frac{1}{m^2n^{-6}}$$

$$= \frac{5}{1} \cdot \frac{x^{-2}}{1} \cdot \frac{y^3}{1} \cdot \frac{1}{m^2} \cdot \frac{1}{n^{-6}}$$

Change the two negative exponents that appear.

$$= \frac{5}{1} \cdot \frac{1}{x^2} \cdot \frac{y^3}{1} \cdot \frac{1}{m^2} \cdot \frac{n^6}{1}$$

$$= \frac{5y^3n^6}{x^2m^2} \quad \text{or} \quad \frac{5n^6y^3}{m^2x^2} \quad \checkmark$$

The final form leaves the letters in alphabetical order.

Compare the final form containing only positive exponents with the original form, which contained negative exponents. Notice that a negative exponent in the numerator became a positive exponent in the denominator and that a negative exponent in the denominator became a positive exponent in the numerator. That is what our two boxed definitions of negative exponents accomplish.

EXAMPLE 18 Write $\dfrac{-5x^{-3}}{y^{-4}}$ without using negative exponents.

Solution

$$\frac{-5x^{-3}}{y^{-4}} = \frac{-5x^{-3}}{1} \cdot \frac{1}{y^{-4}}$$

$$= \frac{-5}{1} \cdot \frac{x^{-3}}{1} \cdot \frac{1}{y^{-4}}$$

$$= \frac{-5}{1} \cdot \frac{1}{x^3} \cdot \frac{y^4}{1}$$

$$= \frac{-5y^4}{x^3} \quad \checkmark \quad \text{or} \quad -\frac{5y^4}{x^3} \quad \checkmark$$

Notice that the -5 in the numerator *does not* become a $+5$ in the denominator. It is *not an exponent* and should not be expected to behave like one.

EXAMPLE 19 Simplify $(3x^5y^4z^{-8})^{-2}$ and write it without using negative exponents.

Solution Since this is a product raised to a power, the power, -2, is distributed. The result is

$$(3x^5y^4z^{-8})^{-2} = 3^{-2}x^{-10}y^{-8}z^{16}$$

$$= \frac{1}{3^2} \cdot \frac{1}{x^{10}} \cdot \frac{1}{y^8} \cdot \frac{z^{16}}{1}$$

$$= \frac{z^{16}}{9x^{10}y^8} \quad \checkmark$$

EXAMPLE 20 Simplify $7x^{-5}x^{-3}$ and write it without using negative exponents.

Solution Since the bases are the same and the numbers are being multiplied, add the exponents.

$$7x^{-5}x^{-3} = 7 \cdot x^{-5-3}$$

$$= 7 \cdot x^{-8}$$

$$= 7 \cdot \frac{1}{x^8}$$

$$= \frac{7}{x^8} \quad \checkmark$$

EXERCISES 20.2

1. Evaluate. Write each as one number.
   *(a) 2^{-3} (b) 2^{-4} *(c) 3^{-3} (d) 5^{-2}
   *(e) 6^{-1} (f) 4^{-4} *(g) $3 \cdot 5^{-1}$ (h) $5 \cdot 2^{-3}$
   *(i) $7 \cdot 8^{-1}$ (j) $9 \cdot 3^{-2}$ *(k) $(2 \cdot 3)^{-2}$ (l) $(3 \cdot 4)^{-3}$

2. Write without using negative exponents.
   *(a) x^{-5} (b) y^{-9} *(c) $4m^{-6}$
   (d) $3x^{-2}$ *(e) $-7t^{-4}$ (f) $-5b^{-8}$

   *(g) $(3x)^{-2}$ (h) $(5y)^{-3}$ *(i) $\dfrac{n^{-7}}{4}$

   (j) $\dfrac{r^{-5}}{3}$ *(k) $\dfrac{2m^{-4}}{7}$ (l) $\dfrac{3x^{-9}}{2}$

   *(m) $5x^{-7}y^{-4}$ (n) $6m^{-9}n^{-3}$ *(o) $7m^{-6}n^7$
   (p) $19b^{-2}c^6$ *(q) $a^2b^{-3}c^4d^{-5}$ (r) $8w^6x^{-4}y^{-5}z$
   *(s) $-8m^{-7}n^2p^{-9}$ (t) $-5ab^7c^{-8}d^{-2}$

3. Write without using negative exponents.
   *(a) $\dfrac{3x^{-4}y}{z^5}$ (b) $\dfrac{7m^{-3}n}{p^6}$ *(c) $\dfrac{5a^{-4}b^{-6}}{2c}$

   (d) $\dfrac{9wx^{-6}y^{-8}}{5z}$ *(e) $\dfrac{a^{-4}b^{-3}}{c^{-8}}$ (f) $\dfrac{x^{-9}y^{-3}}{z^{-5}}$

   *(g) $\dfrac{7a^{-6}b^4c^{-2}}{x^5y^{-3}}$ (h) $\dfrac{4m^5n^{-9}}{p^{-7}r^3}$ *(i) $\dfrac{-5x^{-4}y^{-5}z}{v^{-3}w^2}$

   (j) $\dfrac{-9a^{-9}b^{-7}c^{-4}}{5d^{-1}e^{-5}}$

4. Simplify each expression and write it without using negative exponents.
   *(a) $x^{-2}x^{-7}$ (b) $a^{-5}a^{-7}$ *(c) $6t^{-4} \cdot t^{-3} \cdot t^{-6}$
   (d) $4x^{-3} \cdot x^{-7} \cdot x^{-8}$ *(e) $(x^{-3}y^{-5})^{-4}$ (f) $(a^{-7}b^{-4})^{-2}$
   *(g) $(2x^{-6}y^{-5})^{-3}$ (h) $(3x^{-2}y^{-8})^{-4}$ *(i) $(5a^{-3}b^{-7}c^{-5})^{-2}$
   (j) $(4a^{-2}b^{-4}c^{-6})^{-2}$

20.3 DIVISION AND EXPONENTS

We already have the property

$$\frac{x^a}{x^b} = x^{a-b}$$

which we have used in simple cases where a was greater than b. For example,

$$\frac{x^{29}}{x^{17}} = x^{29-17} = x^{12}$$

We continue with more complicated cases.

EXAMPLE 21 Simplify $\dfrac{x^5 y^9}{x^2 y^4}$.

Solution

$$\frac{x^5 y^9}{x^2 y^4} = \frac{x^5}{x^2} \cdot \frac{y^9}{y^4}$$

$$= x^{5-2} y^{9-4}$$

$$= x^3 y^5 \quad \checkmark$$

Notice that this procedure did result in the fraction being reduced.

The subtraction of exponents also applies when the power in the denominator is larger than the power in the numerator. To see this fact, notice first that

$$\frac{x^3}{x^7} = \frac{xxx}{xxxxxxx} = \frac{1}{xxxx} = \frac{1}{x^4}$$

If, however, you perform the subtraction of exponents instead, the result is

$$\frac{x^3}{x^7} = x^{3-7} = x^{-4}$$

and we know that x^{-4} is the same as $\dfrac{1}{x^4}$.

EXAMPLE 22 Simplify $\dfrac{4l^5 m^4 n^3}{3m^9 n}$. Do not leave negative exponents in the result.

Solution

$$\frac{4l^5 m^4 n^3}{3m^9 n} = \frac{4}{3} \cdot \frac{l^5}{1} \cdot \frac{m^4}{m^9} \cdot \frac{n^3}{n^1}$$

$$= \frac{4}{3} l^5 m^{4-9} n^{3-1}$$

$$= \frac{4}{3} l^5 m^{-5} n^2$$

$$= \frac{4}{3} \cdot l^5 \cdot \frac{1}{m^5} \cdot n^2 = \frac{4l^5 n^2}{3m^5} \quad \checkmark$$

EXAMPLE 23 Simplify $\dfrac{w^{-6}}{w^2}$. Do not leave any negative exponent in the result.

Solution Shown below are two different ways of obtaining the final result,

$$\frac{1}{w^8} \quad \checkmark$$

1. $\dfrac{w^{-6}}{w^2} = w^{-6-2} = w^{-8} = \dfrac{1}{w^8}$

2. $\dfrac{w^{-6}}{w^2} = \dfrac{w^{-6}}{1} \cdot \dfrac{1}{w^2} = \dfrac{1}{w^6} \cdot \dfrac{1}{w^2} = \dfrac{1}{w^6 w^2} = \dfrac{1}{w^8}$

The result of Example 23 will appear as part of the next example, which is somewhat more involved.

EXAMPLE 24 Simplify and reduce completely. Write without negative exponents.

$$\frac{7w^{-6}x^{10}y^{-9}z^{13}}{21w^2x^{-4}z^5}$$

Solution

$$\frac{7w^{-6}x^{10}y^{-9}z^{13}}{21w^2x^{-4}z^5} = \frac{7}{21} \cdot \frac{w^{-6}}{w^2} \cdot \frac{x^{10}}{x^{-4}} \cdot \frac{y^{-9}}{1} \cdot \frac{z^{13}}{z^5}$$

$$= \frac{1}{3} \cdot \frac{1}{w^8} \cdot \frac{x^{14}}{1} \cdot \frac{1}{y^9} \cdot \frac{z^8}{1}$$

$$= \frac{x^{14}z^8}{3w^8y^9} \quad \checkmark$$

EXERCISES 20.3

1. Simplify each term, leaving positive exponents only. Be sure each fraction is completely reduced.

*(a) $\dfrac{x^{40}}{x^{10}}$ (b) $\dfrac{m^{17}}{m^2}$ *(c) $\dfrac{y^{18}}{y^{15}}$ (d) $\dfrac{n^{58}}{n^{31}}$

*(e) $\dfrac{m^6}{m^9}$ (f) $\dfrac{t^{15}}{t^{19}}$ *(g) $\dfrac{x^6y^{10}}{x^2y^7}$ (h) $\dfrac{m^5n^{12}}{m^3n^{10}}$

*(i) $\dfrac{x^5y^7}{xy^4}$ (j) $\dfrac{x^7y^2}{x^5y^6}$ *(k) $\dfrac{a^3b^8}{a^7b^{19}}$ (l) $\dfrac{m^6n^2}{m^{10}n^9}$

*(m) $\dfrac{p^2q^3r^4}{p^7q^6r^5}$ (n) $\dfrac{w^3x^9y^5}{w^{10}xy^{23}}$

2. Simplify each term, leaving positive exponents only. Be sure each fraction is completely reduced.

*(a) $\dfrac{4x^9y^2}{x^2y^9}$ (b) $\dfrac{-8x^2y^{11}}{x^{24}y^{87}}$ *(c) $\dfrac{4x^9y^2z}{5xy^3z^2}$

$$\text{(d) } \frac{7a^3b^9c^5}{2a^4c} \qquad \text{*(e) } \frac{4w^{-4}x^5}{5w^3x^{-2}} \qquad \text{(f) } \frac{7a^8b^{-3}c^3}{9a^{-2}b^5}$$

$$\text{*(g) } \frac{2x^2y^{-3}z^8}{x^{-5}yz^4} \qquad \text{(h) } \frac{25p^{-3}q^{-7}r^{-2}}{p^{-4}q^6r} \qquad \text{*(i) } \frac{m^{-7}n^8p^{-9}}{m^{-9}n^9p^{-4}}$$

$$\text{(j) } \frac{a^7b^{-6}c^{12}}{a^{-1}bc^{-4}d^{-2}} \qquad \text{*(k) } \frac{w^6x^{-9}y^{-2}z^{-5}}{7w^{10}x^{-3}y^7} \qquad \text{(l) } \frac{r^{-14}s^{-11}t^{-5}}{9r^{-2}st^{-23}}$$

20.4 ZERO EXPONENT

Zero is the only integer we had not previously used as an exponent. The zero exponent is introduced for completeness. *Any nonzero number raised to the zero power is 1*, as justified by the work shown next.

Let's attempt to simplify $\dfrac{x^5}{x^5}$ by using $\dfrac{x^m}{x^n} = x^{m-n}$.

$$\frac{x^5}{x^5} = x^{5-5} = x^0$$

We might also note that $\dfrac{x^5}{x^5} = 1$, since a number divided by itself is 1. As a result, we define the exponent zero as follows.

$$\boxed{x^0 = 1}$$

EXAMPLE 25   Evaluate $8^0 + 253^0$.

Solution
$$8^0 + 253^0 = 1 + 1 \qquad \text{since } 8^0 = 1 \text{ and } 253^0 = 1$$
$$= 2 \ \checkmark$$

EXAMPLE 26   Evaluate $(2 + 4)^0 + 3 \cdot 7^0 + (4 \cdot 2)^0$.

Solution
$$(2 + 4)^0 + 3 \cdot 7^0 + (4 \cdot 2)^0 = (6)^0 + 3 \cdot 7^0 + (8)^0$$
$$= 1 + 3 \cdot 1 + 1$$
$$= 1 + 3 + 1$$
$$= 5 \ \checkmark$$

EXAMPLE 27   Simplify $(3x + 2)^0$.

Solution   Once a value is supplied for x, the entire expression $3x + 2$ will be one number. (For example, if x is 5, then $3x + 2$ is 17.) Thus

$$(3x + 2)^0 = 1 \ \checkmark$$

EXERCISES 20.4

1. Evaluate each expression.

*(a) 5^0 *(b) $3 \cdot 4^0$
*(c) $(3 \cdot 4)^0$ (d) $(abc)^0$
*(e) $5x^0$ (f) $(5x)^0$
*(g) $6 \cdot 2^0$ *(h) $4^0 + (-3)^0$
*(i) $2^0 \cdot 7$ (j) $4^0 + 3^0 - 7^0$
*(k) $-3x^0$ (l) $(3m)^0 + 3m^0$
*(m) $(5x - 1)^0$ *(n) $5x - 1^0$
*(o) $9x^0 - 8 - (9x)^0$ (p) $7 \cdot y^0 \cdot 3$

20.5 FRACTIONAL EXPONENTS

Next, we proceed to obtain an exponent notation for radicals. From the definition of square root we know that for any nonnegative number x,

$$\sqrt{x} \cdot \sqrt{x} = x$$

That is, the square root of x is a number that when squared gives x.

If we examine $\sqrt{x}\sqrt{x} = x$ in terms of exponents by letting $\sqrt{x}$ be x^m, we obtain

$$x^m x^m = x \qquad \text{after replacing } \sqrt{x} \text{ by } x^m \text{ in } \sqrt{x}\sqrt{x} = x$$
$$x^m x^m = x^1 \qquad \text{since } x^1 \text{ is the same as } x$$
$$x^{m+m} = x^1 \qquad \text{by a rule for exponents}$$
$$x^{2m} = x^1 \qquad m + m = 2m$$

Then since $2m = 1$, m must be equal to $\frac{1}{2}$. Thus

$$x^{\frac{1}{2}} \cdot x^{\frac{1}{2}} = x^{\frac{1}{2}+\frac{1}{2}} = x^1 = x$$

So

$$\boxed{\sqrt{x} = x^{\frac{1}{2}}}$$

EXAMPLE 28   Evaluate $25^{\frac{1}{2}}$.

Solution
$$25^{\frac{1}{2}} = \sqrt{25}$$
$$= 5 \quad \checkmark$$

EXAMPLE 29   Evaluate $3 \cdot 16^{\frac{1}{2}}$.

Solution
$$3 \cdot 16^{\frac{1}{2}} = 3 \cdot \sqrt{16}$$
$$= 3 \cdot 4$$
$$= 12 \quad \checkmark$$

If you have forgotten the concept of cube root, note that the *cube root* of 8 is 2, since 2 cubed is 8. In symbols, $\sqrt[3]{8} = 2$, since $2^3 = 8$. Here is a brief table of some common cube roots.

$$\begin{array}{lll}
\sqrt[3]{0} = 0 & \text{since} & 0^3 = 0 \\
\sqrt[3]{1} = 1 & \text{since} & 1^3 = 1 \\
\sqrt[3]{8} = 2 & \text{since} & 2^3 = 8 \\
\sqrt[3]{27} = 3 & \text{since} & 3^3 = 27 \\
\sqrt[3]{64} = 4 & \text{since} & 4^3 = 64 \\
\sqrt[3]{125} = 5 & \text{since} & 5^3 = 125 \\
\sqrt[3]{216} = 6 & \text{since} & 6^3 = 216
\end{array}$$

We can determine the exponent form of $\sqrt[3]{x}$ by using the same reasoning as with square root. The cube root of x is a number $\sqrt[3]{x}$ such that $\sqrt[3]{x}\,\sqrt[3]{x}\,\sqrt[3]{x} = x$. So $\sqrt[3]{x} = x^{\frac{1}{3}}$, since $x^{\frac{1}{3}} \cdot x^{\frac{1}{3}} \cdot x^{\frac{1}{3}} = x^{\frac{1}{3}+\frac{1}{3}+\frac{1}{3}} = x^1 = x$.

$$\boxed{\sqrt[3]{x} = x^{\frac{1}{3}}}$$

EXAMPLE 30 Evaluate $64^{\frac{1}{2}} + 8^{\frac{1}{3}}$.

Solution
$$64^{\frac{1}{2}} = \sqrt{64} = 8$$
$$8^{\frac{1}{3}} = \sqrt[3]{8} = 2$$

so
$$64^{\frac{1}{2}} + 8^{\frac{1}{3}} = 8 + 2 = 10 \quad \checkmark$$

Similarly, for fourth root, $\sqrt[4]{x} = x^{\frac{1}{4}}$. Similarly, for fifth root, $\sqrt[5]{x} = x^{\frac{1}{5}}$. In general,

$$\boxed{\sqrt[n]{x} = x^{\frac{1}{n}}}$$

Next, let's consider fractional exponents in which the numerator is not 1. For instance, let's look at $x^{\frac{3}{4}}$.

$$x^{\frac{3}{4}} = x^{\frac{1}{4}\cdot 3} = (x^{\frac{1}{4}})^3 \text{ or } (x^3)^{\frac{1}{4}}$$

Similarly, for $x^{\frac{3}{2}}$,

$$x^{\frac{3}{2}} = (x^{\frac{1}{2}})^3 \text{ or } (x^3)^{\frac{1}{2}}$$

In general,

$$\boxed{x^{\frac{m}{n}} = (x^{\frac{1}{n}})^m \quad \text{or} \quad (x^m)^{\frac{1}{n}}}$$

Keep in mind that x must be nonnegative when the root is even (square root, fourth root, etc.).

EXAMPLE 31   Evaluate (a) $8^{\frac{2}{3}}$, (b) $4^{\frac{5}{2}}$, and (c) $27^{\frac{1}{3}}$.

Solution   (a) $8^{\frac{2}{3}} = (8^{\frac{1}{3}})^2 = (\sqrt[3]{8})^2 = (2)^2 = 4$ ✓

                  (b) $4^{\frac{5}{2}} = (4^{\frac{1}{2}})^5 = (\sqrt[2]{4})^5 = (2)^5 = 32$ ✓

                  (c) $27^{\frac{1}{3}} = (27)^{\frac{1}{3}} = \sqrt[3]{27} = 3$ ✓

Generally *it is easier to simplify by taking the root first and then apply-ing the power,* as done above, because the root is easier to determine when the number is smaller. Compare.

$$\text{Root first:} \quad 8^{\frac{5}{3}} = (8^{\frac{1}{3}})^5 = (2)^5 = 32$$

$$\text{Power first:} \quad 8^{\frac{5}{3}} = (8^5)^{\frac{1}{3}} = (32768)^{\frac{1}{3}} = 32$$

EXAMPLE 32   Evaluate $27^{\frac{2}{3}} + 16^{\frac{5}{4}} - 4 \cdot 5^0$.

Solution $27^{\frac{2}{3}} + 16^{\frac{5}{4}} - 4 \cdot 5^0 = (27^{\frac{1}{3}})^2 + (16^{\frac{1}{4}})^5 - 4 \cdot 5^0$
$$= (3)^2 + (2)^5 - 4 \cdot 1$$
$$= 9 + 32 - 4$$
$$= 37 \quad ✓$$

EXAMPLE 33   Determine the value of $9^{1.5}$.

Solution   Begin by writing the exponent 1.5 as a mixed number, $1\frac{1}{2}$. Then change the mixed number to an improper fraction, namely, $\frac{3}{2}$.

$$9^{1.5} = 9^{1\frac{1}{2}} = 9^{\frac{3}{2}} = (9^{\frac{1}{2}})^3 = (3)^3 = 27 \quad ✓$$

EXAMPLE 34   Simplify $\left(\dfrac{49}{4}\right)^{\frac{3}{2}}$.

Solution
$$\left(\frac{49}{4}\right)^{\frac{3}{2}} = \left[\left(\frac{49}{4}\right)^{\frac{1}{2}}\right]^3$$

$$= \left[\frac{49^{\frac{1}{2}}}{4^{\frac{1}{2}}}\right]^3$$

$$= \left(\frac{7}{2}\right)^3$$

$$= \frac{7^3}{2^3}$$

$$= \frac{343}{8} \quad ✓$$

EXAMPLE 35   Evaluate $9^{-\frac{1}{2}}$.

Solution
$$9^{-\frac{1}{2}} = \frac{1}{9^{\frac{1}{2}}}$$

$$= \frac{1}{\sqrt{9}}$$

$$= \frac{1}{3} \quad \checkmark$$

EXAMPLE 36   Evaluate $8^{-\frac{5}{3}}$.

Solution

$$8^{-\frac{5}{3}} = \frac{1}{8^{\frac{5}{3}}}$$

$$= \frac{1}{(8^{\frac{1}{3}})^5}$$

$$= \frac{1}{(2)^5}$$

$$= \frac{1}{32} \quad \checkmark$$

EXAMPLE 37   Simplify $(x^{\frac{2}{3}})^{12}$.

Solution

$$(x^{\frac{2}{3}})^{12} = x^{\frac{2}{3} \cdot 12}$$

$$= x^8 \quad \checkmark$$

EXAMPLE 38   Simplify $(x^2 y^{10} z)^{\frac{1}{2}}$.

Solution

$$(x^2 y^{10} z)^{\frac{1}{2}} = x^{2 \cdot \frac{1}{2}} y^{10 \cdot \frac{1}{2}} z^{\frac{1}{2}}$$

$$= x^1 \cdot y^5 \cdot z^{\frac{1}{2}}$$

$$= xy^5 z^{\frac{1}{2}} \quad \checkmark$$

EXAMPLE 39   Simplify $x^{\frac{1}{2}} \cdot x^{\frac{1}{3}}$.

Solution

$$x^{\frac{1}{2}} \cdot x^{\frac{1}{3}} = x^{\frac{1}{2}+\frac{1}{3}}$$

$$= x^{\frac{3}{6}+\frac{2}{6}}$$

$$= x^{\frac{5}{6}} \quad \checkmark$$

EXAMPLE 40   Evaluate $\sqrt{9x^{10}y^6}$.

Solution

$$\sqrt{9x^{10}y^6} = (9x^{10}y^6)^{\frac{1}{2}}$$

$$= 9^{\frac{1}{2}} x^5 y^3$$

$$= 3x^5 y^3 \quad \checkmark$$

EXAMPLE 41   Write $\sqrt{x^5}$, using fractional exponents.

Solution

$$\sqrt{x^5} = (x^5)^{\frac{1}{2}}$$

$$= x^{\frac{5}{2}} \quad \checkmark$$

EXAMPLE 42 Write $x^{\frac{2}{3}}$, using a radical.

Solution
$$x^{\frac{2}{3}} = (x^2)^{\frac{1}{3}}$$
$$= \sqrt[3]{x^2} \quad \checkmark$$

EXERCISES 20.5

1. Evaluate each expression.
   *(a) $25^{\frac{1}{2}}$ (b) $81^{\frac{1}{2}}$
   *(c) $100^{\frac{1}{2}}$ (d) $49^{\frac{1}{2}}$

   *(e) $8^{\frac{1}{3}}$ (f) $1^{\frac{1}{3}}$
   *(g) $27^{\frac{1}{3}}$ (h) $64^{\frac{1}{3}}$
   *(i) $125^{\frac{1}{3}}$ *(j) $5 \cdot 4^{\frac{1}{2}}$
   *(k) $9 \cdot 49^{\frac{1}{2}}$ (l) $10 \cdot 144^{\frac{1}{2}}$

   *(m) $5 \cdot 27^{\frac{1}{3}}$ (n) $7 \cdot 9^{\frac{1}{2}} + 3 \cdot 125^{\frac{1}{3}}$

   *(o) $(64^{\frac{1}{2}} + 8^{\frac{1}{3}})^0$ (p) $(27^{\frac{1}{3}} + 49^{\frac{1}{2}})^0$

2. Evaluate each expression.
   *(a) $\left(\dfrac{9}{25}\right)^{\frac{1}{2}}$ (b) $\left(\dfrac{4}{9}\right)^{\frac{1}{2}}$

   *(c) $\left(\dfrac{8}{27}\right)^{\frac{1}{3}}$ (d) $\left(\dfrac{64}{125}\right)^{\frac{1}{3}}$

   *(e) $\left(\dfrac{81}{4}\right)^{\frac{1}{2}} + \left(\dfrac{27}{8}\right)^{\frac{1}{3}}$ (f) $\left(\dfrac{9}{4}\right)^{\frac{1}{2}} + \left(\dfrac{1}{8}\right)^{\frac{1}{3}}$

3. Evaluate each expression.
   *(a) $4^{\frac{3}{2}}$ (b) $8^{\frac{5}{3}}$
   *(c) $16^{\frac{3}{2}}$ (d) $9^{\frac{3}{2}}$
   *(e) $125^{\frac{2}{3}}$ (f) $8^{\frac{2}{3}} + 8^0$

   *(g) $5 \cdot 8^{\frac{2}{3}}$ *(h) $25^{\frac{3}{2}} + 4^{\frac{3}{2}}$
   *(i) $16^{\frac{1}{4}} + 16^{\frac{3}{4}}$ *(j) $4^{\frac{3}{2}} + 1^{\frac{1}{2}} + 3^0$
   *(k) $27^{\frac{5}{3}} + 81^{\frac{3}{4}}$ *(l) $64^{\frac{4}{3}} + 16^{\frac{3}{4}}$

4. Evaluate each expression.
   *(a) $49^{-\frac{1}{2}}$ (b) $64^{-\frac{1}{2}}$ *(c) $27^{-\frac{1}{3}}$
   (d) $16^{-\frac{1}{4}}$ *(e) $4^{-\frac{5}{2}}$ (f) $64^{-\frac{2}{3}}$

   *(g) $8^{\frac{1}{3}} + 8^{-\frac{1}{3}}$ (h) $25^{-\frac{3}{2}}$ *(i) $1^{-\frac{5}{2}} + 9^{\frac{3}{2}}$
   (j) $16^{-\frac{3}{4}} + 4^{-\frac{3}{2}}$

5. Evaluate each expression.
   *(a) $4^{2.5}$ (b) $16^{2.5}$ *(c) $9^{2.5}$ (d) $25^{1.5}$ *(e) $1^{3.5}$
   (f) $4^{3.5}$ *(g) $9^{-1.5}$ (h) $4^{-2.5}$ *(i) $1^{-2.5}$ (j) $16^{-1.5}$

6. Simplify.
   *(a) $(x^8 y^4)^{\frac{1}{2}}$ (b) $(16x^2)^{\frac{1}{2}}$ *(c) $(49a^2 b^6)^{\frac{1}{2}}$
   *(d) $(-8x^3)^{\frac{1}{3}}$ (e) $(27b^3)^{\frac{1}{3}}$ (f) $(64x^3 y^6)^{\frac{1}{3}}$

*(g) $(x^{\frac{1}{4}})^8$ (h) $(m^{\frac{3}{4}})^{16}$ *(i) $(-125r^3s^9)^{\frac{2}{3}}$

(j) $(27m^{12}n^6)^{\frac{4}{3}}$ *(k) $(121x^{10}y^{16})^{\frac{3}{2}}$ *(l) $\sqrt{25a^6y^{10}}$

(m) $\sqrt{81x^4}$ *(n) $(16x^4y^8z^{12})^{\frac{1}{4}}$ *(o) $\sqrt{144a^{10}b^6}$

(p) $\sqrt{169n^{50}r^{20}}$ *(q) $\sqrt[3]{216m^3}$ *(r) $\sqrt[3]{-64x^3y^6}$

7. Simplify.

*(a) $\sqrt[4]{16x^8}$ *(b) $\sqrt[5]{-32y^5}$ (c) $\sqrt[3]{-8x^6y^{21}z^3}$

(d) $\sqrt[3]{-m^{99}n^{36}}$ *(e) $\sqrt[4]{81a^{12}b^{28}}$ (f) $\sqrt[4]{256m^{84}n^{36}}$

(g) $\sqrt[4]{625x^4y^{100}}$ (h) $\sqrt[5]{32x^{10}z^{20}}$

8. Simplify.

*(a) $x^{\frac{1}{4}}x^{\frac{1}{2}}$ (b) $x^{\frac{1}{3}}x^{\frac{1}{6}}$ *(c) $x^{\frac{3}{4}}x^{\frac{1}{4}}$

(d) $x^{\frac{1}{2}}x^{\frac{1}{5}}$ *(e) $x^{\frac{2}{3}}x^{\frac{1}{6}}$ (f) $x^{\frac{3}{8}}x^{\frac{1}{4}}$

9. Write with fractional exponents.

*(a) $\sqrt{x^3}$ (b) $\sqrt{m^9}$ *(c) $\sqrt[3]{x^5}$ (d) $\sqrt[3]{n^4}$ *(e) $\sqrt[4]{y^7}$

(f) $\sqrt[4]{y^3}$

10. Write using a radical.

*(a) $x^{\frac{5}{2}}$ (b) $m^{\frac{3}{5}}$ *(c) $n^{\frac{7}{3}}$ (d) $y^{\frac{7}{2}}$ *(e) $t^{\frac{4}{5}}$

(f) $c^{\frac{5}{4}}$

20.6 SCIENTIFIC NOTATION (OPTIONAL)

Large numbers like

$$6720000000000000000000$$

and small numbers like

$$.00000000000000000000053$$

can be represented with fewer digits by using *scientific notation* (also called *standard notation*).

The large number 6720000000000000000000 can be written as 6.72×10^{22}. We use a power of 10 to eliminate the writing of a long string of zeros. Observe.

$$8400 = 840 \times 10 \quad \text{or} \quad 840 \times 10^1$$
$$= 84 \times 100 \quad \text{or} \quad 84 \times 10^2$$
$$= 8.4 \times 1000 \quad \text{or} \quad 8.4 \times 10^3$$

The most common form of numbers in **scientific notation** is *a number between 1 and 10, multiplied by an appropriate power of 10*. We will call this the "standard form" of scientific notation. For example,

$$8.5 \times 10^{15}$$
$$9.213 \times 10^2$$
$$1.0037 \times 10^{-12}$$
$$4.7 \times 10^{29}$$

EXAMPLE 43   Represent 765000000000000 by using scientific notation.

Solution   The original number 765000000000000 has 14 more digits before the decimal point than does 7.65. So it is 10^{14} times 7.65. Thus 7.65 must be multiplied by 10^{14} to be equal to 765000000000000. So we conclude that

$$765000000000000 = 7.65 \times 10^{14} \quad \checkmark$$

Multiplication by 10^{14} has the effect of moving the decimal 14 places to the right.

$$7\!\wedge\!\underbrace{65000000000000}_{14 \text{ places}}.$$

EXAMPLE 44   Represent 89000000, using scientific notation.

Solution   Since 8.9 must be multiplied by 10^7 in order to move the decimal 7 places to the right, we have

$$89000000 = 8.9 \times 10^7 \quad \checkmark$$

EXAMPLE 45   Represent .00000034, using scientific notation.

Solution   The number 3.4 would have to be divided by 10^7 in order to move the decimal point 7 places to the left. In other words,

$$.00000034 = \frac{3.4}{10^7}$$

Now, using

$$\frac{3.4}{10^7} = 3.4 \times 10^{-7}$$

we obtain

$$.00000034 = 3.4 \times 10^{-7} \quad \checkmark$$

In general, to move the decimal point n places to the left, multiply by 10^{-n}.

EXAMPLE 46   Represent .00000000000001745, using scientific notation.

Solution   For 1.745 to be changed to .00000000000001745, the decimal point would have to be moved 14 places to the left. Thus

$$.00000000000001745 = 1.745 \times 10^{-14} \quad \checkmark$$

EXAMPLE 47   Change 3.6×10^7 to a number without a power of 10.

Solution   Multiplication by 10^7 shifts the decimal point 7 places to the right. Thus

$$3.6 \times 10^7 = 36,000,000 \quad \checkmark$$

EXAMPLE 48   Write 3500×10^8 in scientific notation.

Solution   Standard form is 3.5 times the appropriate power of 10. So

$$3_\wedge 500 \times 10^8 = 3.5 \times 10^3 \times 10^8$$
$$= 3.5 \times 10^{3+8}$$
$$= 3.5 \times 10^{11} \quad \checkmark$$

EXAMPLE 49   Change $.043 \times 10^{20}$ to standard form.

Solution   Standard form is 4.3 times a power of 10.

$$.04_\wedge 3 \times 10^{20} = 4.3 \times 10^{-2} \times 10^{20}$$
$$= 4.3 \times 10^{18} \quad \checkmark$$

EXAMPLE 50   Change 5270×10^{-5} to standard form.

Solution $$5_\wedge 270 \times 10^{-5} = 5.27 \times 10^{-5 \pm 3}$$
$$= 5.27 \times 10^{-2} \quad \checkmark$$

EXAMPLE 51   Change $.00062 \times 10^{-7}$ to standard form.

Solution $$.0006_\wedge 2 \times 10^{-7} = 6.2 \times 10^{-7 - 4}$$
$$= 6.2 \times 10^{-11} \quad \checkmark$$

EXAMPLE 52   Simplify the expression.

$$\frac{30,000,000 \times 200,000,000}{.000004 \times 500,000}$$

Solution   Using scientific notation,

$$30,000,000 = 3 \times 10^7$$
$$200,000,000 = 2 \times 10^8$$
$$.000004 = 4 \times 10^{-6}$$
$$500,000 = 5 \times 10^5$$

Now the expression can be changed to scientific notation and simplified in the steps shown next.

$$\frac{30,000,000 \times 200,000,000}{.000004 \times 500,000} = \frac{3 \times 10^7 \times 2 \times 10^8}{4 \times 10^{-6} \times 5 \times 10^5}$$
$$= \frac{3 \times 2 \times 10^7 \times 10^8}{4 \times 5 \times 10^{-6} \times 10^5}$$
$$= \frac{6 \times 10^{15}}{20 \times 10^{-1}}$$

$$= \frac{6}{20} \times 10^{16}$$
$$= .3 \times 10^{16}$$
$$= 3.0 \times 10^{15} \quad \checkmark$$

EXERCISES 20.6

1. Write each number in scientific notation.
    *(a) 4700000
    (b) 810000
    *(c) 350000000000
    (d) 2734000000000000
    *(e) 453000000000000000
    (f) 600000000000000000
    *(g) .00023
    (h) .000005
    *(i) .0000007
    (j) .000000000142
    *(k) .0000000000013
    (l) .00000000000000095
    *(m) 12,000,000,000,000
    (n) 5,900,000,000,000,000

2. Change each standard form to a number without a power of 10.
    *(a) 3.4×10^8
    (b) 2.0×10^{15}
    *(c) 9.42×10^{11}
    (d) 1.829×10^{12}
    *(e) 1.3×10^{-9}
    (f) 2.81×10^{-13}
    *(g) 5.0×10^{-10}
    (h) 4.734×10^{-11}
    *(i) 7.563×10
    (j) 3.8×10^{-8}

*3. Change each number below to standard form.
    (a) 450×10^7
    (b) 1200×10^{16}
    (c) $.62 \times 10^{23}$
    (d) $.00731 \times 10^{15}$
    (e) 473×10^{-12}
    (f) 841×10^{-16}
    (g) $.621 \times 10^{19}$
    (h) $.456 \times 10^{-21}$
    (i) $3,000,000,000 \times 10^{15}$
    (j) $.000052 \times 10^{-17}$

*4. Simplify each expression in the manner of Example 52.

(a) $\dfrac{.0008 \times 50,000,000,000}{20,000 \times .0000004}$

(b) $\dfrac{.0000009 \times 2.0 \times 10^8 \times 1,000,000}{3.0 \times 10^{-3} \times 30,000,000}$

(c) $\dfrac{800,000,000 \times 3.0 \times 10^{-5}}{12,000,000 \times 10^8}$

(d) $\dfrac{20,000 \times 30,000 \times 50,000,000}{6,000,000 \times 5.0 \times 10^{-2}}$

CHAPTER 20 REVIEW EXERCISES

1. Simplify each expression.
    *(a) $x^7 \cdot x^5$
    (b) $x^8 \cdot x^9$
    *(c) $(x^{12})^4$
    (d) $(y^{15})^6$
    *(e) $3x^4 \cdot 5x^{12}$
    (f) $6x^5 \cdot 4x^{13}$
    *(g) $(2m^4y^5)^3$
    (h) $(2a^6b^{11})^4$
    *(i) $(5xy^3)(3x^7y^{10})^2$
    (j) $(3u^2v)(7u^8v^5)^2$

*(k) $\left(\dfrac{3x^2}{y^5}\right)^4$ (l) $\left(\dfrac{2y^6}{x^3}\right)^6$

*(m) $t^3 t^6 t^{12}$ (n) $x^5 x^{11} x^7$
*(o) $(-2x^2 y)^5$ (p) $(-3x^7 y)^3$
*(q) $(-5a^3 b^6)(7ab^5 c^4)$ (r) $(-6a^4 b^9)(8a^6 bc^7)$

2. Write each expression without using negative exponents.
   *(a) x^{-7} (b) c^{-3} *(c) $4m^{-2}$ (d) $6y^{-7}$

   *(e) $-7x^{-3}$ (f) $-2w^{-4}$ *(g) $\dfrac{y^{-6}}{3}$ (h) $\dfrac{m^{-3}}{8}$

3. Simplify each expression and write it without using negative exponents.

   *(a) $\dfrac{x^5 y^9}{x^2 y^2}$ (b) $\dfrac{a^4 b^7}{a^3 b^5}$ *(c) $\dfrac{x^6 y^8}{x^9 y^7}$

   (d) $\dfrac{a^5 b^6}{a^2 b^8}$ *(e) $(4x)^{-2}$ (f) $(2y)^{-3}$

   *(g) $5x^{-3} y^4 z^{-1}$ (h) $4x^7 y^{-3} z^{-9}$ *(i) $-3x^{-7} y^{-8}$

   (j) $-7a^{-4} b^{-5}$ *(k) $\dfrac{4r^{-6} s^{12}}{t^3 u^{-5}}$ (l) $\dfrac{5a^{-7} b^8}{c^{-4} d^4}$

   *(m) $x^{-7} x^{-3}$ (n) $y^{-8} y^{-5}$ *(o) $3x^{-4} x^{-3} x^{-2}$
   (p) $7t^{-8} t^{-2} t^{-7}$ *(q) $(b^{-4} c^{-5})^{-3}$ (r) $(x^{-9} y^{-4})^{-2}$

   *(s) $\dfrac{m^5 n^{-3} p^{-7} q^6}{m^{-8} n^7 p^{-3} q^9}$ (t) $\dfrac{w^{-4} x^5 y^{-8} z^{12}}{w^7 x^{-2} y^{-1} z^9}$

*4. Evaluate each expression.
   (a) $3 \cdot 2^3$ (b) 7^0 (c) $8 \cdot 3^0$ (d) $9^0 - 3$

   (e) $100^{\frac{1}{2}}$ (f) $8^{\frac{1}{3}}$ (g) $16^{\frac{1}{4}}$ (h) $\left(\dfrac{49}{64}\right)^{\frac{1}{2}}$

   (i) $4 \cdot 9^{\frac{1}{2}}$ (j) $4^{\frac{5}{2}}$ (k) $27^{\frac{2}{3}}$ (l) $3 \cdot 16^{\frac{3}{4}}$
   (m) 2^{-4} (n) $5 \cdot 3^{-2}$ (o) $(2x)^0$ (p) $4^{3.5}$

5. Simplify each expression.
   *(a) $\sqrt{49m^{10} n^{24}}$ (b) $\sqrt{81x^6 y^{18}}$ *(c) $(64x^3 y^{12} z^{30})^{\frac{2}{3}}$
   (d) $(27u^6 v^3 w^{21})^{\frac{2}{3}}$

CUMULATIVE REVIEW EXERCISES (CHAPTERS 10–20)

*1. Simplify the expression. $-6 + 13 - (-5)(-3)(-2) + (-3)^3$
*2. Perform the long division. $x - 4\overline{\smash{)}x^3 - x^2 + 17x - 8}$
*3. Solve the linear equation. $12x + 19 = 2x - 32$
*4. The sum of the three interior angles of a triangle is 180°. Suppose that one angle is 23° more than the smallest angle and the third angle is 13° less than twice the smallest angle. Use algebra to find the number of degrees in each angle.

*5. Graph the lines $y = 6x - 2$ and $y = -2x + 6$ and determine approximately where they meet.

*6. Solve each system of linear equations. Use addition or substitution as appropriate.

(a) $\begin{cases} 2x - 3y = 4 \\ 3x - 7y = 1 \end{cases}$ (b) $\begin{cases} x = y + 13 \\ 7x + 2y = 10 \end{cases}$

*7. Factor each expression, if possible.

(a) $y^2 - 18y - 63$ (b) $3x^2 - 2x - 10$

*8. Simplify the complex fraction.

$$\frac{10 + \dfrac{3}{2x}}{\dfrac{1}{x} - 5}$$

*9. Solve the equation.

$$\frac{5}{x} + 12 = \frac{1}{7x}$$

*10. Solve each formula for y.

(a) $x = w - (1 - 3y)$ (b) $t = \dfrac{a}{x} + \dfrac{b}{y}$

*11. Simplify each expression. Leave no negative exponents.

(a) $\dfrac{x^4 y^7}{x^9 y^3}$ (b) $49^{-\frac{1}{2}}$ (c) $14x^0 - 27^{\frac{4}{3}}$

RADICALS
AND ROOTS

21.1 REVIEW

This section offers a brief review of some of the ideas presented in Chapters 4 and 20. We'll be using these concepts throughout the chapter.

The **square root** of a quantity is the number that when squared (that is, multiplied by itself) produces the original quantity. For instance, the square root of 49 is 7, since $7^2 = 49$—that is, $7 \cdot 7 = 49$. This can be written $\sqrt{49} = 7$. The symbol $\sqrt{}$ is called a **radical.** In general,

$$\boxed{\sqrt{x} \cdot \sqrt{x} = x}$$

For example,

$$\sqrt{5}\sqrt{5} = 5$$
$$\sqrt{9}\sqrt{9} = 9$$

The numbers 0, 1, 4, 9, 16, 25, 36, etc. are called **perfect squares.** A whole number is a perfect square if its square root is a whole number. To illustrate, 49 is a perfect square because $\sqrt{49} = 7$, and 7 is a whole number. But 18 is not a perfect square because $\sqrt{18}$ is not equal to a whole number. Here is a short table of squares roots of perfect squares.

$\sqrt{0} = 0$	$\sqrt{49} = 7$	$\sqrt{196} = 14$
$\sqrt{1} = 1$	$\sqrt{64} = 8$	$\sqrt{225} = 15$
$\sqrt{4} = 2$	$\sqrt{81} = 9$	$\sqrt{256} = 16$
$\sqrt{9} = 3$	$\sqrt{100} = 10$	$\sqrt{289} = 17$
$\sqrt{16} = 4$	$\sqrt{121} = 11$	$\sqrt{324} = 18$
$\sqrt{25} = 5$	$\sqrt{144} = 12$	$\sqrt{361} = 19$
$\sqrt{36} = 6$	$\sqrt{169} = 13$	$\sqrt{400} = 20$

Numbers that can be written as the quotient of two integers are called **rational numbers.** Integers, fractions, and decimals are rational numbers. What isn't? The square root of any whole number that is not a perfect square is an example of a number that is not rational. So $\sqrt{2}$, $\sqrt{3}$, $\sqrt{5}$, $\sqrt{6}$, and $\sqrt{52}$ are examples of such **irrational numbers.**

21.2 SIMPLIFICATION OF RADICALS

In many algebra problems it is acceptable to leave results that contain radicals. For instance, $5\sqrt{2}$ would be an acceptable final form in many cases. However, when we do leave an answer that contains a radical, we insist that it be *simplified.* In other words, in the end *the whole number left under the radical must be as small as possible.* The technique used to simplify radicals is demonstrated in the examples that follow, in which the following property is always used:

$$\boxed{\sqrt{x \cdot y} = \sqrt{x}\sqrt{y}}$$

As an example of this property, consider that $\sqrt{9 \cdot 2} = \sqrt{9}\sqrt{2}$.

Using exponent notation, we can readily show why this property is true.

$$\sqrt{xy} = (xy)^{\frac{1}{2}} = x^{\frac{1}{2}}y^{\frac{1}{2}} = \sqrt{x}\sqrt{y}$$

EXAMPLE 1   Simplify $\sqrt{18}$.

Solution   We must write 18 as the *product* of two whole numbers, *one of which is a perfect square.* Then we can apply the property of radicals boxed above. So consider 18 as $9 \cdot 2$, because 9 is a perfect square.

$$\sqrt{18} = \sqrt{9 \cdot 2} \qquad \text{since } 9 \cdot 2 = 18$$
$$= \sqrt{9}\sqrt{2} \qquad \text{since } \sqrt{x \cdot y} = \sqrt{x}\sqrt{y}$$
$$= 3\sqrt{2} \ \checkmark \qquad \text{replacing } \sqrt{9} \text{ by } 3$$

So $\sqrt{18} = 3\sqrt{2}$. We factored 18 into $9 \cdot 2$ because 9 is a perfect square; that is, $\sqrt{9} = 3$. Had we chosen to consider 18 as $6 \cdot 3$, we could not have been able to proceed as smoothly. *When simplifying radicals, always look for the perfect square.*

EXAMPLE 2   Simplify $\sqrt{75}$.

Solution
$$\sqrt{75} = \sqrt{25 \cdot 3} \qquad \text{Note that 25 is a perfect square.}$$
$$= \sqrt{25}\sqrt{3} \qquad \text{since } \sqrt{x \cdot y} = \sqrt{x}\sqrt{y}$$
$$= 5\sqrt{3} \ \checkmark \qquad \text{replacing } \sqrt{25} \text{ by } 5$$

You are less likely to make an error if you *place the perfect square first,* as we have been doing. In other words, write $\sqrt{25 \cdot 3}$, not $\sqrt{3 \cdot 25}$; write $\sqrt{9 \cdot 2}$, not $\sqrt{2 \cdot 9}$.

EXAMPLE 3   Simplify $\sqrt{72}$.

Solution

$$\sqrt{72} = \sqrt{36 \cdot 2} \qquad \text{36 is a perfect square.}$$
$$= \sqrt{36}\sqrt{2} \qquad \text{since } \sqrt{x \cdot y} = \sqrt{x}\sqrt{y}$$
$$= 6\sqrt{2} \quad \checkmark \qquad \text{replacing } \sqrt{36} \text{ by 6}$$

EXAMPLE 4   Suppose that in the last example you considered 72 as $9 \cdot 8$ rather than as $36 \cdot 2$. After all, 9 is also a perfect square.

Solution

$$\sqrt{72} = \sqrt{9 \cdot 8}$$
$$= \sqrt{9}\sqrt{8}$$
$$= 3\sqrt{8}$$

Is this answer wrong? No, it's not "wrong," but it is not a completely reduced form. It can be simplified further. Remember, the whole number left under the radical must be made as small as possible. So continue the process.

$$3\sqrt{8} = 3 \cdot \sqrt{8}$$
$$= 3 \cdot \sqrt{4 \cdot 2} \qquad \text{by writing } \sqrt{8} \text{ as } \sqrt{4 \cdot 2}$$
$$= 3 \cdot \sqrt{4} \cdot \sqrt{2}$$
$$= 3 \cdot 2 \cdot \sqrt{2} \qquad \text{since } \sqrt{4} \text{ is 2}$$
$$= 6\sqrt{2} \quad \checkmark \qquad \text{since } 3 \cdot 2 \text{ is 6}$$

As demonstrated by Examples 3 and 4, you should always try to find the *largest* perfect square factor. It simplifies the process.

EXAMPLE 5   Simplify $\sqrt{400}$.

Solution   You may recognize 400 as a perfect square and realize that $\sqrt{400} = 20$. If so, that's it. If not, you proceed as in the other examples, perhaps considering 400 as $4 \cdot 100$.

$$\sqrt{400} = \sqrt{4 \cdot 100}$$
$$= \sqrt{4}\sqrt{100}$$
$$= 2 \cdot 10$$
$$= 20 \quad \checkmark$$

EXAMPLE 6   Simplify $5\sqrt{24}$.

Solution

$$5\sqrt{24} = 5 \cdot \sqrt{24}$$
$$= 5 \cdot \sqrt{4 \cdot 6}$$
$$= 5 \cdot \sqrt{4}\sqrt{6}$$
$$= 5 \cdot 2 \cdot \sqrt{6}$$
$$= 10\sqrt{6} \quad \checkmark$$

In Example 2 we simplified $\sqrt{75}$ to $5\sqrt{3}$. One reason the form $5\sqrt{3}$ is considered "simpler" is that it can lead to simplification of a more complicated expression in which it appears. Consider the following example, which shows how a fraction can be reduced once the radical has been simplified.

EXAMPLE 7   Reduce the fraction $\dfrac{\sqrt{75}}{5}$.

Solution

$$\frac{\sqrt{75}}{5} = \frac{5\sqrt{3}}{5} = \frac{5 \cdot \sqrt{3}}{5 \cdot 1} = \frac{5}{5} \cdot \frac{\sqrt{3}}{1} = 1 \cdot \frac{\sqrt{3}}{1} = \frac{\sqrt{3}}{1} = \sqrt{3} \quad \checkmark$$

EXAMPLE 8   Reduce the fraction $\dfrac{2 + \sqrt{24}}{2}$.

Solution   If you are tempted to simply divide out the 2's that appear in the numerator and denominator, then you had better reread Example 10 of Chapter 18. In fact, perhaps you should read the entire section on reducing fractions. In order to reduce a fraction, you must have a common *factor* in both the numerator and denominator. So to reduce this fraction, we'll need to factor a 2 out of the numerator to go with the 2 in the denominator. And this can only be done if 2 can be factored out of $\sqrt{24}$. So simplify $\sqrt{24}$ in order to find out.

$$\sqrt{24} = \sqrt{4}\sqrt{6}$$
$$= 2\sqrt{6}$$

Thus the original fraction will simplify, as shown in the steps below.

$$\frac{2 + \sqrt{24}}{2} = \frac{2 + 2\sqrt{6}}{2} \qquad \text{since } \sqrt{24} = 2\sqrt{6}$$

$$= \frac{2(1 + \sqrt{6})}{2} \qquad \text{after 2 is factored out of numerator}$$

$$= 1 + \sqrt{6} \quad \checkmark \qquad \text{after dividing out the 2's}$$

Sometimes it can be hard to spot a perfect square factor within the number under the radical. For example, $\sqrt{351}$ has a 9 in it and simplifies as

$\sqrt{9 \cdot 39} = \sqrt{9} \cdot \sqrt{39} = 3\sqrt{39}$. Keep in mind that *it is arithmetically easy to test for the presence of a factor of 9 or 4 in a number.* You can divide the number by 9 or 4.

$$\begin{array}{r} 39 \\ 9\overline{)351} \end{array} \quad \text{shows that } 351 = 9 \cdot 39$$

$$\begin{array}{r} 29 \\ 4\overline{)116} \end{array} \quad \text{shows that } 116 = 4 \cdot 29$$

You can also use the divisibility tests for 9 and 4 as explained in Section 6.4. They take even less time than does the actual division process.

If you cannot see a perfect square factor within the number, and there's no factor of 4 or 9 in it, then consider dividing the number by 2 or 3 to see if that makes it easier to find a perfect square factor. For example, $\sqrt{242}$ simplifies readily when considered as $\sqrt{121 \cdot 2}$ or $\sqrt{121}\sqrt{2}$, or $11\sqrt{2}$. Dividing by 2 makes the perfect square factor 121 visible.

EXERCISES 21.2

Answers to starred exercises are given in the back of the book.

*1. Simplify each radical.
  (a) $\sqrt{8}$ (b) $\sqrt{12}$ (c) $\sqrt{27}$ (d) $\sqrt{36}$ (e) $\sqrt{50}$
  (f) $\sqrt{60}$ (g) $\sqrt{28}$ (h) $\sqrt{44}$ (i) $\sqrt{45}$ (j) $\sqrt{100}$
  (k) $\sqrt{125}$ (l) $\sqrt{63}$ (m) $\sqrt{32}$ (n) $\sqrt{48}$ (o) $\sqrt{52}$
  (p) $\sqrt{56}$ (q) $\sqrt{200}$ (r) $\sqrt{900}$

2. Simplify each radical, if possible.
  *(a) $\sqrt{68}$ (b) $\sqrt{84}$ *(c) $\sqrt{90}$ (d) $\sqrt{88}$
  *(e) $\sqrt{83}$ (f) $\sqrt{96}$ *(g) $\sqrt{112}$ (h) $\sqrt{150}$
  *(i) $\sqrt{128}$ (j) $\sqrt{115}$ *(k) $\sqrt{2500}$ (l) $\sqrt{180}$

3. Use the idea that $\sqrt{x}\sqrt{x} = x$ to simplify each product.
  *(a) $\sqrt{7}\sqrt{7}$ (b) $\sqrt{c}\sqrt{c}$
  *(c) $\sqrt{5}\sqrt{5}\sqrt{5}$ (d) $\sqrt{2}\sqrt{2}\sqrt{2}$
  *(e) $\sqrt{3}\sqrt{3}\sqrt{3}\sqrt{3}$ (f) $\sqrt{7}\sqrt{7}\sqrt{7}\sqrt{7}$
  *(g) $\sqrt{2}\sqrt{10}$ (h) $\sqrt{3}\sqrt{21}$
  *(i) $\sqrt{50}\sqrt{2}$ (j) $\sqrt{12}\sqrt{3}$

4. Simplify.
  *(a) $5\sqrt{8}$ *(b) $3\sqrt{25}$ (c) $4\sqrt{27}$ (d) $2\sqrt{20}$
  *(e) $10\sqrt{68}$ (f) $8\sqrt{44}$ *(g) $2\sqrt{45}$ (h) $7\sqrt{32}$
  *(i) $5\sqrt{300}$ (j) $3\sqrt{40}$

5. Simplify the radical, and then reduce the fraction.
  *(a) $\dfrac{\sqrt{18}}{3}$ (b) $\dfrac{\sqrt{44}}{2}$ *(c) $\dfrac{\sqrt{45}}{18}$ (d) $\dfrac{\sqrt{72}}{2}$

  *(e) $\dfrac{\sqrt{80}}{4}$ (f) $\dfrac{\sqrt{200}}{10}$

6. Reduce each fraction, if possible.

*(a) $\dfrac{2 + \sqrt{20}}{2}$ (b) $\dfrac{3 + \sqrt{63}}{3}$

*(c) $\dfrac{6 + \sqrt{63}}{3}$ (d) $\dfrac{4 - \sqrt{20}}{4}$

*(e) $\dfrac{5 + \sqrt{150}}{15}$ (f) $\dfrac{2 - \sqrt{24}}{3}$

7. Simplify each radical, if possible. Consider using the methods suggested in the last two paragraphs of the section.

*(a) $\sqrt{117}$ (b) $\sqrt{92}$ *(c) $\sqrt{98}$ (d) $\sqrt{124}$
*(e) $\sqrt{104}$ (f) $\sqrt{108}$ *(g) $\sqrt{162}$ (h) $\sqrt{240}$
*(i) $\sqrt{143}$ (j) $\sqrt{140}$ *(k) $\sqrt{147}$ (l) $\sqrt{228}$
*(m) $\sqrt{153}$ (n) $\sqrt{221}$ *(o) $\sqrt{323}$ (p) $\sqrt{261}$
*(q) $\sqrt{279}$ (r) $\sqrt{356}$

21.3 COMBINING LIKE RADICALS

The distributive property can be used to combine like radicals.

EXAMPLE 9   Combine $5\sqrt{2} + 6\sqrt{2}$.

Solution
$$5\sqrt{2} + 6\sqrt{2} = (5 + 6)\sqrt{2}$$
$$= 11\sqrt{2} \;\; \checkmark$$

EXAMPLE 10   Combine $8\sqrt{3} + 5\sqrt{2}$.

Solution $\sqrt{3}$ and $\sqrt{2}$ are not like radicals; they cannot be combined. The distributive property does not apply.

EXAMPLE 11   Combine $3\sqrt{8} + 7\sqrt{2}$.

Solution $\sqrt{8}$ and $\sqrt{2}$ are not like radicals as written, but $\sqrt{8}$ can be simplified.
$$\sqrt{8} = \sqrt{4 \cdot 2} = \sqrt{4}\sqrt{2} = 2\sqrt{2}$$

So
$$3\sqrt{8} + 7\sqrt{2}$$

becomes
$$3 \cdot 2\sqrt{2} + 7\sqrt{2}$$

or
$$6\sqrt{2} + 7\sqrt{2}$$
which simplifies to
$$13\sqrt{2} \;\; \checkmark$$

EXAMPLE 12   Combine $7\sqrt{3} + \sqrt{27} + \sqrt{3}$.

Solution　Since $\sqrt{3}$ means $1\sqrt{3}$ and $\sqrt{27} = 3\sqrt{3}$ when simplified,

$$7\sqrt{3} + \sqrt{27} + \sqrt{3} = 7\sqrt{3} + 3\sqrt{3} + 1\sqrt{3}$$
$$= (7 + 3 + 1)\sqrt{3}$$
$$= 11\sqrt{3} \quad \checkmark$$

EXAMPLE 13　Combine $5 + 3\sqrt{2} + \sqrt{3} + 4\sqrt{2}$.

Solution　The only like radicals are $3\sqrt{2}$ and $4\sqrt{2}$, which combine to yield $7\sqrt{2}$. The other numbers cannot be combined, so the final result is

$$5 + 7\sqrt{2} + \sqrt{3} \quad \checkmark$$

EXERCISES 21.3

1. Combine when possible.
   *(a)　$4\sqrt{7} - 2\sqrt{7}$　　　　　(b)　$5\sqrt{6} + \sqrt{6}$
   *(c)　$7\sqrt{2} + 10\sqrt{2} - 3\sqrt{2}$　(d)　$8\sqrt{5} + \sqrt{7} - \sqrt{5}$
   *(e)　$\sqrt{8} + 9\sqrt{2}$　　　　　*(f)　$\sqrt{16} + \sqrt{4}$
   *(g)　$\sqrt{32} + 6\sqrt{2}$　　　　(h)　$2\sqrt{3} + 5\sqrt{27}$
   *(i)　$\sqrt{72} - 2\sqrt{2} + \sqrt{2}$　(j)　$\sqrt{54} + 3\sqrt{24}$
   *(k)　$5\sqrt{41} - 3\sqrt{17}$　　　*(l)　$5\sqrt{100} + 6\sqrt{81}$
   *(m)　$2\sqrt{3} + 6\sqrt{75}$　　　(n)　$\sqrt{28} + 2\sqrt{63}$
   *(o)　$\sqrt{45} + 8\sqrt{5}$　　　　*(p)　$7 + 6\sqrt{2} + 5\sqrt{3}$
   *(q)　$\sqrt{90} + \sqrt{40}$　　　　*(r)　$5\sqrt{12} + 2\sqrt{27} - \sqrt{48}$
   *(s)　$2\sqrt{20} - \sqrt{5} + \sqrt{10}$　(t)　$\sqrt{52} + 2\sqrt{117}$

21.4 RATIONALIZING THE DENOMINATOR OR NUMERATOR

In some applications, it is undesirable to have a fraction that contains a radical in the denominator. Our problem, then, is how to remove a radical from the denominator of a fraction. In other applications, it is desirable to remove a radical from the numerator of a fraction. The examples below demonstrate how it can be done. Removing the radical from the denominator removes the irrational number from the denominator and hence **rationalizes** the denominator. Removing the radical from the numerator removes the irrational number from the numerator and hence rationalizes the numerator. The procedure is based on $\sqrt{x}\sqrt{x} = x$.

EXAMPLE 14　Remove the radical from the denominator of $\dfrac{5}{\sqrt{3}}$.

Solution　Multiplying the denominator $\sqrt{3}$ by $\sqrt{3}$ eliminates the radical from the denominator, since $\sqrt{3} \cdot \sqrt{3} = 3$. The numerator must also be multiplied by $\sqrt{3}$ so that the overall effect is multiplication of the fraction by 1.

$$\frac{5}{\sqrt{3}} = \frac{5}{\sqrt{3}} \cdot \frac{\sqrt{3}}{\sqrt{3}}$$

$$= \frac{5\sqrt{3}}{3} \quad \checkmark \qquad \text{since } \sqrt{3}\sqrt{3} = 3$$

EXAMPLE 15 Remove the irrational number from the numerator of $\dfrac{\sqrt{5}}{3}$.

Solution Multiply both numerator and denominator of $\dfrac{\sqrt{5}}{3}$ by $\sqrt{5}$. Doing so will eliminate the radical from the numerator, and yet the value of the fraction will be the same, since we are merely multiplying the fraction by an appropriate form of the number 1.

$$\frac{\sqrt{5}}{3} = \frac{\sqrt{5}}{3} \cdot \frac{\sqrt{5}}{\sqrt{5}}$$

$$= \frac{5}{3\sqrt{5}} \quad \checkmark$$

EXAMPLE 16 Rationalize the denominator of $\dfrac{\sqrt{6}}{\sqrt{2}}$.

Solution As before,

$$\frac{\sqrt{6}}{\sqrt{2}} = \frac{\sqrt{6}}{\sqrt{2}} \cdot \frac{\sqrt{2}}{\sqrt{2}} = \frac{\sqrt{12}}{2}$$

But $\sqrt{12}$ can be simplified, as shown next.

$$\frac{\sqrt{12}}{2} = \frac{\sqrt{4 \cdot 3}}{2} = \frac{\sqrt{4}\sqrt{3}}{2} = \frac{2\sqrt{3}}{2} = \sqrt{3} \quad \checkmark$$

Another approach:

$$\frac{\sqrt{6}}{\sqrt{2}} = \sqrt{\frac{6}{2}} = \sqrt{3} \quad \checkmark$$

The second approach showed the use of another important property of radicals. The property:

$$\boxed{\sqrt{\frac{x}{y}} = \frac{\sqrt{x}}{\sqrt{y}}}$$

We use it again in the next example.

EXAMPLE 17 Simplify $\sqrt{\dfrac{1}{2}}$ and leave it with a rational denominator.

Solution
$$\sqrt{\frac{1}{2}} = \frac{\sqrt{1}}{\sqrt{2}} = \frac{1}{\sqrt{2}} \qquad \text{since } \sqrt{1} = 1$$

If we want to rationalize the denominator, we continue as

$$= \frac{1}{\sqrt{2}} \cdot \frac{\sqrt{2}}{\sqrt{2}}$$

$$= \frac{\sqrt{2}}{2} \quad \checkmark$$

Historically, people automatically rationalized the denominator of an expression like $\frac{1}{\sqrt{2}}$ because $\frac{1}{\sqrt{2}}$ was a more difficult division than was $\frac{\sqrt{2}}{2}$. Using 1.414 as an approximation for $\sqrt{2}$, you can see that $\frac{1}{1.414}$ is a more difficult division than is $\frac{1.414}{2}$. However, with modern calculators readily available, this no longer matters. Still, sometimes there are algebraic advantages to having a rational denominator. But there are also times when a rational *numerator* is desirable, especially in some calculus situations. So you must be able to rationalize either a denominator or a numerator.

EXAMPLE 18   Rationalize the denominator of $\dfrac{5\sqrt{3}}{\sqrt{8}}$.

Solution   There are several ways of doing so. One approach:

$$\frac{5\sqrt{3}}{\sqrt{8}} = \frac{5\sqrt{3}}{\sqrt{8}} \cdot \frac{\sqrt{8}}{\sqrt{8}} = \frac{5\sqrt{24}}{8} = \frac{5\sqrt{4}\sqrt{6}}{8} = \frac{5 \cdot 2\sqrt{6}}{8} = \frac{5\sqrt{6}}{4} \quad \checkmark$$

Another:

$$\frac{5\sqrt{3}}{\sqrt{8}} = \frac{5\sqrt{3}}{2\sqrt{2}} = \frac{5\sqrt{3}}{2\sqrt{2}} \cdot \frac{\sqrt{2}}{\sqrt{2}} = \frac{5\sqrt{6}}{2 \cdot 2} = \frac{5\sqrt{6}}{4} \quad \checkmark$$

Another:

$$\frac{5\sqrt{3}}{\sqrt{8}} = \frac{5\sqrt{3}}{\sqrt{8}} \cdot \frac{\sqrt{2}}{\sqrt{2}} = \frac{5\sqrt{6}}{\sqrt{16}} = \frac{5\sqrt{6}}{4} \quad \checkmark$$

Next, three multiplication examples are presented as background for a different kind of rationalizing example.

EXAMPLE 19   Multiply $\sqrt{3}\sqrt{7}$.

Solution
$$\sqrt{3}\sqrt{7} = \sqrt{3 \cdot 7}$$

$$= \sqrt{21} \quad \checkmark$$

EXAMPLE 20   Multiply $(2 + \sqrt{3})(5 - \sqrt{3})$.

Solution   The form of this multiplication is the same as that of multiplication of binomials (see Section 11.3).

$$(2 + \sqrt{3})(5 - \sqrt{3})$$

$$= 2 \cdot 5 - 2\sqrt{3} + \sqrt{3} \cdot 5 - \sqrt{3} \cdot \sqrt{3}$$

$$= 10 - 2\sqrt{3} + 5\sqrt{3} - 3 \qquad \begin{cases} \text{since } 2 \cdot 5 = 10 \\ \text{and } \sqrt{3} \cdot \sqrt{3} = 3 \end{cases}$$

$$= (10 - 3) - 2\sqrt{3} + 5\sqrt{3} \qquad \text{when regrouped}$$

$$= 7 + 3\sqrt{3} \quad \checkmark \qquad \text{after combining like terms}$$

EXAMPLE 21 Multiply $(5 + \sqrt{7})(2 + \sqrt{2})$.

Solution $(5 + \sqrt{7})(2 + \sqrt{2}) = 5 \cdot 2 + 5\sqrt{2} + \sqrt{7} \cdot 2 + \sqrt{7} \cdot \sqrt{2}$

$$= 10 + 5\sqrt{2} + 2\sqrt{7} + \sqrt{14} \quad \checkmark$$

which cannot be simplified.

EXAMPLE 22 Remove the radical from the denominator of $\dfrac{1}{\sqrt{3} + 1}$.

Solution If the denominator $\sqrt{3} + 1$ is multiplied by $\sqrt{3} - 1$, the result will be the difference of two squares, $(\sqrt{3})^2 - (1)^2$, which is 2. The radical will be eliminated. Of course, the numerator must also be multiplied by $\sqrt{3} - 1$. Here is the work.

$$\frac{1}{\sqrt{3} + 1} = \frac{1}{\sqrt{3} + 1} \cdot \frac{\sqrt{3} - 1}{\sqrt{3} - 1}$$

$$= \frac{\sqrt{3} - 1}{3 - 1}$$

$$= \frac{\sqrt{3} - 1}{2} \quad \checkmark$$

As another example, if the denominator were $\sqrt{11} - 3$, then you would multiply by $\sqrt{11} + 3$ to create the difference of two squares.

EXERCISES 21.4

1. Rationalize the denominator of each fraction and simplify as much as possible.

    *(a) $\dfrac{3}{\sqrt{2}}$ (b) $\dfrac{5}{\sqrt{7}}$ *(c) $\dfrac{2}{\sqrt{5}}$ (d) $\dfrac{4}{\sqrt{3}}$

    *(e) $\dfrac{6}{\sqrt{3}}$ (f) $\dfrac{10}{\sqrt{2}}$ *(g) $\dfrac{10}{\sqrt{5}}$ (h) $\dfrac{8}{\sqrt{2}}$

2. Simplify. Do not leave a radical in the denominator.

    *(a) $\dfrac{\sqrt{10}}{\sqrt{2}}$ (b) $\dfrac{\sqrt{21}}{\sqrt{3}}$ *(c) $\dfrac{5\sqrt{27}}{\sqrt{3}}$ (d) $\dfrac{\sqrt{98}}{7\sqrt{2}}$

*(e) $\dfrac{4\sqrt{6}}{\sqrt{200}}$　(f) $\dfrac{5\sqrt{3}}{\sqrt{128}}$　*(g) $\dfrac{6\sqrt{5}}{\sqrt{8}}$　(h) $\dfrac{3\sqrt{15}}{\sqrt{32}}$

3. Rationalize the numerator of each fraction and simplify as much as possible.

*(a) $\dfrac{\sqrt{2}}{5}$　(b) $\dfrac{\sqrt{7}}{2}$　*(c) $\dfrac{\sqrt{8}}{3}$　(d) $\dfrac{\sqrt{27}}{4}$

*(e) $\dfrac{\sqrt{24}}{12}$　(f) $\dfrac{\sqrt{54}}{3}$　*(g) $\dfrac{\sqrt{32}}{8}$　(h) $\dfrac{\sqrt{18}}{6}$

4. Simplify and remove the radical from the denominator when it appears.

*(a) $\sqrt{\dfrac{1}{3}}$　(b) $\sqrt{\dfrac{1}{5}}$　*(c) $\sqrt{\dfrac{8}{7}}$　(d) $\sqrt{\dfrac{4}{13}}$

*(e) $\sqrt{\dfrac{1}{9}}$　(f) $\sqrt{\dfrac{27}{32}}$　*(g) $\sqrt{\dfrac{80}{70}}$　(h) $\sqrt{\dfrac{80}{90}}$

5. Multiply.

*(a) $\sqrt{5}\,\sqrt{6}$　　　　　　(b) $\sqrt{7}\,\sqrt{6}$

*(c) $(1+\sqrt{2})(1+\sqrt{2})$　　(d) $(5+\sqrt{3})(4+\sqrt{3})$

*(e) $(5+\sqrt{7})(5-\sqrt{7})$　　(f) $(3+\sqrt{5})(2-\sqrt{5})$

*(g) $(6-\sqrt{13})(8-\sqrt{3})$　　(h) $(12-\sqrt{17})(5-\sqrt{3})$

6. Remove the radical from the denominator.

*(a) $\dfrac{1}{\sqrt{5}+2}$　　*(b) $\dfrac{3}{1+\sqrt{2}}$

*(c) $\dfrac{6}{\sqrt{7}-2}$　　*(d) $\dfrac{7}{\sqrt{10}+3}$

(e) $\dfrac{13}{\sqrt{17}-4}$　　(f) $\dfrac{4}{\sqrt{6}+2}$

21.5　EQUATIONS CONTAINING RADICALS

Here is a series of examples that shows how to solve equations that contain radicals. We shall be using a new operation called **squaring both sides of an equation.** If you have an equation, then the left- and right-hand sides are equal. For instance, in the equation

$$\sqrt{x}=6$$

$\sqrt{x}$ and 6 are equal. So if we multiply each side by itself (that is, square both sides)

$$\underline{\sqrt{x}\,\sqrt{x}}=\underline{6\cdot 6}$$

we are actually just multiplying both sides by equal quantities. This "new operation" produces the result

$$x=36\quad\checkmark$$

since $\sqrt{x}\,\sqrt{x}=x$ and $6\cdot 6=36$.

Check:

$$\sqrt{x} = 6$$
$$\sqrt{36} \overset{?}{=} 6 \qquad \text{after substituting 36 for } x$$
$$6 = 6 \quad \checkmark$$

EXAMPLE 23 Solve $\sqrt{x + 1} = 11$.

Solution Begin by squaring both sides of the equation. Squaring removes the radical, since by the definition of square root, $\sqrt{x + 1} \sqrt{x + 1} = x + 1$.

$$(\sqrt{x + 1})^2 = (11)^2$$
$$x + 1 = 121$$
$$x = 120 \quad \checkmark$$

Check:

$$\sqrt{x + 1} = 11$$
$$\sqrt{120 + 1} \overset{?}{=} 11 \qquad \text{after substituting 120 for } x$$
$$\sqrt{121} \overset{?}{=} 11$$
$$11 = 11 \quad \checkmark$$

EXAMPLE 24 Solve $3\sqrt{7n - 3} = 4$.

Solution On squaring,

$$(3\sqrt{7n - 3})^2 = (4)^2$$

or

$$3^2(\sqrt{7n - 3})^2 = 4^2$$
$$9(7n - 3) = 16 \qquad \text{since } (\sqrt{7n - 3})^2 = 7n - 3$$
$$63n - 27 = 16 \qquad \text{after distributing the 9}$$
$$63n = 43$$
$$n = \frac{43}{63} \quad \checkmark$$

Check:

$$3\sqrt{7n - 3} = 4$$
$$3\sqrt{7 \cdot \frac{43}{63} - 3} \overset{?}{=} 4 \qquad \text{after substituting } \frac{43}{63} \text{ for } n$$

$$3\sqrt{\frac{43}{9} - 3} \overset{?}{=} 4$$

$$3\sqrt{\frac{43}{9} - \frac{27}{9}} \overset{?}{=} 4$$

$$3\sqrt{\frac{16}{9}} \overset{?}{=} 4$$

$$3 \cdot \frac{4}{3} \overset{?}{=} 4$$

$$4 = 4 \quad \checkmark$$

EXAMPLE 25   Solve $\sqrt{6x - 1} = 2\sqrt{x}$.

Solution

$$(\sqrt{6x - 1})^2 = (2\sqrt{x})^2$$

$$6x - 1 = 4x$$

$$2x - 1 = 0$$

$$2x = 1$$

$$x = \frac{1}{2} \quad \checkmark$$

Check:

$$\sqrt{6x - 1} = 2\sqrt{x}$$

$$\sqrt{6 \cdot \frac{1}{2} - 1} \overset{?}{=} 2\sqrt{\frac{1}{2}}$$

$$\sqrt{3 - 1} \overset{?}{=} 2 \cdot \frac{\sqrt{1}}{\sqrt{2}}$$

$$\sqrt{2} \overset{?}{=} 2 \cdot \frac{\sqrt{1}}{\sqrt{2}} \cdot \frac{\sqrt{2}}{\sqrt{2}}$$

$$\sqrt{2} \overset{?}{=} 2 \cdot \frac{\sqrt{2}}{2}$$

$$\sqrt{2} = \sqrt{2} \quad \checkmark$$

EXAMPLE 26   Solve $\sqrt{2x} + 3 = 9$.

Solution   You *could* immediately square both sides, as we have done in the other examples. But doing so will not remove the radical. Observe.

$$(\sqrt{2x} + 3)^2 = (\sqrt{2x} + 3)(\sqrt{2x} + 3)$$

$$= 2x + 3\sqrt{2x} + 3\sqrt{2x} + 9$$

$$= 2x + 6\sqrt{2x} + 9$$

So instead of squaring immediately, simply add -3 to both sides of the original equation, $\sqrt{2x} + 3 = 9$. The result is an equation with the radical alone on one side.

$$\sqrt{2x} = 6$$

Now, on squaring both sides, we obtain a simple equation with no radical in it.

$$2x = 36$$

which yields

$$x = 18 \quad \checkmark$$

Be sure to check 18 in the original equation.

EXAMPLE 27 Solve $\sqrt{x} + 1 = 0$.

Solution
$$\sqrt{x} + 1 = 0$$
$$\sqrt{x} = -1 \qquad \text{after adding } -1 \text{ to both sides}$$
$$x = 1 \qquad \text{after squaring both sides of the equation}$$

But $x = 1$ *will not check in the original equation.* Try it. No number will check in the original equation; there is no solution. Squaring the -1 removed the minus and introduced an error. If you look at the original equation, you will notice that it suggests that the sum of two positive numbers must be equal to zero. And that cannot be true.

In the last example a solution was determined that did not check; that is, it was not a solution of the original equation. Such a "solution" is called an **extraneous root.** Whenever both sides of an equation are squared, all solutions must be checked in order to determine which values are indeed solutions and which are extraneous. This means that *all equations containing radicals should be checked.*

As a simple example, consider the equation $x = 3$, which has only one solution, 3. Now square both sides of this equation to obtain $x^2 = 9$. This new equation has two solutions, -3 and 3, yet only one of them (3) is a solution of the original equation. In this example -3 is an extraneous root.

EXERCISES 21.5

1. Solve each equation.
   *(a) $\sqrt{x} = 9$ (b) $\sqrt{m} = 11$
   *(c) $\sqrt{x} = 7$ (d) $\sqrt{y} = 4$
   *(e) $\sqrt{x - 1} = 4$ (f) $\sqrt{2x + 1} = 5$
   *(g) $\sqrt{1 + 7x} = 9$ (h) $\sqrt{4x - 5} = 8$
   *(i) $3\sqrt{x + 2} = 7$ (j) $2\sqrt{x - 3} = 11$
   *(k) $6\sqrt{5y - 2} = 9$ (l) $10\sqrt{1 - 3x} = 12$
   *(m) $\sqrt{3t - 1} = \sqrt{t + 7}$ (n) $\sqrt{2n + 7} = \sqrt{5n + 1}$
   *(o) $\sqrt{5x + 10} = 4\sqrt{x}$ (p) $\sqrt{2x - 3} = 3\sqrt{x - 5}$

2. Solve each equation, if possible. Check your answer.
   *(a) $\sqrt{3x - 5} + 2 = 6$ (b) $\sqrt{2n} + 5 = 19$
   *(c) $\sqrt{x + 1} + 4 = 0$ (d) $3 + \sqrt{x} = 1$
   *(e) $6 + \sqrt{2x - 7} = 10$ (f) $8 + \sqrt{5x + 1} = 15$

*(g) $\sqrt{4n + 3} = \sqrt{6}$ *(h) $\sqrt{2x + 3} - 4 = 1$
*(i) $\sqrt{10 - 3x} = \sqrt{13}$ *(j) $\sqrt{7t - 6} - 8 = 13$

CHAPTER 21 REVIEW EXERCISES

1. Simplify, if possible.
     *(a) $\sqrt{24}$ (b) $\sqrt{80}$ *(c) $\sqrt{500}$ (d) $\sqrt{64}$
     *(e) $\sqrt{54}$ *(f) $\sqrt{76}$ *(g) $\sqrt{160}$ (h) $\sqrt{256}$
     *(i) $\sqrt{102}$ *(j) $\sqrt{171}$ *(k) $\sqrt{243}$

2. Simplify.
     *(a) $6\sqrt{2} + 5\sqrt{32}$ (b) $\sqrt{27} + \sqrt{75}$

3. Rationalize the denominator of each fraction.
     *(a) $\dfrac{7}{\sqrt{5}}$ (b) $\dfrac{3}{5 - \sqrt{7}}$

4. Simplify. Leave with a rational denominator.
     *(a) $\sqrt{\dfrac{1}{11}}$ (b) $\sqrt{\dfrac{9}{17}}$

5. Solve each equation.
     *(a) $\sqrt{3x + 2} = 11$ (b) $\sqrt{x + 3} - 2 = 7$

CUMULATIVE REVIEW EXERCISES (CHAPTERS 10–21)

*1. Multiply $(2x + 5)(3x - 4)$.

*2. Solve the equation.
$$4x - 8(5x - 2) = 34$$

*3. Write an algebraic expression to represent each phrase.
     (a) The number of months in r years.
     (b) The perimeter of a triangle which has smallest side x, another side $x + 3$, and the third side is twice the smallest side.
     (c) The profit obtained by buying an antique at $3x$ dollars and selling it for $5y$ dollars.

*4. Determine the equation of the line with slope -3 that passes through the point $(7, -2)$.

*5. Solve the system.
$$\begin{cases} 2m + 3n = 21 \\ 5m - 4n = 18 \end{cases}$$

*6. Factor each expression, if possible.
     (a) $5x^2 - 10x$ (b) $2x^2 - 162$
     (c) $3x^2 + 3x - 60$ (d) $4x^2 + 13x - 12$

*7. Perform the indicated operation. Reduce when possible.
     (a) $\dfrac{x}{3x - 6} + \dfrac{7}{2x - 4}$ (b) $\dfrac{4m^2n^3}{3m^5p} \cdot \dfrac{mp^3}{n^2t^7}$

*8. Solve the formula for x.
$$y = mx + nx - t$$

*9. Simplify each expression. Leave no negative exponents.

(a) $\dfrac{a^{-9}b^5}{a^2b^{-3}}$ (b) $(x^3y^{-5})(x^{14}y^2)$ (c) $9^{\frac{1}{2}}x^0$

*10. Simplify.

(a) $\sqrt{120}$ (b) $\sqrt{90}$

22

MORE QUADRATIC EQUATIONS

22.1 INTRODUCTION

In Chapter 17 you learned how to solve quadratic equations by factoring. In this chapter you will see how to solve quadratic equations which cannot be solved by factoring. The methods presented are *completing the square* (Section 22.2) and the *quadratic formula* (Section 22.3).

We proceed now with some special examples that will prepare you for the process of completing the square. The examples are based on solving quadratic equations in which the factoring for the difference of two squares can be applied.

Here are a few examples of quadratic equations that can be solved by using the factoring for the difference of two squares.

$$x^2 = 9 \longrightarrow x^2 - 9 = 0 \longrightarrow (x + 3)(x - 3) = 0 \longrightarrow x = -3, 3$$
$$x^2 = 4 \longrightarrow x^2 - 4 = 0 \longrightarrow (x + 2)(x - 2) = 0 \longrightarrow x = -2, 2$$
$$x^2 = 16 \longrightarrow x^2 - 16 = 0 \longrightarrow (x + 4)(x - 4) = 0 \longrightarrow x = -4, 4$$

If we use the notation $\pm$ to mean **plus or minus,** then $x = -3, 3$ can be written $x = \pm 3$; that is, $+3$ is a solution and -3 is also a solution. The preceding work can be summarized as

$$x^2 = 9 \longrightarrow x = \pm 3$$
$$x^2 = 4 \longrightarrow x = \pm 2$$
$$x^2 = 16 \longrightarrow x = \pm 4$$

Of course, it is also true that

$$\text{If } x^2 = 25, \text{ then } x = \pm 5.$$
$$\text{If } x^2 = 81, \text{ then } x = \pm 9.$$

In general,

$$\boxed{\text{If } x^2 = n, \text{ then } x = \pm\sqrt{n}.}\quad\text{("root taking")}$$

A few more examples.

$$x^2 = 49 \longrightarrow x = \pm 7$$
$$x^2 = 21 \longrightarrow x = \pm\sqrt{21}$$
$$x^2 = 3 \ \longrightarrow x = \pm\sqrt{3}$$

EXAMPLE 1   Solve the quadratic equation $(x + 3)^2 = 17$.

Solution   Following the preceding pattern, if

$$(x + 3)^2 = 17$$
then $$x + 3 = \pm\sqrt{17}$$

To solve for x, add -3 to both sides. The result is

$$x = -3 \pm \sqrt{17} \ \ \checkmark$$

Note that $-3 \pm \sqrt{17}$ is read "minus 3 plus or minus the square root of 17." It indicates that both $-3 + \sqrt{17}$ and $-3 - \sqrt{17}$ are solutions.

EXAMPLE 2   Solve the equation $(x - 5)^2 = 12$.

Solution
$$(x - 5)^2 = 12$$
$$x - 5 = \pm\sqrt{12}$$
$$x = 5 \pm\sqrt{12}$$

And since $\sqrt{12} = \sqrt{4 \cdot 3} = \sqrt{4}\sqrt{3} = 2\sqrt{3}$, the solutions simplify to

$$x = 5 \pm 2\sqrt{3} \ \ \checkmark$$

NOTE
In an expression such as $5 + 2\sqrt{3}$, you *cannot* combine the 5 and the 2. The normal order of operations tells us that the 2 must be multiplied by $\sqrt{3}$ before the addition can be considered. After the multiplication, the terms you would consider adding are 5 and $2\sqrt{3}$, and they are not like terms.

EXERCISES 22.1

Answers to starred exercises are given in the back of the book.
*1. Solve each quadratic equation.
    (a) $x^2 = 64$ (b) $m^2 = 1$
    (c) $y^2 = 36$ (d) $c^2 = 100$
    (e) $x^2 = 7$ (f) $x^2 = 11$
    (g) $a^2 = 15$ (h) $t^2 = 47$

2. Solve each quadratic equation.
*(a) $(x + 1)^2 = 11$ (b) $(x - 1)^2 = 13$
*(c) $(x - 3)^2 = 19$ (d) $(m + 5)^2 = 3$
*(e) $(y - 4)^2 = 8$ (f) $(t + 2)^2 = 24$
*(g) $(p + 1)^2 = 28$ (h) $(x + 6)^2 = 50$

3. Solve each quadratic equation.
*(a) $(x - 2)^2 - 3 = 0$ (b) $(x + 5)^2 - 7 = 0$
*(c) $(x + 1)^2 + 1 = 8$ (d) $(x - 2)^2 + 5 = 19$

22.2 COMPLETING THE SQUARE

You should now have the appropriate orientation for studying the process called **completing the square.** The equation $x^2 + 4x + 1 = 0$ cannot be solved by factoring. (Try it and you will see.) So another approach is needed. The expression $x^2 + 4x + 1$ is almost the square of $(x + 2)$, since

$$(x + 2)^2 = x^2 + 4x + 4$$

Because we need only add a constant to $x^2 + 4x + 1$ to make it the square of $(x + 2)$, the process of completing the square can be used here. Let's continue. We have

$$x^2 + 4x + 1 = 0$$

and we want

$$x^2 + 4x + 4 = \cdots$$

So add 3 to both sides of the original equation.

$$
\begin{array}{rcc}
x^2 + 4x + 1 = & 0 \\
+3 & +3 \\
\hline
x^2 + 4x + 4 = & 3
\end{array}
$$

Writing $x^2 + 4x + 4$ as $(x + 2)^2$, we obtain

$$(x + 2)^2 = 3$$

which resembles the form of Examples 1 and 2 and should be easy for you to solve. The work is shown next.

$$
\begin{aligned}
(x + 2)^2 &= 3 \\
x + 2 &= \pm\sqrt{3} \\
x &= -2 \pm \sqrt{3} \quad \checkmark
\end{aligned}
$$

The roots $-2 \pm \sqrt{3}$ can be written separately as $-2 + \sqrt{3}$ and $-2 - \sqrt{3}$, but there is no advantage in doing so in this instance.

EXAMPLE 3   Solve the equation $x^2 + 6x - 2 = 0$ by completing the square.

Solution   The square that includes $x^2 + 6x$ is $(x + 3)^2$, since $(x + 3)(x + 3) = x^2 + 6x + 9$. So add $+11$ to both sides of $x^2 + 6x - 2 = 0$ to get the square $x^2 + 6x + 9$.

$$\begin{array}{rcl} x^2 + 6x - 2 &=& 0 \\ + 11 && +11 \\ \hline x^2 + 6x + 9 &=& 11 \end{array}$$

Since $x^2 + 6x + 9$ is $(x + 3)^2$, our equation can be written as

$$(x + 3)^2 = 11$$

Now
$$x + 3 = \pm\sqrt{11}$$

So
$$x = -3 \pm \sqrt{11} \quad \checkmark$$

You can always complete a square of the form $x^2 + nx + \cdots$ by noting that the square is $\left(x + \dfrac{n}{2}\right)^2$ and working from there. For instance,

$$\begin{array}{ll} x^2 + 8x \cdots \longrightarrow (x + 4)^2 & +4 \text{ is half of } +8 \\ m^2 - 2m \cdots \longrightarrow (m - 1)^2 & -1 \text{ is half of } -2 \\ y^2 + 3y \cdots \longrightarrow \left(y + \dfrac{3}{2}\right)^2 & +\dfrac{3}{2} \text{ is half of } +3 \\ x^2 - 50x \cdots \longrightarrow (x - 25)^2 & -25 \text{ is half of } -50 \end{array}$$

EXAMPLE 4   Solve the equation $y^2 - 12y + 12 = 0$ by completing the square.

Solution   You can see that $y^2 - 12y + 12$ differs from the square $(y - 6)^2$ by a constant, since $(y - 6)^2 = y^2 - 12y + 36$. Add the constant 24 to both sides to make the expression on the left side equal to the desired square.

$$\begin{array}{rcl} y^2 - 12y + 12 &=& 0 \\ + 24 &=& +24 \\ \hline y^2 - 12y + 36 &=& 24 \end{array}$$

$$(y - 6)^2 = 24$$

$$y - 6 = \pm\sqrt{24}$$

$$y = 6 \pm \sqrt{24}$$

And since $\sqrt{24} = \sqrt{4}\sqrt{6} = 2\sqrt{6}$, replace $\sqrt{24}$ with the simpler form $2\sqrt{6}$ to get

$$y = 6 \pm 2\sqrt{6} \quad \checkmark$$

The next two examples involve fractions and are considerably more involved. They should serve as an incentive to pursue another approach (which we do in the next section). On the other hand, there are situations in which you must be able to complete the square even when fractions arise.

EXAMPLE 5   Solve the equation $x^2 + 7x - 5 = 0$ by completing the square.

Solution   The $x^2 + 7x$ portion suggests, since $\frac{7}{2}$ is half of the 7, that we use the square

$$\left(x + \frac{7}{2}\right)^2$$

or $$x^2 + 7x + \frac{49}{4}$$

Since we are beginning with the equation $x^2 + 7x - 5 = 0$, we need to add $+5$ and $+\frac{49}{4}$ to it (to both sides) in order to obtain the desired square $x^2 + 7x + \frac{49}{4}$.

$$
\begin{array}{rcl}
x^2 + 7x - 5 &=& 0 \\
+5 & & +5 \\
+\dfrac{49}{4} & & +\dfrac{49}{4} \\
\hline
x^2 + 7x + \dfrac{49}{4} &=& 5 + \dfrac{49}{4}
\end{array}
$$

Since $x^2 + 7x + \frac{49}{4}$ is $\left(x + \frac{7}{2}\right)^2$ and since $5 + \frac{49}{4} = \frac{20}{4} + \frac{49}{4} = \frac{69}{4}$, we have

$$\left(x + \frac{7}{2}\right)^2 = \frac{69}{4}$$

It follows that

$$x + \frac{7}{2} = \pm \sqrt{\frac{69}{4}}$$

On solving for x, we have

$$x = -\frac{7}{2} \pm \sqrt{\frac{69}{4}}$$

Also, $\sqrt{\dfrac{69}{4}} = \dfrac{\sqrt{69}}{\sqrt{4}} = \dfrac{\sqrt{69}}{2}$. So

$$x = -\frac{7}{2} \pm \frac{\sqrt{69}}{2}$$

or $$x = \frac{-7 \pm \sqrt{69}}{2} \quad \checkmark$$

EXAMPLE 6   Solve the equation $5n^2 - 8n + 2 = 0$ by completing the square.

Solution   In order to complete the square, we must first get a coefficient of 1 for the n^2 term. So multiply both sides of $5n^2 - 8n + 2 = 0$ by $\frac{1}{5}$ to get a coefficient of 1 for n^2.

$$n^2 - \frac{8}{5}n + \frac{2}{5} = 0$$

Since half of $-\frac{8}{5}$ is $-\frac{4}{5}$, our square is

$$\left(n - \frac{4}{5}\right)^2$$

or

$$n^2 - \frac{8}{5}n + \frac{16}{25}$$

We must change $n^2 - \frac{8}{5}n + \frac{2}{5} = 0$ in order to make the left side equal to $n^2 - \frac{8}{5}n + \frac{16}{25}$. To do so, add $-\frac{2}{5}$ and $+\frac{16}{25}$ so as to change the $\frac{2}{5}$ term to $\frac{16}{25}$.

$$n^2 - \frac{8}{5}n + \frac{2}{5} - \frac{2}{5} + \frac{16}{25} = 0 - \frac{2}{5} + \frac{16}{25}$$

or

$$n^2 - \frac{8}{5}n + \frac{16}{25} = -\frac{10}{25} + \frac{16}{25}$$

or

$$n^2 - \frac{8}{5}n + \frac{16}{25} = \frac{6}{25}$$

which can be put in the form

$$\left(n - \frac{4}{5}\right)^2 = \frac{6}{25}$$

and then

$$n - \frac{4}{5} = \pm \sqrt{\frac{6}{25}}$$

or

$$n = \frac{4}{5} \pm \sqrt{\frac{6}{25}}$$

and since $\sqrt{\frac{6}{25}} = \frac{\sqrt{6}}{5}$, we get

$$n = \frac{4}{5} \pm \frac{\sqrt{6}}{5} = \frac{4 \pm \sqrt{6}}{5} \quad \checkmark$$

EXERCISES 22.2

1. Solve each quadratic equation by completing the square.
   *(a) $x^2 + 8x + 1 = 0$ (b) $x^2 + 4x + 1 = 0$
   *(c) $m^2 + 6m + 7 = 0$ (d) $x^2 + 6x + 1 = 0$
   *(e) $y^2 + 12y + 15 = 0$ (f) $t^2 + 12t + 33 = 0$
   *(g) $x^2 - 8x + 8 = 0$ (h) $x^2 + 8x + 5 = 0$
   *(i) $x^2 - 10x + 1 = 0$ (j) $x^2 - 6x + 2 = 0$
   *(k) $x^2 - 12x - 3 = 0$ (l) $x^2 - 4x - 1 = 0$
   *(m) $x^2 + 10x = 4$ (n) $x^2 - 4x = 10$
   *(o) $y^2 + 4y = 1$ (p) $y^2 + 10y = 3$

2. Solve each quadratic equation by completing the square.
   *(a) $x^2 - 3x + 1 = 0$ (b) $n^2 + 7n - 4 = 0$
   *(c) $x^2 + 5x + 1 = 0$ (d) $y^2 - 5y - 2 = 0$

*(e) $x^2 + 3x - 8 = 0$ (f) $m^2 + 9m + 2 = 0$
*(g) $2x^2 + 6x - 1 = 0$ (h) $2x^2 - 8x - 9 = 0$
*(i) $5x^2 + 3x = 1$ (j) $3x^2 - 5x = 6$

22.3 THE QUADRATIC FORMULA

A useful formula, called the **quadratic formula,** has been derived for solving any quadratic equation—that is, any equation of the form

$$ax^2 + bx + c = 0 \qquad a \neq 0$$

The formula was obtained by completing the square for the general equation $ax^2 + bx + c = 0$. The derivation is given as Section 22.4. The result is

> *Quadratic Formula*
>
> $$x = \frac{-b \pm \sqrt{b^2 - 4ac}}{2a}$$

To use the formula, obtain a, b, and c from the equation and substitute them into the formula.

- a is the coefficient of x^2.
- b is the coefficient of x.
- c is the *constant* term.

Here are some examples specifying the a, b, and c.

- In $x^2 + 7x - 10 = 0$, $a = 1$, $b = 7$, $c = -10$.
- In $2x^2 - 3x + 2 = 0$, $a = 2$, $b = -3$, $c = 2$.
- In $5x^2 - 3 = 0$, $a = 5$, $b = 0$, $c = -3$.
- In $x^2 + 2x = 0$, $a = 1$, $b = 2$, $c = 0$.
- In $5x^2 = 6x + 1$, we must transform the original equation into $5x^2 - 6x - 1 = 0$ to see that $a = 5$, $b = -6$, $c = -1$.

NOTE
a, the coefficient of x^2, must be positive. So if the coefficient of x^2, as given, is negative, you can multiply the entire equation (all terms, both sides) by -1 in order to make it positive. For example,

$$-x^2 + 4x - 2 = 0 \xrightarrow{-1} x^2 - 4x + 2 = 0$$

Now a is positive. In fact,

$$a = 1$$
$$b = -4$$
$$c = 2$$

EXAMPLE 7   Use the quadratic formula to solve the equation $x^2 + 7x + 4 = 0$.

Solution
$$a = 1, b = 7, c = 4$$

$$x = \frac{-b \pm \sqrt{b^2 - 4ac}}{2a}$$

$$= \frac{-7 \pm \sqrt{(7)^2 - 4(1)(4)}}{2(1)}$$

$$= \frac{-7 \pm \sqrt{49 - 16}}{2}$$

$$= \frac{-7 \pm \sqrt{33}}{2} \quad \checkmark$$

NOTE
You *cannot* split $\sqrt{49 - 16}$ into $\sqrt{49} - \sqrt{16}$. After all, $\sqrt{49 - 16}$ is $\sqrt{33}$, and $\sqrt{49} - \sqrt{16}$ is $7 - 4$, or 3. Surely $\sqrt{33} \neq 3$. Similarly, an expression such as $\sqrt{9 + 16}$ *cannot* be split into $\sqrt{9} + \sqrt{16}$. Observe that $\sqrt{9 + 16} = \sqrt{25} = 5$, yet $\sqrt{9} + \sqrt{16} = 3 + 4 = 7$. And $5 \neq 7$.

EXAMPLE 8   Use the quadratic formula to solve the equation $2x^2 + 4x - 3 = 0$.

Solution
$$a = 2, b = 4, c = -3$$

$$x = \frac{-b \pm \sqrt{b^2 - 4ac}}{2a}$$

$$= \frac{-4 \pm \sqrt{(4)^2 - 4(2)(-3)}}{2(2)}$$

$$= \frac{-4 \pm \sqrt{16 + 24}}{4}$$

$$= \frac{-4 \pm \sqrt{40}}{4}$$

$$= \frac{-4 \pm 2\sqrt{10}}{4} \qquad \text{since } \sqrt{40} = \sqrt{4 \cdot 10} = \sqrt{4}\sqrt{10} = 2\sqrt{10}$$

$$= \frac{\mathbf{2}(-2 \pm \sqrt{10})}{\mathbf{2} \cdot 2} \qquad \text{after factoring out } \mathbf{2}\text{'s}$$

$$= \frac{-2 \pm \sqrt{10}}{2} \quad \checkmark \qquad \text{since the } \mathbf{2}\text{'s divide out}$$

EXAMPLE 9   Solve the equation $x^2 = 5 + 2x$ by using the quadratic formula.

Solution   Change $x^2 = 5 + 2x$ to $ax^2 + bx + c = 0$ form in order to select a, b, and c for use in the quadratic formula. First, add $-2x$ to each side.

$$
\begin{array}{rl}
x^2 & = 5 + 2x \\
-2x & \quad -2x \\
\hline
x^2 - 2x & = 5
\end{array}
$$

Next, add -5 to each side.

$$
\begin{array}{rl}
x^2 - 2x = & 5 \\
-5 & -5 \\
\hline
x^2 - 2x - 5 = & 0
\end{array}
$$

From $x^2 - 2x - 5 = 0$ we can see that $a = 1$, $b = -2$, and $c = -5$. These values can be substituted in the quadratic formula to find the roots of the equation.

$$x = \frac{-b \pm \sqrt{b^2 - 4ac}}{2a}$$

$$= \frac{-(-2) \pm \sqrt{(-2)^2 - 4(1)(-5)}}{2(1)}$$

$$= \frac{2 \pm \sqrt{4 + 20}}{2}$$

$$= \frac{2 \pm \sqrt{24}}{2}$$

$$= \frac{2 \pm 2\sqrt{6}}{2} \qquad \text{since } \sqrt{24} = \sqrt{4 \cdot 6} = 2\sqrt{6}$$

$$= \frac{2(1 \pm \sqrt{6})}{2}$$

$$x = 1 \pm \sqrt{6} \quad \checkmark \qquad \text{since } \frac{2}{2} = 1$$

EXAMPLE 10   Solve the equation $x^2 + 7x + 12 = 0$ by using the quadratic formula.

Solution   This equation can be solved by factoring and actually should be solved that way. Suppose, however, that you are careless at some time and use the quadratic formula because you do not realize that the quadratic expression could be factored. Let's see what sort of result you can expect. Keep in mind, of course, that *you should, in fact, use the formula only when the quadratic expression cannot be factored by ordinary means.*

$$a = 1, b = 7, c = 12$$

$$x = \frac{-b \pm \sqrt{b^2 - 4ac}}{2a} = \frac{-7 \pm \sqrt{(7)^2 - 4(1)(12)}}{2(1)}$$

$$= \frac{-7 \pm \sqrt{49 - 48}}{2}$$

$$= \frac{-7 \pm \sqrt{1}}{2}$$

$$= \frac{-7 \pm 1}{2}$$

Since no radical is left at this point, we consider both cases (+ and −) and continue the computations.

$$x = \frac{-7 + 1}{2} \quad \text{and} \quad \frac{-7 - 1}{2}$$

$$= \frac{-6}{2} \quad \text{and} \quad \frac{-8}{2}$$

$$= -3 \quad \text{and} \quad -4 \quad \checkmark$$

So the solutions are $x = -3$ and $x = -4$.

EXERCISES 22.3

*1. Solve each equation by using the quadratic formula.
    (a) $x^2 + 3x + 1 = 0$ (b) $x^2 + 5x + 2 = 0$
    (c) $x^2 - 5x - 3 = 0$ (d) $y^2 - 3y + 1 = 0$
    (e) $3t^2 + 5t - 6 = 0$ (f) $2x^2 + 3x - 3 = 0$
    (g) $x^2 - 7x + 2 = 0$ (h) $x^2 - 11x + 2 = 0$
    (i) $2x^2 - 9x - 1 = 0$ (j) $3y^2 - 7y - 4 = 0$

2. Solve each equation by using the quadratic formula.
    *(a) $x^2 - 4x + 1 = 0$ (b) $x^2 + 4x - 1 = 0$
    *(c) $x^2 + 6x + 1 = 0$ (d) $x^2 - 4x + 1 = 0$
    *(e) $x^2 - 2x - 1 = 0$ (f) $m^2 - 8m + 2 = 0$
    *(g) $x^2 + 8x - 4 = 0$ (h) $x^2 + 8x + 4 = 0$
    *(i) $x^2 - 10x + 3 = 0$ (j) $x^2 - 6x - 2 = 0$
    *(k) $3x^2 - 2x - 7 = 0$ (l) $3x^2 + 8x + 3 = 0$
    *(m) $4x^2 - 8x + 1 = 0$ (n) $4y^2 - 6y - 3 = 0$
    *(o) $2x^2 + 6x - 1 = 0$ (p) $2x^2 - 10x + 1 = 0$

3. Solve each equation by using the quadratic formula.
    *(a) $x^2 = 5 - 7x$ (b) $x^2 = 2 - 9x$
    *(c) $x^2 - 8 = 3x$ (d) $m = m^2 - 10$
    *(e) $-t^2 - 9t + 2 = 0$ (f) $-2x^2 + 9x + 3 = 0$
    *(g) $x = 3x^2 - 3$ (h) $x = 2x^2 - 5$
    *(i) $10 = y^2 + y$ (j) $7 = t^2 + t$

4. Solve each quadratic equation by using the quadratic formula, even though they can be solved by factoring or "root taking."
    *(a) $5y^2 - 7y - 6 = 0$ (b) $2x^2 + 5x + 2 = 0$

*(c) $x^2 + 6x = 0$ (d) $x^2 = 3x$
*(e) $x^2 - 3 = 0$ (f) $x^2 - 10 = 0$
*(g) $x = 2x^2 - 1$ (h) $x^2 = 12 - x$
*(i) $x = 3x^2 - 2$ (j) $3x = 2x^2 - 5$

22.4 DERIVATION OF THE QUADRATIC FORMULA (OPTIONAL)

In this section we begin with the general form of the quadratic equation, $ax^2 + bx + c = 0$, and find the general solution for x—namely,

$$x = \frac{-b \pm \sqrt{b^2 - 4ac}}{2a}$$

First, begin with

$$ax^2 + bx + c = 0 \qquad a \neq 0$$

and divide both sides by a. (This is why a cannot be zero.)

$$\frac{ax^2}{a} + \frac{bx}{a} + \frac{c}{a} = \frac{0}{a}$$

The result is

$$x^2 + \frac{b}{a}x + \frac{c}{a} = 0$$

Next, add $-\dfrac{c}{a}$ to both sides.

$$x^2 + \frac{b}{a}x = -\frac{c}{a}$$

The left-hand side is almost the square $\left(x + \dfrac{b}{2a}\right)^2$. But $\left(x + \dfrac{b}{2a}\right)^2 =$ $x^2 + \dfrac{b}{a}x + \dfrac{b^2}{4a^2}$; so add $\dfrac{b^2}{4a^2}$ to the left side (and to the right side), in order to complete the square.

$$x^2 + \frac{b}{a}x + \frac{b^2}{4a^2} = \frac{b^2}{4a^2} - \frac{c}{a}$$

or

$$\left(x + \frac{b}{2a}\right)^2 = \frac{b^2}{4a^2} - \frac{c}{a}$$

Now let's combine the fractions that are on the right. Their common denominator is $4a^2$. Thus

$$\frac{b^2}{4a^2} - \frac{c}{a}$$

can be written

$$\frac{b^2}{4a^2} - \frac{c}{a} \cdot \frac{4a}{4a}$$

or

$$\frac{b^2}{4a^2} - \frac{4ac}{4a^2}$$

or

$$\frac{b^2 - 4ac}{4a^2}$$

So far we have

$$\left(x + \frac{b}{2a}\right)^2 = \frac{b^2 - 4ac}{4a^2}$$

It follows that

$$x + \frac{b}{2a} = \pm \sqrt{\frac{b^2 - 4ac}{4a^2}}$$

This equation can easily be solved for x by adding $-\dfrac{b}{2a}$ to both sides.

$$x = -\frac{b}{2a} \pm \sqrt{\frac{b^2 - 4ac}{4a^2}}$$

The radical can be simplified.

$$\sqrt{\frac{b^2 - 4ac}{4a^2}} = \frac{\sqrt{b^2 - 4ac}}{\sqrt{4a^2}} = \frac{\sqrt{b^2 - 4ac}}{2a}$$

Now x can be written

$$x = -\frac{b}{2a} \pm \frac{\sqrt{b^2 - 4ac}}{2a}$$

Finally, we can combine the two fractions to obtain

$$x = \frac{-b \pm \sqrt{b^2 - 4ac}}{2a}$$

And this is the quadratic formula, the formula used to solve equations of the form $ax^2 + bx + c = 0$.

22.5 IMAGINARY NUMBERS (OPTIONAL)

All the numbers used so far in this book, and probably all the numbers you have ever used, are classified as **real numbers**. Real numbers possess an important property: The square of any real number is either positive or zero. If a real number is squared, the result is never a negative number. Are there any numbers whose squares are negative numbers? Solving the quadratic equation $x^2 + 1 = 0$ leads to such a number, for here x must be a number whose square is -1, since x^2 and 1 must add up to zero. The number whose square is -1 is an example of a so-called **imaginary number**. If we solve $x^2 + 1 = 0$ mechanically, we first get $x^2 = -1$ and then $x = \pm \sqrt{-1}$. The number $\sqrt{-1}$ is called i, for imaginary. The roots of the equation $x^2 + 1 = 0$ are $x = \pm i$.

The number $\sqrt{-1}$ is written as i. In general, $\sqrt{-a}$ for positive numbers a is written $i\sqrt{a}$. Consider some other examples of imaginary numbers: $\sqrt{-4}$, $\sqrt{-5}$, $\sqrt{-75}$, and observe their simplification.

$$\sqrt{-4} = i\sqrt{4}$$
$$= i \cdot 2$$
$$= 2i \quad \checkmark$$

$$\sqrt{-5} = i\sqrt{5} \quad \checkmark$$

$$\sqrt{-75} = i\sqrt{75}$$
$$= i\sqrt{25}\sqrt{3}$$
$$= i \cdot 5\sqrt{3}$$
$$= 5i\sqrt{3} \quad \checkmark$$

EXAMPLE 11   Solve the equation $x^2 + 2x + 5 = 0$.

Solution   Using the quadratic formula with $a = 1$, $b = 2$, and $c = 5$, we obtain

$$x = \frac{-b \pm \sqrt{b^2 - 4ac}}{2a}$$

$$= \frac{-2 \pm \sqrt{(2)^2 - 4(1)(5)}}{2(1)}$$

$$= \frac{-2 \pm \sqrt{4 - 20}}{2}$$

$$= \frac{-2 \pm \sqrt{-16}}{2}$$

$$= \frac{-2 \pm 4i}{2} \qquad \text{since } \sqrt{-16} = i\sqrt{16} = i \cdot 4 = 4i$$

$$= \frac{2(-1 \pm 2i)}{2}$$

$$= -1 \pm 2i \quad \checkmark$$

EXAMPLE 12   Solve the equation $x^2 - 4x + 7 = 0$.

Solution   Using the quadratic formula with $a = 1$, $b = -4$, and $c = 7$, we have

$$x = \frac{-b \pm \sqrt{b^2 - 4ac}}{2a}$$

$$= \frac{-(-4) \pm \sqrt{(-4)^2 - 4(1)(7)}}{2(1)}$$

$$= \frac{4 \pm \sqrt{16 - 28}}{2}$$

$$= \frac{4 \pm \sqrt{-12}}{2}$$

$$= \frac{4 \pm 2i\sqrt{3}}{2} \qquad \begin{cases} \text{since } \sqrt{-12} = i\sqrt{12} \\ = i\sqrt{4}\sqrt{3} = 2i\sqrt{3} \end{cases}$$

$$= \frac{2(2 \pm i\sqrt{3})}{2}$$

$$= 2 \pm i\sqrt{3} \quad \checkmark$$

EXERCISES 22.5

1. Express each number in terms of i. For example, $\sqrt{-9} = 3i$.
   *(a) $\sqrt{-16}$ (b) $\sqrt{-25}$ *(c) $\sqrt{-3}$ (d) $\sqrt{-7}$
   *(e) $\sqrt{-24}$ (f) $\sqrt{-20}$ *(g) $\sqrt{-27}$ (h) $\sqrt{-45}$
   *(i) $\sqrt{-80}$ (j) $\sqrt{-63}$

2. Solve each quadratic equation.
   *(a) $x^2 + 4 = 0$ (b) $x^2 + 9 = 0$
   *(c) $x^2 - 2x + 5 = 0$ *(d) $x^2 + x + 1 = 0$
   *(e) $m^2 + m + 2 = 0$ (f) $x^2 + 4x + 7 = 0$
   *(g) $2x^2 - 3x + 4 = 0$ (h) $2x^2 + 6x + 8 = 0$
   *(i) $x^2 = 2x - 9$ (j) $-x^2 + 2x - 10 = 0$

22.6 NATURE OF THE ROOTS (OPTIONAL)

We can determine the types of roots (that is, the nature of the roots) of a quadratic equation by examining $b^2 - 4ac$. The expression $b^2 - 4ac$ is called the **discriminant**.

$$x = \frac{-b \pm \sqrt{\boxed{b^2 - 4ac}}}{2a} \qquad \text{discriminant}$$

Case I. Discriminant $= 0$

If $b^2 - 4ac = 0$, then $\sqrt{} = 0$ and the roots are $x = \frac{-b}{2a}$; that is, both roots are the same. The roots are therefore *real and equal.* As an example, in $x^2 + 6x + 9 = 0$, $a = 1$, $b = 6$, and $c = 9$. So $b^2 - 4ac$ is $6^2 - 4 \cdot 1 \cdot 9 = 36 - 36 = 0$. The roots of $x^2 + 6x + 9 = 0$ are real and equal. Note

$$x = \frac{-b \pm \sqrt{\text{discriminant}}}{2a}$$

$$= \frac{-6 \pm \sqrt{0}}{2(1)} = \frac{-6 \pm 0}{2} = -\frac{6}{2} = -3 \quad \checkmark$$

Case II. Discriminant > 0

If $b^2 - 4ac > 0$, then the number under the $\sqrt{}$ is positive. This means that the roots are real. Because of the $\pm$, the roots are different. Thus the roots are *real and unequal*. To illustrate, the roots of $x^2 + 7x + 12 = 0$ are real and unequal, since $b^2 - 4ac$ is $7^2 - 4 \cdot 1 \cdot 12 = 1$.

$$x = \frac{-b \pm \sqrt{\text{discriminant}}}{2a}$$

$$= \frac{-7 \pm \sqrt{1}}{2(1)} = \frac{-7 \pm 1}{2} = -3 \text{ and } -4 \quad \checkmark$$

Case III. Discriminant < 0

If $b^2 - 4ac < 0$, then the number under the $\sqrt{}$ is negative. This means that the roots are imaginary. Because of the $\pm$, the roots are different. Thus the roots are *imaginary and unequal.* As an example, the roots of $x^2 + 3x + 7 = 0$ are imaginary and unequal, because $b^2 - 4ac = 3^2 - 4 \cdot 1 \cdot 7 = -19$.

$$x = \frac{-b \pm \sqrt{\text{discriminant}}}{2a} = \frac{-3 \pm \sqrt{-19}}{2(1)} = \frac{-3 \pm i\sqrt{19}}{2} \quad \checkmark$$

EXERCISES 22.6

1. Check the discriminant to determine the nature of the roots of each quadratic equation below. Do not actually solve the equations.
   *(a) $x^2 + 5x + 2 = 0$ (b) $m^2 - 3m + 1 = 0$
   *(c) $y^2 + 2y + 6 = 0$ (d) $x^2 - 3x + 2 = 0$
   *(e) $x^2 - 6x + 9 = 0$ (f) $x^2 - 6x - 9 = 0$
   *(g) $x^2 - 5 = 0$ *(h) $n^2 + 5 = 0$
   *(i) $x^2 + 2x = 0$ (j) $3x^2 + 7x + 1 = 0$
   (k) $5x^2 - 10x + 6 = 0$ *(l) $x^2 = 3x + 10$
   *(m) $x^2 - 1 = x$ *(n) $-x^2 + 7x + 2 = 0$
   (o) $-x^2 + 6x - 9 = 0$ (p) $x^2 = 12x$

22.7 PARABOLAS

In Chapter 14 we saw that graphs of straight lines are associated with linear equations. Here we'll see an example of a graph that is associated with a quadratic equation.

EXAMPLE 13   Sketch the graph of $y = x^2 + 3x + 2$.

Solution   Obtain points by substituting different values for x into $x^2 + 3x + 2$ in order to get corresponding values for y. Since it is not a straight line, several points should be obtained to see how the curve is supposed to look. The work below shows the substitution of 0, 1, 2, −1, −2, −3, −4, and −5 for x to obtain eight (x, y) points.

$$x = \underline{0}: \qquad y = \underline{0}^2 + 3 \cdot \underline{0} + 2 = 2 \qquad \longrightarrow (0, 2)$$
$$x = \underline{1}: \qquad y = \underline{1}^2 + 3 \cdot \underline{1} + 2 = 6 \qquad \longrightarrow (1, 6)$$
$$x = \underline{2}: \qquad y = \underline{2}^2 + 3 \cdot \underline{2} + 2 = 12 \qquad \longrightarrow (2, 12)$$
$$x = \underline{-1}: \qquad y = (\underline{-1})^2 + 3 \cdot (\underline{-1}) + 2 = 0 \longrightarrow (-1, 0)$$
$$x = \underline{-2}: \qquad y = (\underline{-2})^2 + 3(\underline{-2}) + 2 = 0 \longrightarrow (-2, 0)$$
$$x = \underline{-3}: \qquad y = (\underline{-3})^2 + 3(\underline{-3}) + 2 = 2 \longrightarrow (-3, 2)$$
$$x = \underline{-4}: \qquad y = (\underline{-4})^2 + 3(\underline{-4}) + 2 = 6 \longrightarrow (-4, 6)$$
$$x = \underline{-5}: \qquad y = (\underline{-5})^2 + 3(\underline{-5}) + 2 = 12 \longrightarrow (-5, 12)$$

The graph (shown next) is obtained by passing a smooth curve through the points.

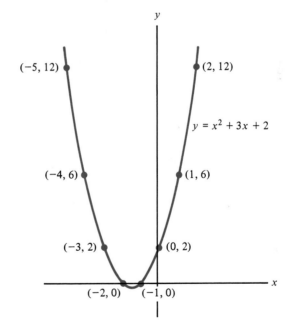

The curve is an example of a **parabola.** Equations of the form

$$y = ax^2 + bx + c$$

are parabolas when graphed.

The parabola $y = x^2 + 3x + 2$ crosses the x axis at $x = -2$ and at $x = -1$. The y value on the x axis is 0; so when $x = -2$ or $x = -1$, then $y = 0$. In other words, the roots of $x^2 + 3x + 2 = 0$ (that is, when y is 0) are −2 and −1 *by graphing*. The roots can also be determined by factoring, as shown next.

$$x^2 + 3x + 2 = 0$$
$$(x + 2)(x + 1) = 0$$

$x + 2 = 0$	$x + 1 = 0$
$x = -2$	$x = -1$

If the coefficient of x^2 is negative, then the graph will appear "upside down," as shown in the next example.

EXAMPLE 14 Sketch the graph of $y = -x^2 - 2x + 3$.

Solution We shall use the following values for x: $-4, -3, -2, -1, 0, 1, 2, 3$. Note that in evaluating $-x^2 - 2x + 3$, the $-x^2$ means $-(x^2)$, *not* $(-x)^2$.

x	$y = -x^2 - 2x + 3$	(x, y)
-4	$y = -(16) - 2(-4) + 3 = -5$	$(-4, -5)$
-3	$y = -(9) - 2(-3) + 3 = 0$	$(-3, 0)$
-2	$y = -(4) - 2(-2) + 3 = 3$	$(-2, 3)$
-1	$y = -(1) - 2(-1) + 3 = 4$	$(-1, 4)$
0	$y = -(0) - 2(0) + 3 = 3$	$(0, 3)$
1	$y = -(1) - 2(1) + 3 = 0$	$(1, 0)$
2	$y = -(4) - 2(2) + 3 = -5$	$(2, -5)$
3	$y = -(9) - 2(3) + 3 = -12$	$(3, -12)$

The graph is shown below. Note that the roots of $-x^2 - 2x + 3 = 0$ are -3 and 1.

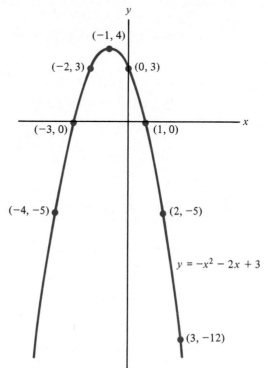

EXERCISES 22.7

*1. Sketch the graph of each parabola.
   (a) $y = x^2$ (b) $y = x^2 - 1$
   (c) $y = -x^2$ (d) $y = x^2 - 4x + 3$
   (e) $y = 2x^2$ (f) $y = -x^2 + 3x + 2$

2. Sketch the graph of each parabola.
   *(a) $y = x^2 - 5x + 6$ (b) $y = x^2 + 4x + 3$
   *(c) $y = x^2 + 3x - 4$ (d) $y = x^2 + 1$
   *(e) $y = 1 - x^2$ (f) $y = x^2 - 9$
   *(g) $y = 3x^2 + 1$ (h) $y = 3x^2$
   *(i) $y = x^2 + 3x$ (j) $y = x^2 - 2x$
   *(k) $y = -x^2 + 2x - 3$ (l) $y = -x^2 + 1$

CHAPTER 22 REVIEW EXERCISES

*1. Solve each equation. Do not use the quadratic formula *unless* the expression will not factor.
   (a) $3x^2 - 2x - 1 = 0$ (b) $3x^2 - 6x + 2 = 0$
   (c) $2t^2 + 9t + 1 = 0$ (d) $4y^2 + 16y + 7 = 0$

2. Solve by completing the square.
   *(a) $m^2 + 10m + 1 = 0$ (b) $x^2 + 6x + 2 = 0$

*3. Determine the nature of the roots of each quadratic equation.
   (a) $x^2 - 2x + 4 = 0$ (b) $2x^2 + x - 3 = 0$
   (c) $x^2 - 8x + 16 = 0$

*4. Sketch the graph of the parabola $y = x^2 - 3x - 4$.

CUMULATIVE REVIEW EXERCISES (CHAPTERS 10–22)

*1. Simplify each expression.

   (a) $(0)(-6)(+1) - (-3)(-4)^2$ (b) $\dfrac{-8}{+4} - (-3)(-6)(-1) - 7^2$

   (c) $3x - (x - 7) + 2x - 18$ (d) $5 + 8(1 - 4x) + 3x$

*2. Perform the division.
$$x - 1\overline{)x^3 - 3x^2 + 3x - 1}$$

*3. Solve the equation.
$$5(2 + x) - 3 = 17x$$

*4. Solve each word problem.
   (a) 854 meters of fencing will be used to enclose a rectangular field. The length of the field is 1 meter longer than twice the width. Determine the length and width of the field to be enclosed.
   (b) Marcy is 7 years older than Janice. Thirty-nine years ago she was twice as old as Janice was then. How old is each person today?

*5. Determine the equation of the line that passes through $(0, 3)$ and $(1, 5)$.

*6. Solve each system of equations.

(a) $\begin{cases} x + 6y = 4 \\ 3x + 9y = 21 \end{cases}$ (b) $\begin{cases} x = 3y - 10 \\ 5x + 11y = 2 \end{cases}$

*7. Solve the inequality.

$$8(2x + 4) - 3 \geq 0$$

*8. Perform the indicated operation. Reduce if possible.

(a) $\dfrac{12xy}{7y} \cdot \dfrac{5}{3x}$ (b) $\dfrac{5x}{6x^2} \div \dfrac{x^2 - 12x + 36}{2x - 12}$

(c) $\dfrac{5}{6xy} + \dfrac{3y}{8x}$ (d) $\dfrac{5}{x - 3} - \dfrac{3}{x + 2}$

*9. Solve each equation.

(a) $\dfrac{4}{x} + 9 = \dfrac{x + 1}{x}$ (b) $x^2 + 7x + 1 = 0$

*10. Simplify each expression. Write with positive exponents only.

(a) $(2x^5y^3)^6$ (b) $(2m^8)(-3m^6)^3$ (c) $-5x^{-19}y^6$

(d) $\sqrt{225x^4y^{100}z^{14}}$

*11. Evaluate each expression.

(a) $2^{-1} + 2^0$ (b) $(7 \cdot 6)^0$ (c) $81^{3/4}$

(d) $32^{3/5}$

*12. Solve each formula for the letter indicated.

(a) $y = mx + b$ for m (b) $5x + 9y + 7 = 0$ for y

(c) $x = \dfrac{ab}{c + d}$ for a (d) $x = \dfrac{ab}{c + d}$ for d

(e) $a = b + c(d - e)$ for d

*13. Simplify. Leave the denominator rational.

$$\sqrt{\dfrac{13}{25}}$$

*14. Solve $\sqrt{2 - 2x} + 5 = 13$.

*15. Solve each equation.

(a) $2x^2 - 17x + 21 = 0$ (b) $x^2 + 4x + 5 = 0$

(c) $x^2 = 7x - 6$ (d) $2x^2 - 9x - 3 = 0$

*16. Solve $x^2 - 6x - 1 = 0$ by completing the square.

23

SETS AND FUNCTIONS

23.1 INTRODUCTION

Sets were introduced about a hundred years ago by the German mathematician Georg Cantor. The original theory has been expanded over the years and now serves as a basic tool in many areas of advanced mathematics.

To begin, a **set** is a collection of things. In mathematics, the things are often numbers. However, a set can contain anything. Here are some examples of sets.

- The set of all students in the classroom
- The set of all people who have walked on the moon
- The set of integers between one and ten inclusive
- The set of manufacturers of automobiles
- The set of geometric figures of eight sides or less
- The set of all books written by Edgar Allan Poe

The things that make up the set are called **elements** or *members* of the set. The elements of the set of integers between one and ten inclusive are 1, 2, 3, 4, 5, 6, 7, 8, 9, 10.

This last set is much easier to discuss and work with if written as {1, 2, 3, 4, 5, 6, 7, 8, 9, 10}. *Braces* { } are used to indicate a set. The elements are placed inside the braces and separated by commas. It does not matter in which order the elements are written. This means that the sets {1, 2, 3, 4, 5} and {3, 2, 1, 5, 4} are the same.

The symbol $\in$ is used to mean "is an element of" or "is in." For instance, $5 \in \{2, 3, 5, 9, 17\}$ is read as 5 *is in* the set containing 2, 3, 5, 9, 17 or as 5 *is an element of* that set.

Indicate whether the following are true or false.

$$3 \in \{1, 2, 3, 4, 5\}$$
$$3 \in \{1, 3\}$$
$$7 \in \{1, 2, 3, 6\}$$
$$x \in \{a, b, c, x, y, z\}$$
$$\$ \in \{*, /, +, -, \&\}$$

The answers are: true, true, false, true, false.

The statement that 7 *is not an element of* $\{1, 2, 3, 6\}$ can be written $7 \notin \{1, 2, 3, 6\}$. The symbol $\notin$ means *is not an element of* or *is not in*.

Indicate whether the following are true or false.

$$5 \in \{1, 2, 3, 4, 6, 7, 8\}$$
$$5 \notin \{1, 2, 3, 4, 6, 7, 8\}$$
$$3 \notin \{1, 5, 9, 17\}$$
$$a \notin \{a, b, c, d, q\}$$
$$x \notin \{u, v, w, x, y, z\}$$
$$e \in \{a, x, y, e, z\}$$

The answers are: false, true, true, false, false, true.

We sometimes use *three dots* (. . .) to indicate a "continuation in the same manner." This notation is used to avoid writing all the members of a set explicitly. Thus the set of whole numbers

$$\{1, 2, 3, 4, 5, 6, 7, 8, 9, 10, 11, 12, 13\}$$

can be written

$$\{1, 2, 3, 4, \ldots, 13\}$$

The set $\{14, 16, 18, 20, 22, 24, 26, 28, 30, 32, 34, 36\}$ of even whole numbers between 14 and 36 can be written $\{14, 16, 18, \ldots, 36\}$.

The set of all the whole numbers greater than 13 can be written as $\{14, 15, 16, 17, \ldots\}$.

EXERCISES 23.1

Answers to starred exercises are given in the back of the book.

1. Write each using set notation.
    *(a) The set of all even numbers from 2 through 24.
    (b) The set of all vowels in the alphabet.
    *(c) The set of integers from 7 through 15.
    (d) The set of all positive integers.
    *(e) The set of all months beginning with the letter M.

2. Write each using set notation.
    *(a) 5 is in the set of all odd numbers between 1 and 15 inclusive.
    *(b) 7 is not in the set of even numbers less than 20.

*(c) x is not in the set of vowels of the alphabet.

(d) June is in the set of months of the year.

(e) Sunday is not in the set of months of the year.

3. Indicate whether the following statements are true or false.

*(a) $x \in \{a, b, c, x, y, z\}$ *(b) $x \notin \{a, b, c, x, y, z\}$

*(c) $5 \in \{2, 4, 6, 8, 10\}$ *(d) $5 \in \{1, 3, 15, 17, 19\}$

*(e) $d \in \{a, e, q, r\}$ *(f) $* \notin \{\square, \triangle, *, \$\}$

*(g) $*** \in \{+, -, *, /, **\}$ (h) $-19 \in \{integers\}$

(i) $t \notin \{R, S, T, U, V\}$ (j) $341 \notin \{odd\ numbers\}$

23.2 SET-BUILDER NOTATION

The set $\{2, 4, 6, 8, 10, \ldots\}$ is the set of all positive, even integers. It can also be written as $\{x | x \text{ positive, even integer}\}$, which is read "the set of all numbers x such that x is a positive, even integer." The vertical line $|$ is read as *"such that."* This is particularly useful notation for some statements about sets. For example, how could we represent the set of all numbers that are greater than zero (including integers, decimals, fractions, etc.)? Using the new notation, we simply write this as

$$\{x | x > 0\}$$

the set of all numbers x such that x is greater than zero.† Set notation that uses the "such that" vertical line is called **set-builder notation.**

EXAMPLE 1 Denote the set of all numbers less than 17 except 0. Call the set Q.

Solution
$$Q = \{m | m < 17, m \neq 0\}$$

EXAMPLE 2 Denote the set of all odd numbers greater than or equal to 7. Call the set R.

Solution
$$R = \{x | x \text{ odd}, x \geq 7\}$$

EXAMPLE 3 Denote the set of all numbers that are between 3 and 5—that is, all that are both greater than 3 and less than 5. Call the set P.

Solution
$$P = \{n | 3 < n < 5\}$$

The notation $3 < n$ indicates n *greater than* 3. Also, $n < 5$ indicates n *less than* 5. They are combined as $3 < n < 5$ to mean all numbers n that are both greater than 3 *and* less than 5.

EXAMPLE 4 Explain what is in the set $\{t | 1 < t \leq 7\}$.

---

† The set $\{x | x > 0\}$ is an example of an **infinite set,** since it has an unlimited number of elements. Most sets that we discuss will be **finite sets;** they will have a limited (1 or 5 or 1000, etc.) number of elements.

Solution   The set consists of all numbers that are between 1 and 7, excluding 1 and including 7. In other words, it is the set of all numbers that are greater than 1 and less than or equal to 7.

EXERCISES 23.2

1. Use the notation of this section to denote the following sets. The set of:
   *(a) All numbers greater than 13.
   *(b) All numbers less than $\frac{1}{2}$.
   (c) All numbers greater than or equal to -27.
   *(d) All numbers less than or equal to 2.
   *(e) All numbers not less than 0.
   (f) All odd numbers less than 10.
   *(g) All numbers between (and including) -5 and 16.
   (h) All numbers between (but not including) 5 and 16.
   *(i) All even numbers between 3 and 95.
   (j) All even numbers between (and including) 2 and 500.
   *(k) All nonzero numbers between -25 and 25 inclusive.
   (l) All integers between (but not including) -100 and $+200$.

2. Explain what is in each set.
   *(a) $\{x \mid x > 0\}$ *(b) $\{t \mid t \neq 0\}$
   (c) $\{m \mid m < -9\}$ *(d) $\{x \mid x \text{ even}, x \geq 6\}$
   (e) $\{x \mid x \text{ odd}, x \neq 1\}$ *(f) $\{y \mid y \text{ integer}, y \geq 0\}$
   (g) $\{y \mid y \text{ integer}, y > 0\}$ (h) $\{y \mid y \text{ integer}, y < 0\}$
   *(i) $\{x \mid 3 < x < 9\}$ (j) $\{x \mid 3 \leq x \leq 9\}$
   *(k) $\{x \mid 3 \leq x < 9\}$ (l) $\{x \mid 3 < x \leq 9\}$

23.3 UNION AND INTERSECTION

The **union** of two sets is computed in several examples below. See if you can determine just what we mean by the union of two sets. The operation of union is denoted by the U-shaped symbol ∪.

$$\{1, 2, 3\} \cup \{4, 5\} = \{1, 2, 3, 4, 5\}$$
$$\{a, b\} \cup \{x, y\} = \{a, b, x, y\}$$
$$\{5, 6, 17\} \cup \{5, 9\} = \{5, 6, 9, 17\}$$
$$\{1, 2\} \cup \{1, 2\} = \{1, 2\}$$
$$\{2, 4, 6, 8\} \cup \{1, 2, 3\} = \{1, 2, 3, 4, 6, 8\}$$

The examples show that the union of two sets is formed by taking all the elements involved (from both sets) and placing them in one new set. The new set is called their union. More technically, *the union of two sets is a set that contains all elements that are in either of the two original sets.*

The **intersection** of two sets is computed in several examples below. Read them and determine what we mean by the intersection of two sets. The symbol ∩ is used to indicate intersection.

$${1, 2, 3, 4, 5} \cap {1, 2, 3} = {1, 2, 3}$$
$${b, g, x, z} \cap {b, c, q} = {b}$$
$${2, 4, 6, 8} \cap {1, 2, 3, 4} = {2, 4}$$
$${1, 2, 3} \cap {4, 5, 6} = {\ \ }$$

It should be clear from the examples that *the intersection of two sets is a set that contains the elements common to both sets.*

If you are wondering about the example

$${1, 2, 3} \cap {4, 5, 6} = {\ \ }$$

you should note that the two sets have no elements in common, so their intersection is a set with no elements. The set with no elements can be denoted $\{\ \}$ or $\emptyset$. It is called the **empty set** or *null set*. For example, ${a, b, c} \cap {5, 9} = \emptyset$. When two sets have no elements in common, they are called **disjoint** sets.

EXERCISES 23.3

1. Determine each union.
   *(a) ${5, 6, 19, 35} \cup {8, 12, 19}$
   (b) ${1, 5, 6, 9, 2} \cup {3, 4, 8, 7, 10}$
   *(c) ${4, 8, 7, 3} \cup {14, 3, 7, 12}$
   (d) ${a, b, c, d, e} \cup {e, c, b, a, d}$
   *(e) ${1, 2, 3, 4, 5, 6} \cup {5, 2, 4}$

2. Determine each intersection.
   *(a) ${6, 4, 19, 35} \cap {19, 12, 8}$
   (b) ${a, b, c, d, e} \cap {b, e, a, c, d}$
   *(c) ${3, 4, 8, 7, 10} \cap {1, 5, 6, 9, 2}$
   (d) ${7, 3, 12, 14} \cap {4, 8, 7, 3}$
   *(e) ${4, 3, 1} \cap {3, 5, 1, 2, 4}$

3. Determine each union or intersection.
   *(a) ${x, y, z} \cup {x, y, z}$
   (b) ${x, y, z} \cap {x, y, z}$
   *(c) ${a, b, c, d} \cup {a, b, c, d, e, f}$
   (d) ${a, b, c} \cap {d, e, f}$

23.4 SUBSETS

${1, 2}$ is a **subset** of ${1, 2, 19, 50}$. This means that every element of ${1, 2}$ is also in the set ${1, 2, 19, 50}$. Similarly,

${a, b, c}$ is a subset of ${a, b, c, d, e, f}$.
${1, 4}$ is a subset of ${1, 2, 3, 4, \ldots, 100}$
${50}$ is a subset of ${2, 4, 6, \ldots, 200}$.
${3, 4, 5}$ is a subset of ${3, 4, 5}$.

Any set is a subset of itself.

{ } is a subset of {1, 2, 3}. ← The empty set is a subset of every set.

The words "is a subset of" are often replaced by the symbol ⊆, which means the same thing. For instance,

$$\{3, 5\} \subseteq \{1, 3, 5, 7, 9, 11\}$$

For convenience, we often name sets by using capital letters. If we name the set {1, 2, 3} A and call B the set {1, 2, 3, 4, 5, 6}, then

$$A = \{1, 2, 3\}$$
$$B = \{1, 2, 3, 4, 5, 6\}$$

and

$$A \subseteq B, \quad A \text{ is a subset of } B.$$

EXAMPLE 5 Let $A = \{2, 5\}$, $B = \{1, 2, 3, 4, 5\}$, $C = \{1, 3, 6\}$.

Solution Then

$A \subseteq B$ (A is a subset of B, since both its elements, 2 and 5, are in B.)

and

$C \not\subseteq B$ (C is *not* a subset of B, since 6 is in C but not in B.)

Also,
$$A \cap B = \{2, 5\}$$
$$A \cap C = \varnothing$$
$$A \cup C = \{1, 2, 3, 5, 6\}$$
$$A \cup B = \{1, 2, 3, 4, 5\}$$
$$B \cap C = \{1, 3\}$$
$$C \cup B = \{1, 2, 3, 4, 5, 6\}$$

EXERCISES 23.4

1. If $A = \{7, 9, 3, 8, 2\}$, $B = \{3, 8, 10\}$, $C = \{2, 7\}$, which of the following are true statements?

*(a) $A \subseteq B$ *(b) $A \subseteq C$ (c) $B \subseteq C$
*(d) $B \not\subseteq A$ (e) $C \subseteq A$ *(f) $C \cap A \subseteq A$
(g) $C \not\subseteq B$ (h) $A \cap B \subseteq C$ *(i) $B \cap C \subseteq A$
*(j) $A \cup B \not\subseteq C$

2. List all the subsets of each set.

*(a) $\{a,b\}$ *(b) $\{x, y, z\}$ (c) $\{p, q, r, s\}$

23.5 COMPLEMENTS

The elements that we select for sets are taken from another set—a **universe**. The universe can change from example to example and from discussion to discussion, and it is determined by the nature of the problem. In each example below a universe is selected, and in each case it is denoted by $\mathcal{U}$.

EXAMPLE 6   Let $\mathcal{U} = \{1, 2, 3, 4, 5, 6, 7, 8, 9, 10, 11, 12\}$.

Solution   We will use sets containing numbers selected from 1, 2, 3, 4, 5, 6, 7, 8, 9, 10, 11, 12; for instance, $X = \{1, 2, 3\}$, $Y = \{2, 8, 11, 12\}$, and $Z = \{1, 2, 3, 4, 5, 6, 7, 8, 9, 10, 11, 12\}$. Sets discussed must be subsets of the universe.

$$X \subseteq \mathcal{U}, \qquad Y \subseteq \mathcal{U}, \qquad Z \subseteq \mathcal{U}$$

It was necessary to introduce the concept of "universe" in order to discuss complements. The **complement** of a set X will be denoted by $\bar{X}$ (read as X-bar). *The complement of a set X is the set $\bar{X}$ containing all elements of the universe that are not in X.*

EXAMPLE 7   Let $\mathcal{U} = \{a, b, c, d, e, f, g\}$.

Solution   If $X = \{a, b, c, d\}$, then $\bar{X} = \{e, f, g\}$.
If $Y = \{b, f\}$, then $\bar{Y} = \{a, c, d, e, g\}$.
The sets X and $\bar{X}$ have no elements in common; that is, $X \cap \bar{X} = \varnothing$. Also, X and $\bar{X}$ together contain all elements in the universe; that is, $X \cup \bar{X} = \mathcal{U}$. The same is true for Y and $\bar{Y}$ as well as for any other set and its complement.

EXERCISES 23.5

*1. If $\mathcal{U} = \{1, 3, 5, 7, 9, 11, 13, 15, 17\}$, find the complement of each set below.
   (a) $A = \{1, 3, 5\}$ (b) $B = \{7, 15\}$
   (c) $C = \{\ \}$ (d) $D = \{1, 3, 5, 7, 9, 11, 13, 15, 17\}$
   (e) $E = \{13\}$

2. If $\mathcal{U} = \{m, a, t, h, i, s, o, k\}$, find the complement of each set below.
   (a) $F = \{m, a, t, h\}$ (b) $G = \{i, s\}$
   (c) $H = \{o, k\}$ (d) $I = \{\ \}$
   (e) $J = \{m, a, t, h, i, s, o, k\}$ (f) $K = \{m, i, o\}$
   (g) $L = \{h, o, t\}$

23.6 VENN DIAGRAMS

Venn diagrams are useful for visualizing relationships between sets. The diagrams are named for John Venn, who used them in 1881 to illustrate logic. They were first used by Leonhard Euler and hence are also called *Euler's circles*. In Venn diagrams, a rectangle represents the universe.

Circles (and their interiors) represent sets.

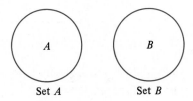

Set *A* Set *B*

And since the sets are made up of elements from the universe, the circles are placed within the rectangular universe.

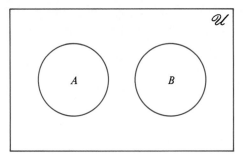

The circles in the rectangle shown are separated, indicating that the sets have no elements in common. If two sets have elements in common, then the circles that represent them overlap, as shown below.

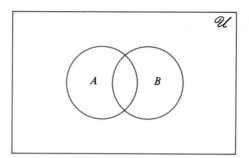

EXAMPLE 8   Represent $A \cap B$ in the case that A and B overlap. Shade the intersection.

Solution

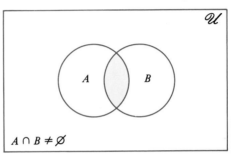

The shaded area represents the intersection of A and B and gives a visual picture of $A \cap B$. Note that $A \cap B \neq \varnothing$ means that the intersection is *not* the empty set; that is, the sets have at least one element in common.

EXAMPLE 9   Represent $A \cap B$ for the case $A \cap B = \varnothing$ (that is, the sets have no elements in common).

Solution

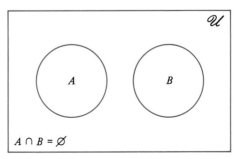

There is nothing to shade, because there are no elements in the intersection.

EXAMPLE 10   Represent $A \cup B$ in the case where A and B have elements in common (that is, when $A \cap B \neq \varnothing$).

Solution

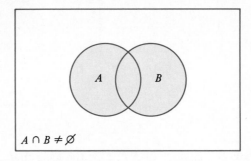

The shaded area is $A \cup B$.

EXAMPLE 11   Represent $A \cup B$ in the case $A \cap B = \varnothing$.

Solution

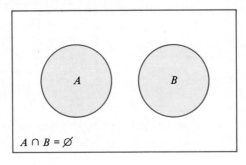

The shaded areas represent $A \cup B$.

EXAMPLE 12   Represent $\overline{A}$ for some set A in the universe $\mathcal{U}$.

Solution

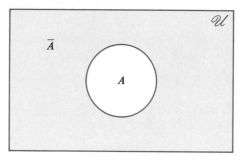

The shaded area is $\overline{A}$, since $\overline{A}$ is all of $\mathcal{U}$ that is not in A.

EXAMPLE 13 Represent $A \cap B$ for sets A and B such that B is a subset of A.

Solution

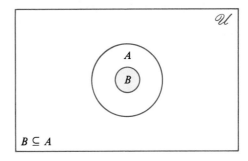

The shaded area is $A \cap B$.

EXAMPLE 14 Represent $A \cup B$ for A and B such that $B \subseteq A$.

Solution

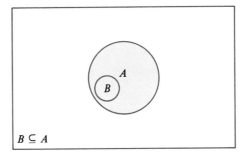

The shaded area is $A \cup B$.

EXAMPLE 15 Represent $A \cup B$ for $A = B$.

Solution

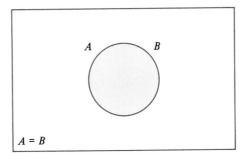

The shaded area is $A \cup B$.

EXAMPLE 16 Represent $A \cap B \cap C$ for three sets A, B, and C that overlap each other.

Solution

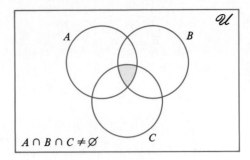

The shaded area is $A \cap B \cap C$.

EXERCISES 23.6

*1. Consider the Venn diagram shown below. The numbers 1, 2, 3, and 4 are elements. List the elements that are in the region requested.

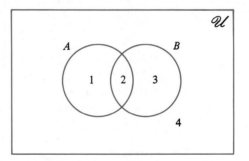

(a) A (b) B (c) $A \cup B$ (d) $A \cap B$ (e) $\mathcal{U}$

(f) $\bar{A}$ (g) $\bar{B}$ (h) $\overline{A \cup B}$ (i) $\overline{A \cap B}$

2. Consider the Venn diagram shown below. The numbers 1, 2, 3, 4, 5, and 6 are elements. List the elements that are in the region requested.

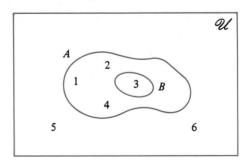

*(a) A *(b) B (c) $A \cup B$ (d) $A \cap B$

(e) $\mathcal{U}$ *(f) $\bar{A}$ *(g) $\bar{B}$ (h) $\overline{A \cup B}$

(i) $\overline{A \cap B}$

3. Consider the Venn diagram. List the elements that are in the region requested.

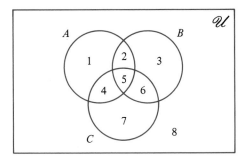

*(a) A (b) B (c) C
*(d) $\mathcal{U}$ *(e) $A \cap B$ (f) $B \cap C$
(g) $A \cap C$ *(h) $A \cap B \cap C$ *(i) $A \cup B$
(j) $B \cup C$ (k) $A \cup C$ *(l) $A \cup B \cup C$

4. Draw Venn diagrams for the indicated sets.
   *(a) $X \cap Y$ for two sets X and Y that do not overlap.
   (b) $X \cup Y$ for two sets X and Y that do not overlap.
   *(c) $A \cup B \cup C$ for three overlapping sets.
   (d) $A \cap B \cap C$ for three overlapping sets.
   *(e) $A \cap B$ for two identical sets A and B.
   (f) $A \cup B$ for two identical sets A and B.
   *(g) $A \cap B$ for $A \subseteq B$.
   (h) $A \cap B$ for $B \subseteq A$.
   (i) $A \cup B$ for $A \subseteq B$.
   *(j) $A \cup B$ for $B \subseteq A$.

5. Assuming that A and B are two overlapping sets, represent

   (a) $\overline{A \cap B}$ (b) $\bar{A} \cap \bar{B}$ (c) $\overline{A \cup B}$ (d) $\bar{A} \cup \bar{B}$

   Discuss any apparent relationships.

6. Use the Venn diagram of Exercise 3 to verify that the following statements are true.
   (a) $A \cap B = B \cap A$
   (b) $A \cup B = B \cup A$
   (c) $A \cup (B \cap C) = (A \cup B) \cap (A \cup C)$
   (d) $A \cap (B \cup C) = (A \cap B) \cup (A \cap C)$

23.7 RELATIONS AND FUNCTIONS (OPTIONAL)

The concept of set can be used to define special sets called relations and functions. Since relations and functions are important in mathematics, we now proceed to have a brief look at them.

Consider the equation $y = 2x + 1$. For every value of x that you supply, you obtain a corresponding y value. Informally, the equation $y = 2x + 1$ is an example of a *relation*, a correspondence involving two variables. Here are some other relations.

$$x - y = 6$$

$$y = x^2 + 2x + 5$$

$$x^2 + y^2 = 9$$

Each relation includes all x's and y's that satisfy the equation. This means that each relation includes all points that could be graphed by using the equation to determine points. Set notation can be used to formally express the four relations that we have mentioned above.

$$\{(x, y)|y = 2x + 1\}$$

$$\{(x, y)|x - y = 6\}$$

$$\{(x, y)|y = x^2 + 2x + 5\}$$

$$\{(x, y)|x^2 + y^2 = 9\}$$

Recall the notation. $\{(x, y)|y = 2x + 1\}$ is read "the set of all points (x, y) such that $y = 2x + 1$."

We can now formally define a **relation** as a set of ordered pairs, a set of points. Here are a few simple sets of ordered pairs; each is a relation, just like the more complicated sets above.

$$\{(1, 3), (2, 6), (5, 2)\}$$

$$\{(0, 5), (5, -1), (5, 2)\}$$

$$\{(2, 4)\}$$

All relations can be expressed as sets of ordered pairs, or "points." Some relations are also called **functions.** To be a function, a relation cannot have two points in which x (the first coordinate) is the same and y (the second coordinate) is different. For instance, the relation

$$\{(1, 3), (2, 5), (2, 4), (3, -7)\}$$

is not a function because it contains the pairs $(2, 5)$ and $(2, 4)$ in which there are two different y values (5 and 4) for the same x value (2).

The relation

$$\{(1, 3), (2, -2), (4, 3), (7, 5)\}$$

is a function because there are not two y values for a single x value. Note that there are two x values (1 and 4) for the same y value (3), but it does not matter; that is not the test.

You can also examine the graph of a relation to determine whether or not it is a function. An easy test is to pass a vertical line through various parts of the graph. If the line never catches two (or more) points, then the graph is of a function. However, if two (or more) points can be caught, then the relation is not a function. Here are two examples.

Not a function:

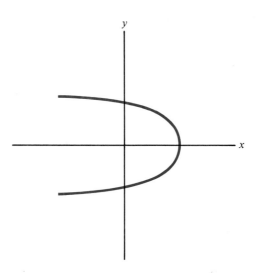

since two points can be caught by a vertical line

Function:

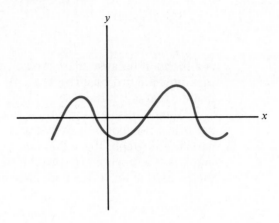

EXERCISES 23.7

*1. Which sketches are graphs of functions and which are not?

(a)

(b)

(c)

(d)

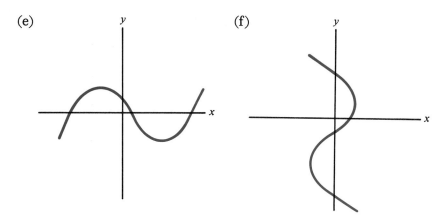

(e)

(f)

*2. Determine which of the following are functions and which are not.
(a) $\{(1, 2), (3, 4)\}$ (b) $\{(0, 1), (1, 0), (3, 4)\}$
(c) $\{(1, 0), (1, 1)\}$ (d) $\{(1, 4), (2, 4), (3, 4)\}$
(e) $y = x + 3$ (f) $y = x^2$
(g) $y = \pm \sqrt{x}$ (h) $y = \pm x$

3. Consider the following statement: "All functions are relations, but not all relations are functions." Use examples in order to explain it.

23.8 FUNCTION NOTATION (OPTIONAL)

The famous Swiss mathematician Leonhard Euler (1707–1783) suggested that functions could be written using a special notation—**function notation**—for some applications. Considering a function such as $y = 2x + 1$, we can say that it represents

$$y \text{ as a function of } x$$

and we can write

$$y = f(x)$$

and

$$f(x) = 2x + 1$$

The difference is that we now use $f(x)$ instead of y. It is only a change in notation. The meaning is the same.

Values of the function $f(x) = 2x + 1$ can be obtained for any x supplied. Since

$$f(\underline{x}) = 2 \cdot \underline{x} + 1$$

we see that

$$f(\underline{3}) = 2 \cdot \underline{3} + 1, \text{ which is } 7$$

Notice that $f(3)$ means the same as *the y value in* $y = 2x + 1$ *when* $x = 3$, but it is an abbreviated form and therefore very simple and useful. Similarly,

$$f(\underline{6}) = 2 \cdot \underline{6} + 1 = 13 \quad \checkmark$$

$$f(\underline{0}) = 2 \cdot \underline{0} + 1 = 1 \quad \checkmark$$

$$f(\underline{m}) = 2 \cdot \underline{m} + 1 = 2m + 1 \quad \checkmark$$

$$f(\underline{x + 5}) = 2 \cdot (\underline{x + 5}) + 1 \quad \text{or} \quad 2x + 11 \quad \checkmark$$

$$f\left(\underline{\tfrac{1}{2}}\right) = 2 \cdot \underline{\tfrac{1}{2}} + 1 = 2 \quad \checkmark$$

The symbol $f(x)$ is read "f at x" or "f of x." For example, $f(6)$ is f at 6, that is, the value of f when x is 6, or the value of y when x is 6.

EXAMPLE 17   If $f(x) = 3x^2 + 5x + 2$, determine
(a) $f(0)$
(b) $f(1)$
(c) $f(-1)$
(d) $f(x + 1)$
(e) $f(r - 2)$
(f) $f(x + h)$

Solution

$$f(\underline{x}) = 3(\underline{x})^2 + 5(\underline{x}) + 2$$

(a) $\quad f(\underline{0}) = 3(\underline{0})^2 + 5(\underline{0}) + 2 = 2 \quad \checkmark$

(b) $\quad f(\underline{1}) = 3(\underline{1})^2 + 5(\underline{1}) + 2 = 10 \quad \checkmark$

(c) $\quad f(\underline{-1}) = 3(\underline{-1})^2 + 5(\underline{-1}) + 2 = 0 \quad \checkmark$

(d) $\begin{aligned} f(\underline{x + 1}) &= 3(\underline{x + 1})^2 + 5(\underline{x + 1}) + 2 \\ &= 3(x^2 + 2x + 1) + 5x + 5 + 2 \\ &= 3x^2 + 11x + 10 \quad \checkmark \end{aligned}$

(e) $\begin{aligned} f(\underline{r - 2}) &= 3(\underline{r - 2})^2 + 5(\underline{r - 2}) + 2 \\ &= 3(r^2 - 4r + 4) + 5r - 10 + 2 \\ &= 3r^2 - 7r + 4 \quad \checkmark \end{aligned}$

(f) $\begin{aligned} f(\underline{x + h}) &= 3(\underline{x + h})^2 + 5(\underline{x + h}) + 2 \\ &= 3(x^2 + 2xh + h^2) + 5x + 5h + 2 \\ &= 3x^2 + 6xh + 3h^2 + 5x + 5h + 2 \quad \checkmark \end{aligned}$

EXERCISES 23.8

*1. If $f(x) = 3x - 5$, find the values listed below.

(a) $f(1)$	(b) $f(2)$	(c) $f(0)$	(d) $f(-5)$
(e) $f(a)$	(f) $f(\tfrac{1}{3})$	(g) $f(-\tfrac{1}{3})$	(h) $f(3m)$
(i) $f(x + 1)$	(j) $f(x - 5)$	(k) $f(t - 1)$	(l) $f(x + h)$

*2. If $f(x) = x^2 + 8x - 1$, find the values listed below.
    (a) $f(2)$ (b) $f(3)$ (c) $f(10)$ (d) $f(0)$
    (e) $f(-1)$ (f) $f(-3)$ (g) $f(m)$ (h) $f(\frac{1}{2})$
    (i) $f(-\frac{1}{4})$ (j) $f(x + 2)$ (k) $f(x - 1)$ (l) $f(x + h)$

23.9 RIGHT TRIANGLE TRIGONOMETRY (OPTIONAL)

Right triangle trigonometry was developed in order to provide an indirect means of measuring distances that could not be measured directly—for instance, the distance from the earth to the moon. Hipparchus (about 140 B.C.) and Ptolemy (about A.D. 150) are considered the founders of trigonometry. They devised techniques to catalog the positions of stars relative to the earth. Much of the early interest in trigonometry was due to its natural application to astronomy, map making, and navigation. Before we can consider applications, however, some definitions are needed.

In a right triangle the side opposite the right angle is called the **hypotenuse.** The side opposite a specified nonright angle (called the angle θ) is called the **opposite** *side.* The remaining side next to angle θ is called the **adjacent** *side.* Which side is opposite and which side is adjacent depends on the angle that you choose to refer to, as illustrated by the diagrams shown next.

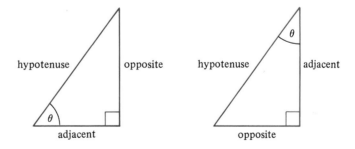

Three *trigonometric functions* are defined in terms of the lengths of the hypotenuse, opposite, and adjacent sides.

$$\text{sine of } \theta = \sin \theta = \frac{\text{opposite}}{\text{hypotenuse}}$$

$$\text{cosine of } \theta = \cos \theta = \frac{\text{adjacent}}{\text{hypotenuse}}$$

$$\text{tangent of } \theta = \tan \theta = \frac{\text{opposite}}{\text{adjacent}}$$

EXAMPLE 18   Find sin θ, cos θ, and tan θ corresponding to the triangle shown.

Solution

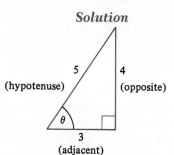

$$\sin \theta = \frac{\text{opposite}}{\text{hypotenuse}} = \frac{4}{5} = .8000$$

$$\cos \theta = \frac{\text{adjacent}}{\text{hypotenuse}} = \frac{3}{5} = .6000$$

$$\tan \theta = \frac{\text{opposite}}{\text{adjacent}} = \frac{4}{3} = 1.333 \qquad \text{approximately}$$

EXAMPLE 19   Find sin x, cos x, and tan x corresponding to the triangle shown.

Solution

$$\sin x = \frac{\text{opposite}}{\text{hypotenuse}} = \frac{5}{13} = .3846 \qquad \text{approximately}$$

$$\cos x = \frac{\text{adjacent}}{\text{hypotenuse}} = \frac{12}{13} = .9231 \qquad \text{approximately}$$

$$\tan x = \frac{\text{opposite}}{\text{adjacent}} = \frac{5}{12} = .4167 \qquad \text{approximately}$$

EXAMPLE 20   Find tan θ in the diagram shown.

Solution

$$\tan \theta = \frac{\text{opposite}}{\text{adjacent}} = \frac{?}{4}$$

The opposite side is not known, but if we apply the Pythagorean theorem (from Chapter 4), we can determine the length of this unknown side. Since

$$c^2 = a^2 + b^2 \qquad \text{the Pythagorean theorem}$$

We have

$$7^2 = 4^2 + b^2 \qquad \text{Call the unknown side } b.$$

or

$$49 = 16 + b^2$$

So

$$b^2 = 33$$

which means
$$b = \sqrt{33}$$
Thus
$$\tan \theta = \frac{\text{opposite}}{\text{adjacent}} = \frac{\sqrt{33}}{4} \quad \checkmark$$

Approximate values for the sine, cosine, and tangent of an angle can be looked up in the table (p. 392). In order to be sure that you can read the table, look up sin 24°, cos 81°, and tan 53°. Then check your results with those given below. The symbol ≈ means approximately equal to.

$$\sin 24° \approx .4067$$
$$\cos 81° \approx .1564$$
$$\tan 53° \approx 1.3270$$

The same table can be used to determine the angle when either the sine, or the cosine, or the tangent is known. Just reverse the table look-up process used above. If sin θ is .9744, how many degrees are in angle θ? If cos θ = .5878, what is the value of θ? If tan θ = .2867, what is θ? When you have finished using Table 1 to determine these values of θ, check your answers with the results given below.

If sin θ = .9744, then $\theta \approx 77°$.

If cos θ = .5878, then $\theta \approx 54°$.

If tan θ = .2867, then $\theta \approx 16°$.

EXAMPLE 21   Tom walks out 35 feet from the side of a building. He then lies down and looks up at the top of the building. He measures his line of vision to be at an angle of 40° with respect to the ground. Approximately how high is the building?

Solution   Here is the situation.

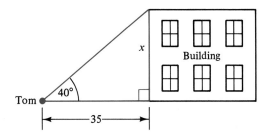

With respect to the angle of 40°, the side of the triangle of length 35 feet is the adjacent side, and the unknown side labeled x is the opposite side. The trigonometric relationship involving these two sides is tangent.

TABLE 1
Values of Trigonometric Functions

Angle	Sin	Cos	Tan	Angle	Sin	Cos	Tan
0°	.0000	1.0000	.0000	46°	.7193	.6947	1.0355
1°	.0175	.9998	.0175	47°	.7314	.6820	1.0724
2°	.0349	.9994	.0349	48°	.7431	.6691	1.1106
3°	.0523	.9986	.0524	49°	.7547	.6561	1.1504
4°	.0698	.9976	.0699	50°	.7660	.6428	1.1918
5°	.0872	.9962	.0875				
				51°	.7771	.6293	1.2349
6°	.1045	.9945	.1051	52°	.7880	.6157	1.2799
7°	.1219	.9925	.1228	53°	.7986	.6018	1.3270
8°	.1392	.9903	.1405	54°	.8090	.5878	1.3764
9°	.1564	.9877	.1584	55°	.8192	.5736	1.4281
10°	.1736	.9848	.1763				
				56°	.8290	.5592	1.4826
11°	.1908	.9816	.1944	57°	.8387	.5446	1.5399
12°	.2079	.9781	.2126	58°	.8480	.5299	1.6003
13°	.2250	.9744	.2309	59°	.8572	.5150	1.6643
14°	.2419	.9703	.2493	60°	.8660	.5000	1.7321
15°	.2588	.9659	.2679				
				61°	.8746	.4848	1.8040
16°	.2756	.9613	.2867	62°	.8829	.4695	1.8807
17°	.2924	.9563	.3057	63°	.8910	.4540	1.9626
18°	.3090	.9511	.3249	64°	.8988	.4384	2.0503
19°	.3256	.9455	.3443	65°	.9063	.4226	2.1445
20°	.3420	.9397	.3640				
				66°	.9135	.4067	2.2460
21°	.3584	.9336	.3839	67°	.9205	.3907	2.3559
22°	.3746	.9272	.4040	68°	.9272	.3746	2.4751
23°	.3907	.9205	.4245	69°	.9336	.3584	2.6051
24°	.4067	.9135	.4452	70°	.9397	.3420	2.7475
25°	.4226	.9063	.4663				
				71°	.9455	.3256	2.9042
26°	.4384	.8988	.4377	72°	.9511	.3090	3.0777
27°	.4540	.8910	.5095	73°	.9563	.2924	3.2709
28°	.4695	.8829	.5317	74°	.9613	.2756	3.4874
29°	.4848	.8746	.5543	75°	.9659	.2588	3.7321
30°	.5000	.8660	.5774				
				76°	.9703	.2419	4.0108
31°	.5150	.8572	.6009	77°	.9744	.2250	4.3315
32°	.5299	.8480	.6249	78°	.9781	.2079	4.7046
33°	.5446	.8387	.6494	79°	.9816	.1908	5.1446
34°	.5592	.8290	.6745	80°	.9848	.1736	5.6713
35°	.5736	.8192	.7002				
				81°	.9877	.1564	6.3138
36°	.5878	.8090	.7265	82°	.9903	.1392	7.1154
37°	.6018	.7986	.7536	83°	.9925	.1219	8.1443
38°	.6157	.7880	.7813	84°	.9945	.1045	9.5144
39°	.6293	.7771	.8098	85°	.9962	.0872	11.4301
40°	.6428	.7660	.8391				
				86°	.9976	.0698	14.3007
41°	.6561	.7547	.8693	87°	.9986	.0523	19.0811
42°	.6691	.7431	.9004	88°	.9994	.0349	28.6363
43°	.6820	.7314	.9325	89°	.9998	.0175	57.2900
44°	.6947	.7193	.9657	90°	1.0000	.0000	
45°	.7071	.7071	1.0000				

$$\tan 40° = \frac{\text{opposite}}{\text{adjacent}} = \frac{x}{35}$$

or

$$.8391 \approx \frac{x}{35}$$

so

$$x \approx 35(.8391) \approx 29.4$$

The building is about 29.4 feet high.

EXERCISES 23.9

Answers to starred exercises are given in the back of the book.

*1. For the triangle at the right, find each of the following.
(a) sin θ
(b) cos θ
(c) tan θ

*2. For the triangle at the right, find each of the following.
(a) sin x
(b) cos x
(c) tan x
(d) sin y
(e) cos y
(f) tan y

*3. For the triangle at the right, find each of the following.
(a) sin y
(b) cos y
(c) tan y
(d) sin x
(e) cos x
(f) tan x

4. For the triangle at the right, find each of the following.
   (a) sin *m*
   (b) cos *m*
   (c) tan *m*
   (d) sin *n*
   (e) cos *n*
   (f) tan *n*

*5. In forestry, right triangle trigonometry is sometimes used to determine the height of a tree. The observer measures his distance (*d*) from the tree and the angle of inclination (*θ*) of his line of sight to the top of the tree. To the nearest foot, what is the height of the tree if the distance of the observer from the tree is 65 feet and the angle of inclination is 53°?

*6. The string of a kite is attached to the ground. The angle formed between the string and the ground is 58°. If the length of the string is 1000 feet (that is, you have let out 1000 feet of string), how high is the kite?

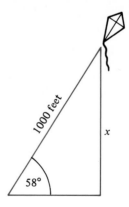

7. A 43-foot ladder is placed against a wall at an angle of 50° with respect to the ground. How high on the wall does the ladder reach?

8. The sun shines on a 25-foot vertical telephone pole, creating a 15-foot shadow on the ground. Determine the angle that the sun's rays make with respect to the ground.

*9. Having climbed to the top of a flagpole, Bob looks downward at a chalk mark that he made 150 feet from the base of the pole. He estimates that the angle between his line of vision to the chalk mark and the flagpole is 65°. About how high is the pole?

CHAPTER 23 REVIEW EXERCISES

*1. Explain what is in the set $\{x \mid 4 \leq x < 9\}$.

2. Let $\mathcal{U} = \{1, 2, 3, 4, 5, 6, 7\}$, $A = \{1, 2, 3\}$, $B = \{4, 5, 6\}$, $C = \{1, 3, 5, 7\}$.
   *(a) $A \cup B =$ *(b) $A \cap B =$
   *(c) $A \cup C =$ *(d) $B \cap C =$
   *(e) $\bar{B} =$ (f) $\bar{C} =$
   *(g) $\bar{\mathcal{U}} =$ *(h) Is $A \subseteq \mathcal{U}$? Explain.
   (i) Is $A \subseteq C$? Explain.

3. Draw a Venn diagram for the indicated sets.
   (a) $A \cup B$ for two sets A and B that overlap
   (b) $A \cap B$ for two sets A and B such that $A \subseteq B$

*4. Is the given relation a function?
   (a) $\{(5, 1), (5, 2), (5, 3)\}$ (b) $y = 7x + 4$

*5. If $f(x) = x^2 + 3x + 5$, determine
   (a) $f(4)$ (b) $f(0)$

*6. For the triangle shown here, determine:
   (a) x
   (b) $\sin \theta$
   (c) $\cos \theta$
   (d) $\tan \theta$

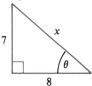

ANSWERS TO SELECTED EXERCISES

EXERCISES 1.1 (page 2)

1. (a) $(1 \times 100) + (4 \times 10) + (3 \times 1)$ (c) $(5 \times 10) + (0 \times 1)$
   (d) $(1 \times 1000) + (6 \times 100) + (0 \times 10) + (3 \times 1)$
   (e) $(7 \times 100) + (7 \times 10) + (7 \times 1)$
   (g) $(6 \times 100) + (0 \times 10) + (0 \times 1)$
   (h) $(1 \times 10,000) + (9 \times 1000) + (7 \times 100) + (3 \times 10) + (4 \times 1)$
   (i) $(3 \times 10,000) + (4 \times 1000) + (8 \times 100) + (5 \times 10) + (1 \times 1)$
   (k) $(6 \times 100,000) + (4 \times 10,000) + (8 \times 1000) + (7 \times 100) +$
       $(1 \times 10) + (3 \times 1)$
   (m) $(4 \times 1,000,000) + (9 \times 100,000) + (8 \times 10,000) + (0 \times 1000) +$
       $(6 \times 100) + (1 \times 10) + (7 \times 1)$

2. (a) seven hundred fifty-four   (c) one thousand three hundred forty-one
   (e) fifty-four thousand six hundred seventeen
   (g) five hundred sixty-three thousand four hundred thirty-two
   (i) one million eight hundred ninety-seven thousand four hundred
   (k) seventy-eight million eight hundred thousand one hundred sixty-seven
   (m) six hundred fifty-four million four hundred fifty-six thousand
       five hundred forty-six
   (o) eight billion ninety-eight million seven hundred sixty-four
       thousand one hundred twenty-three

3. Only 9, 0, 3, and 2 are digits.
4. Only 6, 16, 160, and 10,000,000 are whole numbers.

EXERCISES 1.2 (page 5)

1. (a) 9   (b) 6   (c) 8   (d) 13   (e) 8   (f) 14   (g) 18   (h) 15
   (i) 9   (j) 10   (k) 11   (l) 13   (m) 10   (n) 16   (o) 10
   (p) 13   (q) 12   (r) 14

2. (a) 69   (b) 89   (c) 75   (f) 153   (h) 1302   (i) 1590   (k) 7869   (m) 10,802
(n) 146   (o) 1284   (r) 18,923   (t) 13,858   (u) 18,310   (w) 107,603   (x) 233,574
3. (a) 158.1   (b) 9.57   (d) 169.082   (e) 1629.7   (g) 128.5   (j) 10,000

EXERCISES 1.3 (page 8)

1. (a) 5   (b) 1   (c) 6   (d) 2   (e) 7   (f) 10   (g) 14   (h) 17   (i) 27   (j) 9
(k) 19   (l) 8
2. (a) 24   (c) 29   (e) 275   (g) 599   (i) 5992   (j) 6569   (k) 9317   (m) 12,020
(o) 566,667
3. (a) 14.5   (b) 63.9   (c) 46.94   (e) 6.999   (g) 47.68   (i) 79.9

EXERCISES 1.4 (page 11)

1. (a) 72   (b) 81   (c) 35   (d) 8   (e) 0   (f) 28   (g) 18   (h) 0   (i) 12   (j) 12
(k) 48   (l) 45
2. (a) 1737   (b) 1320   (c) 3588   (d) 11,850   (g) 3,645,158   (i) 22,990,110
(k) 26,499,200   (l) 27,016,000   (n) 5,272,978,200   (p) 768,716,000
3. (a) 492.4668   (b) 75.23988   (e) 261.5459224   (g) 4,711,312.52312   (i) 805.6450008

EXERCISES 1.5 (page 15)

1. (a) 4   (b) 9   (c) 7   (d) 3   (e) 7   (f) 6   (g) 8   (h) 9   (i) 5   (j) 7   (k) 9
(l) 9
2. (a) 711   (b) 8423   (c) 921   (f) 845   (g) 94   (i) 153
3. (a) 8, R: 6   (c) 127, R: 7   (e) 253, R: 8   (g) 232, R: 38   (i) 406, R: 13
4. (a) 311.70   (b) 175.42   (f) 1282.03   (g) 96.89   (i) 402.93
5. (a) 32.810   (c) 126.85   (e) 23,828   (g) 16,450   (i) 11.015

EXERCISES 1.6 (page 17)

1. (a) 73.77   (b) 15.22   (d) 98.14   (f) 765.92   (g) 56.90
2. (a) 17.8   (b) 99.7   (c) 187.2   (d) 40.0   (f) 456.8   (h) 46.0
3. (a) 20   (b) 23   (f) 10   (g) 34
4. (a) 58.644   (b) 46.163   (d) 98.002   (g) 10.000   (i) 5.765
5. (a) 90   (b) 150   (c) 540   (e) 5680   (g) 90
6. (a) 1800   (b) 1900   (c) 5400   (e) 2000   (h) 10,000

EXERCISES 1.7 (page 19)

1. (a) 0   (c) 0   (e) cannot divide   (g) cannot divide   (i) cannot divide   (k) cannot divide

EXERCISES 1.9 (page 20)

1. (a) 6^2 (c) 2^3 (d) 9^7 (f) 3^{10} (h) m^2 (i) x^5 (j) $(x + y)^3$
2. (a) $8 \cdot 8$ (b) $2 \cdot 2 \cdot 2$ (d) $6 \cdot 6 \cdot 6 \cdot 6 \cdot 6 \cdot 6 \cdot 6 \cdot 6$ (g) $x \cdot x$
   (i) $(a + b)(a + b)(a + b)(a + b)$
3. (a) 8 (c) 625 (f) 343 (g) 1 (i) 243 (k) 0

CHAPTER 1 REVIEW EXERCISES (page 21)

1. (a) 2351 (c) 149.93 (e) 223,715
2. (a) 4413 (c) 18,990 (d) 13.088
3. (a) 345,796 (c) 41,090,778 (d) 613.224
4. (a) 164.2 (b) 0 (c) cannot divide (d) 14.71
5. (a) 6.0 (c) 9.9 (e) .9
6. (a) 16 (b) 0 (c) 125 (d) 144 (e) 1.69

EXERCISES 2.1 (page 25)

1. (a) 20 (b) 27 (c) 11 (d) 14 (e) 9 (f) 4 (g) 25 (h) 8 (i) 61 (j) 41
   (k) 11 (l) 32 (m) 28 (n) 19 (o) 30 (p) 27
2. (a) 5 (c) 66 (e) 57 (g) 36
3. (a) 24 (b) 25 (c) 13 (d) 1 (e) 2 (f) 6 (g) 36 (h) 1 (i) 16 (j) 24
   (k) 9 (l) 52 (m) 9 (n) 476

CHAPTER 2 REVIEW EXERCISES (page 26)

1. (a) 24 (b) 25 (c) 9 (d) 1
2. (a) 25 (b) 57 (c) 27 (d) 8 (e) 12 (f) 8
3. (a) 6 (b) 1 (c) 20 (d) 13 (e) 10 (f) 2 (g) 25 (h) 16 (i) 33 (j) 27
   (k) 6 (l) 124

EXERCISES 3.1 (page 31)

1. (a) 1036 (c) 420 (e) 120 (g) 87.92 (i) 249.316 (j) 59
2. (a) 30 (b) 1800
3. (a) 22.8 (b) 153.86 (c) 1017.36 (d) 523.33 (e) 1004.8
4. (a) area 177, perimeter 72 (b) area 128, perimeter 56 (c) area 450, perimeter 100
   (d) area 1104, perimeter 168 (e) area 132, perimeter 74 (f) area 656, perimeter 248

CHAPTER 3 REVIEW EXERCISES (page 33)

1. (a) 204 (b) 85 (c) 125.6 (d) 180 (e) 314

EXERCISES 4.1 (page 37)

3. (a) 3 (b) $\sqrt{5}$ (c) 12 (e) 100 (f) $\sqrt{61}$ (i) 25 (j) $\sqrt{40}$ (k) 8 (l) 25
4. (a) $t = 2$ (c) $x = 4$ (e) $A = 6$
5. 12 feet

CHAPTER 4 REVIEW EXERCISES (page 39)

2. (a) 4   (b) 13   (c) $\sqrt{13}$ (d) $\sqrt{18}$

EXERCISES 5.1 (page 43)

5. (a) commutative property for addition   (b) associative property for addition
     (c) commutative property for addition   (d) distributive property   (e) distributive property
     (f) closure property for addition   (g) associative property for addition
     (h) commutative property for addition   (i) closure property for addition
     (j) commutative property for multiplication   (k) distributive property   (l) distributive property
7. (a) associative property for addition   (b) commutative property for addition
     (c) commutative property for addition   (d) closure property for addition
     (e) closure property for addition
8. (a) distributive property   (b) commutative property for addition   (c) distributive property
     (d) commutative property for multiplication   (e) closure property for multiplication

CHAPTER 5 REVIEW EXERCISES (page 44)

1. (a) associative property for multiplication   (b) commutative property for addition
     (c) distributive property   (d) closure property for multiplication   (e) closure property for addition
     (f) distributive property
2. (a) distributive property   (b) commutative property for multiplication
     (c) closure property for multiplication   (d) commutative property for addition
     (e) closure property for addition

EXERCISES 6.2 (page 47)

1. (a) 3/10   (b) 8/27   (c) 6/35   (e) 15/28   (g) 1/81   (h) 12/25   (j) 60/91
     (k) 180/299   (m) 7/8   (o) 8/9   (p) 3/8   (r) 50/99   (s) 0
2. (a) 10/21   (c) 8/35

EXERCISES 6.3 (page 50)

1. (a) 2/3   (b) 1   (c) 3/4   (d) 3/4   (g) 2/3   (h) 1/2   (k) 3/4   (l) 5/12   (m) 1/3
2. (a) 1/9   (b) 7/20   (c) 1/12   (d) 18/35   (e) 2/3   (f) 1   (h) 4/11   (j) 3   (k) 1/2
     (m) 4/27   (o) 3/10

EXERCISES 6.4 (page 53)

1. (a) 2, 4, 5, 10   (b) 2, 3, 5, 6, 10   (c) 3, 5, 9   (f) 3   (g) none of them   (h) 2
     (j) 2, 3, 6   (l) 3, 5   (m) none of them   (o) 3   (p) 2, 3, 4, 6   (q) 2, 3, 4, 6, 9
     (s) 2, 4
2. (a) 2/5   (b) 7/11   (c) 1/3   (d) 14/15   (e) 4/5   (g) 7/8   (i) 1/4   (j) 1/24
     (m) 5/27   (n) cannot be reduced   (p) 7/8   (r) 13/17
3. (a) 4/7   (b) 1/25   (c) 2/9   (d) 7/8   (f) 7/11   (h) 13/24   (i) 3/16   (k) 8/31
4. (a) 7/31   (b) 6/49   (c) 30/49   (e) 1   (f) 2   (h) 1/40   (j) 5/192   (k) 2/5

EXERCISES 6.5 (page 55)

1. (a) 20/21   (b) 2/3   (d) 5/6   (f) 35/24   (g) 10/7   (h) 24/49   (j) 1   (l) 16/25
   (n) 2/21   (p) 17/24   (q) 15

EXERCISES 6.6 (page 59)

1. (a) 6/13   (b) 4/7   (c) 5/9   (d) 4/5
2. (a) 3/7   (b) 2/11   (c) 1/3   (d) 1/9
3. (a) 1   (c) 1/2   (e) 1/2   (g) 23/20   (i) 67/91   (k) 91/132   (m) 23/16   (o) 73/42
4. (a) 1/3   (c) 16/35   (e) 1/12   (g) 1/4   (i) 1/6
5. (a) 3/4   (b) 22/15   (c) 8/11

EXERCISES 6.7 (page 62)

1. (a) $1\frac{1}{6}$ (c) $4\frac{1}{2}$ (e) $7\frac{1}{4}$ (g) $3\frac{4}{5}$ (i) $25\frac{1}{2}$
2. (a) 5/3   (c) 20/7   (d) 28/9   (e) 27/5   (f) 17/2   (g) 53/12   (h) 65/9
   (i) 12 is not a mixed number, but it can be written as an improper fraction, for example, as 12/1.
   (j) 66/7
3. (a) 147/16 or 9 3/16   (c) 7/3 or 2 1/3   (e) 18/7 or 2 4/7   (g) 9/8 or 1 1/8
4. (a) 17/4 or 4 1/4   (c) 139/12 or 11 7/12   (e) 9/35   (g) 23/12 or 1 11/12
5. (a) 6 1/8   (b) 26 1/8   (c) 109 3/8   (d) 27 5/8

EXERCISES 6.8 (page 65)

1. (a) $3 \cdot 7$ (b) $2 \cdot 5$ (d) $3 \cdot 11$ (e) $2 \cdot 3 \cdot 7$ (f) $3 \cdot 5 \cdot 7$ (h) $2^3 \cdot 5$ (j) $2 \cdot 5^2$
   (l) $2^2 \cdot 3^2$ (m) 2^6 (p) $2 \cdot 3^2 \cdot 5$ (s) $2^4 \cdot 3^2$ (t) $2^3 \cdot 3^3$
2. (a) 30   (b) 210   (c) 100   (d) 480   (e) 105   (g) 60   (h) 18   (i) 120   (l) 180
   (m) 600   (o) 180

EXERCISES 6.9 (page 67)

1. (a) 5/18   (b) 77/120   (c) 29/72   (f) 83/120   (h) 131/420   (j) 67/180   (l) 37/54
2. (a) 73/100   (c) 131/80   (e) 259/180   (g) 49/100   (i) 601/90   (k) 133/60
3. (a) 13/20   (c) 11/48   (e) 5/24
4. 137/90   6. 78/77

CHAPTER 6 REVIEW EXERCISES (page 69)

1. (a) 3/5   (b) 11/40   (c) 3/5   (e) 2/5   (g) 3/14   (i) 56/75
2. (a) 1   (c) 1/2   (e) 8/7   (g) 391/377   (i) 2/35   (k) 65/44   (m) 187/20   (n) 22/7
3. (a) 1   (c) 6/5   (e) 2/3   (g) 76/33   (i) 16/27   (k) 15/124   (m) 23/10   (n) 77/51
4. (a) $3^3 \cdot 5$ (c) $2^3 \cdot 11$ (e) $2^5 \cdot 3$ (g) $2 \cdot 5 \cdot 11$ (i) 3^5
5. (a) 30   (c) 120   (e) 120   (g) 315
6. (a) 13/20   (c) 25/96   (e) 203/280   (g) 113/252   (i) 589/630   (k) 328/45
   (m) 79/8 or $9\frac{7}{8}$ (n) 26/15 or $1\frac{11}{15}$

EXERCISES 7.1 (page 73)

1. (a) 17/100　　(b) 31/100　　(c) 1/2　　(d) 17/50　　(e) 3/4　　(g) 1/5　　(h) 63/200
　(i) 88/125　　(k) 19/20　　(m) 12/25　　(n) 3/10　　(o) 1131/2000　　(q) 33/40　　(s) 19/20
　(t) 999,999/1,000,000
2. (a) .1250　　(b) .6000　　(c) .7777　　(e) .5714　　(f) .6666　　(g) .8461　　(i) .6250
　(k) .1818　　(m) .1111　　(n) .2666

EXERCISES 7.2 (page 77)

1. (a) 7/20　　(c) 7/25　　(e) 1/5　　(g) 9/25　　(h) 1/20　　(i) 3/100　　(j) 16/25　　(k) 21/50
　(m) 17/20　　(o) 2/25
2. (a) .4　　(c) .25　　(e) .15　　(f) .06　　(g) .19　　(i) .54
3. (a) 73%　　(c) 65%　　(e) 7%　　(g) 8%　　(h) 80%　　(i) 80%　　(k) 72.3%　　(m) 6.2%
　(o) 90.1%　　(q) .7%
4. (a) 80%　　(c) 33.3% (approximately)　　(e) 26.6% (approximately)　　(g) 85.7% (approximately)
　(h) 36.3% (approximately)　　(i) 40%　　(k) 58.3% (approximately)　　(m) 69.2% (approximately)
　(o) 21.4% (approximately)　　(p) 37.5%

EXERCISES 7.3 (page 79)

1. (a) 59/1000　　(b) 2　　(c) 153/1000　　(f) 0　　(h) 2/125　　(i) 4　　(k) 9/2000
　(l) 3/10,000　　(m) 12/5　　(o) 269/200　　(q) 313/100　　(s) 10
2. (a) .035　　(c) .0725　　(e) .0275　　(g) .05375
3. (a) 3　　(c) 1.5　　(e) 1.73　　(g) 13.5　　(i) 14.25
4. (a) 230%　　(c) 150%　　(e) 350%　　(g) 1%　　(i) 20%　　(k) .2%　　(m) 7.1%　　(o) 999%
　(q) 740%　　(s) 212.5%

EXERCISES 7.4 (page 81)

1. (a) 32.25　　(b) 46.4　　(c) 164　　(d) 500　　(e) 12.5%　　(f) 14.28% (approximately)
　(g) 263.15 (approximately)
2. (a) 70%　　(c) 6 windows　　(e) $40,000　　(g) 62.5%　　(h) 37.5%　　(j) 85%　　(l) 34 points
　(m) 10%

CHAPTER 7 REVIEW EXERCISES (page 82)

1. (a) .8, 80%　　(b) .1666, 16.66%　　(e) 6, 600%　　(g) 1.6666, 166.66%　　(i) 9.514, 951.4%
2. (a) .55, 11/20　　(c) .72, 18/25　　(e) 7.25, 29/4　　(f) .163, 163/1000　　(h) .0009, 9/10,000
3. (a) 11.1% (approximately)　　(b) 40.6
4. (a) 30%　　(b) 28,000 men　　(c) $11,000

EXERCISES 8.1 (page 90)

1. (a) 123 feet　　(b) 276 inches　　(c) 720 minutes　　(d) 94 hours　　(g) 80 quarts　　(h) 54 feet
2. (a) 10,800 seconds　　(b) 316,800 inches　　(c) 1680 hours　　(d) 256,000 ounces　　(e) 7 hours
　(h) 512 gallons

3. (a) 3 miles per minute   (b) 15,840 feet per minute   (c) 264 feet per second
    (d) 45 miles per hour   (e) 170 feet per minute   (f) 17 inches per second
4. (a) 36 inches   (b) 25 centimeters   (c) 36 meters   (e) 25.12 inches (approximately)
5. (a) 84 square feet   (c) 99 square inches   (e) 1.96 square meters
    (f) 16.61 square centimeters (approximately)
6. (a) 272 cubic centimeters   (c) 1570 cubic inches (approximately)
    (e) 267.95 cubic inches (approximately)   (g) 376.8 cubic centimeters (approximately)
7. (a) 128 feet   (b) 960 square feet   (c) 720 cubic feet   (d) 35.325 cubic inches (approximately)
8. (a) 720 square inches   (b) 16 square feet   (e) 29,376 cubic inches

CHAPTER 8 REVIEW EXERCISES (page 91)

1. (a) 174 yards   (b) 19 minutes   (c) 5,443,200 seconds   (d) 1.4 weeks   (e) 2430 cubic feet
2. (a) 238 square centimeters   (b) 588.68 cubic inches approximately   (c) 14.8 meters
    (d) 1 square inch   (e) 390 cubic centimeters

EXERCISES 9.1 (page 99)

1. (a) 1.34 m (approximately)   (c) 8.83 m (approximately)   (e) 150 km (approximately)
    (g) 48.26 cm (approximately)   (i) 76.2 mm (approximately)
2. (a) 275.59 in. (approximately)   (c) 141.07 ft (approximately)   (e) 9.44 in. (approximately)
    (g) 84 mi (approximately)   (i) 9.25 in. (approximately)
3. (a) 21.69 liters (approximately)   (c) 360 ml (approximately)   (e) 12.84 liters (approximately)
    (g) 67.92 liters (approximately)   (i) 1.13 kl (approximately)
4. (a) 39.22 qt (approximately)   (c) 3 oz (approximately)   (e) 500 oz (approximately)
    (g) 6.09 gal (approximately)   (i) 2385 gal (approximately)
5. (a) 10.45 kg (approximately)   (c) 198.45 g (approximately)   (e) 85,050 mg (approximately)
6. (a) 61.6 lb (approximately)   (c) 3.42 oz (approximately)   (e) .02 oz (approximately)
7. (a) 7000 mm   (c) .158 m   (e) 45,000 ml   (g) .7 liter   (h) 19,000 g   (i) 70,000 mg
    (l) 3,000,000 ml   (m) 700,000 cm   (n) 7,000,000 mm   (o) 4 m   (q) 200,000 g
    (s) 12,000,000 ml   (u) 35,000,000 mm   (w) .5 kl   (y) 1000 cm
8. (a) 300 g   (b) 1900 g   (c) 50 g   (d) 170 g   (e) 25 g   (f) 3.3 g   (g) 740 liters
    (h) 15 dl   (i) 20 m   (j) 3000 dm   (k) 4000 dl   (l) 300 dl
9. (a) 35°C   (c) 5°C   (e) 68°F   (g) 113°F   (i) 24°C   (k) 138°F

CHAPTER 9 REVIEW EXERCISES (page 100)

1. (a) 12.7 cm   (b) 538.65 g   (c) 12.26 liters   (d) 500 km
2. (a) 88 lb   (b) 166.67 fluid ounces   (c) 12.6 in.   (d) 55.77 ft
3. (a) 8520 mm   (b) 85.2 cm   (c) .052 km   (d) .005 kg
4. (a) 11°C   (b) 160°F

REVIEW EXERCISES FOR CHAPTERS 1–9 (page 101)

1. (a) 1632   (c) 978   (e) 786,477   (g) 1196.70   (i) 343
2. (a) 44   (b) 9
3. (a) $A = 20$ (b) $S = 452.16$
4. (a) 13   (c) $\sqrt{2}$

5. (a) 12   (c) $\sqrt{51}$
6. (a) commutative property for addition   (c) closure property for multiplication
    (d) commutative property for multiplication
7. (a) 16/17   (c) 9/5   (e) 393/50   (g) 16/42 or 8/21
8. (a) 105   (c) 360
9. (a) 40%   (b) 87.5%
10. (a) 45/100 or 9/20   (b) 83/100
11. (a) .34   (b) 3.21   (c) .085
12. (a) 18,000 seconds   (c) 4 miles per minute
13. (a) 56.60 liters   (b) 1.10 pounds   (c) 5000 millimeters

EXERCISES 10.1 (page 104)

1. (a) +12   (b) −79   (c) +29,028   (d) −10,918   (g) +2   (i) +5   (j) −14
    (k) −8,000,000,000

EXERCISES 10.2 (page 110)

1. (a) +33   (b) +39   (c) +5   (d) −9   (e) +4   (f) −7   (g) −27   (h) −27   (i) −5
    (j) +5   (k) −77   (l) −18
2. (a) +11   (b) −11   (c) −4   (d) +6   (e) 0   (g) +11   (i) −23   (k) +7   (m) +45
    (o) −9   (q) −31   (s) −31
3. (a) +1/12   (b) −1/10   (c) −19/15   (d) +17/28   (e) +1/8   (f) −4/3   (g) +5/12
    (h) +11/24
4. (a) +16   (b) −3   (c) +6   (d) 0   (e) −6   (g) +11   (i) −5   (k) −27
5. (a) −3.1   (b) +7.7   (c) +2.37   (d) −2.9   (e) +2.36   (f) −8.3

EXERCISES 10.3 (page 113)

1. (a) +25   (b) −35   (c) −60   (d) +2   (e) +51   (f) +29   (g) −98   (h) +79
    (i) −115   (j) −12   (k) −76   (l) +911   (m) −239   (n) −341   (o) +95
2. (a) +117.75   (b) +92.1   (c) +94.72
3. (a) +3   (b) −9   (c) +9   (d) −15   (e) +17   (f) −3   (h) +15   (k) −35
4. (a) −5.12   (b) 113.43   (c) +5/7   (d) −3/4   (e) −4/15   (f) +1/12
5. (a) +4   (c) +9   (e) −13   (g) +4   (i) −8   (k) −40   (m) +4

EXERCISES 10.4 (page 119)

1. (a) −6   (b) −10   (c) +35   (d) −18   (e) +24   (f) +63   (g) +9   (h) −40
    (i) +20   (j) +36   (k) −4   (l) −48   (m) +9   (n) −49   (o) −90   (p) −80
    (q) +84   (r) +76   (s) −120   (t) +64   (u) −57   (v) +99   (w) +240   (x) −105
2. (a) −5.12   (b) +33.28   (c) −74.305   (d) +663.48   (e) −1/21   (f) +6/55   (g) +1/2
    (h) −35/48
3. (a) −24   (b) −6   (c) +216   (d) −60   (e) −96   (g) +16   (h) −1   (i) +360
    (k) −24   (m) +144   (o) +24
4. (a) −1   (b) +16   (c) −27   (d) +32   (e) +81   (f) −243   (g) −128   (h) +1
    (i) −1   (j) −64   (k) +1   (l) +27
5. (a) +15   (b) −7   (c) −64   (d) −82   (e) −1   (f) −983   (g) −3   (h) +18
    (i) +24   (j) −32

EXERCISES 10.5 (page 123)

1. (a) −5 (b) −18 (c) +22 (d) −5 (e) +10 (f) +16 (g) +14 (h) +34
2. (a) −6 (c) +2
3. (a) −13 (b) +15 (c) +10 (d) −15 (e) +17 (f) +3 (g) +6 (h) −2
   (i) +12 (j) −5 (k) −3 (l) +8 (m) +2 (n) −3 (o) +5 (p) 0
4. (a) −17 (b) −2 (c) +22 (e) +78 (g) −31 (i) −7 (j) −4 (k) +16
   (m) −13 (o) +41 (q) +18
5. (a) +32 (b) 0 (c) 0 (e) +6 (g) −103 (h) +20 (i) 0 (j) −2 (k) −13
   (m) −2 (o) +64 (q) +63

EXERCISES 10.6 (page 127)

1. (a) +5 (b) −3 (c) −3 (d) +4 (e) +4 (f) −6 (g) −4 (h) −5 (i) +12
   (j) −1 (k) −6 (l) +3 (m) +27 (n) −6 (o) 0 (p) 0 (q) −1 (r) +4
2. (a) −3/5 (b) −4/7 (c) +7/9 (d) −6/7 (e) +5/8 (f) −1/3 (g) −5/6 (h) +3/4
   (i) +6/7
3. (a) +1 (c) +1 (e) −11 (g) 0
4. (a) −3 (b) 0 (c) not defined (d) 0 (e) −1 (f) −1 (g) not defined
   (h) not defined (i) 0 (j) −1 (k) 0 (l) not defined
5. (a) −12 (b) −2 (c) +3 (d) −1 (g) +30 (h) −11/24 (j) +5

EXERCISES 10.7 (page 129)

1. (a) 21 (b) −7 (c) 3 (d) 47
3. (a) −5 (b) +5 (c) −1 (d) −7
5. (a) +14 (b) −28 (c) +12 (d) −16
7. (a) 24 (c) −8 (e) −1 (g) 21 (i) −15 (k) −9 (m) 7
8. (a) 5 (c) −1 (e) 2 (g) 8 (i) −4
9. (a) 41 (c) 16 (e) −44 (g) 29 (i) 9

EXERCISES 10.8 (page 132)

1. (a) 30 (b) 0 (c) 2 (d) 17 (e) 7 (f) 0 (g) 12 (h) 3
2. (a) true (b) true (c) true (d) false (e) true (f) true (g) false (h) false
   (i) false (j) false (k) true (l) true (m) false (n) true (o) false (p) false
   (q) true (r) false

CHAPTER 10 REVIEW EXERCISES (page 132)

1. (a) +3 (b) −15 (c) −1 (d) +10 (e) −3 (f) −16 (g) +4 (h) +4 (i) −45
   (j) +42 (k) +1 (l) −1 (m) +6 (n) +49 (o) +37 (p) −71 (q) −20
   (r) +1 (s) −32 (t) −3 (u) −27 (v) −5
2. (a) +18 (b) +4 (c) −12 (d) −15 (e) +16 (f) −62 (g) −9 (h) +7
   (i) −37 (j) +36 (k) +9 (l) −17 (m) +54 (n) −22 (o) −24 (p) −2
   (q) −10 (r) +11
3. (a) −16 (b) +20 (c) +17 (d) +10 (e) −21 (f) +9

EXERCISES 11.1 (page 137)

1. (a) $8x$ (b) $17y$ (c) $14m$ (d) $8y$ (e) $5x$ (f) $-2x$ (g) $7x$ (h) $-9t$ (i) $5x$
   (j) 0 (k) n (l) $-3x$ (m) $-6t$ (n) $-5a$ (o) $-2y$ (p) $11t$ (q) b (r) $-3y^2$
   (s) $-3x^2$ (t) $-11t^2$
2. (a) $12x + 7$ (c) $-2y$ (e) $6mn + 5m$ (g) $6x^2 + 5x$ (i) $2x + 3y + 9z$ (k) 7
   (m) $-3x^2$ (o) $8y^2 - 8y$ (q) $1 - 9t$ (r) $-3ab + 5$ (s) $-10a^2 + 16b$
3. (a) 2 (b) 1 (c) -5 (d) -1 (e) y (f) $3y$ (g) $-5y$ (h) $-y$ (i) $25my^2$
   (j) w (k) $-ab$ (l) $-t$
4. (a) $13.2x$ (c) $8.6x^2$ (e) $\frac{17}{12}x$ (g) $\frac{5}{8}x$ (i) $\frac{1}{2}t$

EXERCISES 11.2 (page 142)

1. (a) x^{12} (b) y^{13} (c) m^{14} (d) t^5 (e) cannot simplify   (f) x^{21} (g) a^{40}
   (h) cannot simplify   (i) y^{120} (j) x^{21}
2. (a) $24y^2$ (c) $5m^{10}$ (e) $3x^{11}$ (g) $8x^{11}$ (i) $-6a^3b^7$ (k) $10x^{10}y^7$ (m) $-30u^{10}v$
3. (a) $6x + 7$ (b) $114 - 42x$ (c) $-y - 10$ (e) $6x$ (g) -4 (h) $6 + 16t$ (k) $33x + 3$
   (m) $17x + 6$ (o) 8 (p) $13 - x$ (q) $11 - 11x$ (r) $7y - 2$
4. (a) $7x^2 + 12x - 1$ (c) $6x^2 - 22x + 1$ (d) $-2x^2 - 6x + 20$ (e) $-15x + 11$
5. (a) $-15x + 20$ (b) $-44x + 54$ (c) $53 - 43x$ (e) $5x - 75$ (f) $8x + 4$ (h) $13x + 7$
   (i) $6 - 2x$ (j) $4x + 22$ (l) $-66x - 85$
6. (a) $-x + 14y - 78$ (c) $28x - 26y$ (e) $22x - 45y + 21$ (g) $-2x + 24y + 95$

EXERCISES 11.3 (page 146)

1. (a) $15x^6 + 30x^2$ (c) $56y + 24y^6$ (e) $5x^9 + 8x^7$ (g) $48m^{16} - 96m^7$ (i) $6x^2 + 30$
   (k) $27x^2 + 6x$ (m) $7x^2 + 8x^3$ (o) $25y - 10y^2$
2. (a) $x^2 + 5x + 6$ (c) $x^2 + 2x - 35$ (e) $y^2 - 10y + 16$ (g) $cd - 6c + 5d - 30$
   (i) $ab + 6a + 2b + 12$ (k) $m^2 - 1$ (m) $9 - x^2$ (o) $14 + 5x - x^2$
3. (a) $20x^2 + 17x + 3$ (c) $8x^2 + 2x - 1$ (e) $3x^2 - 11x + 10$ (g) $7m^2 + 30m - 25$
   (i) $6mn + 10m + 21n + 35$ (k) $10 + 23x + 9x^2$ (m) $6x^2 - 24$ (o) $2x^2 + 3xy + y^2$
4. (a) $m^2 + 2mn + n^2$ (c) $a^2 - 2ab + b^2$ (e) $x^2 + 14x + 49$ (g) $y^2 - 10y + 25$
5. (a) $x^3 + 5x^2 + 10x + 12$ (c) $x^3 - 9x^2 + 20x - 12$ (e) $x^3 + 6x^2 + 3x - 10$
   (g) $2x^4 - 11x^3 + 15x^2 - 17x - 5$ (i) $15x^2 - 44xy + 32y^2 + 41x - 64y + 14$

EXERCISES 11.4 (page 151)

1. (a) $x^2 + 6x + 1$ (c) $3y^2 - 2y - 1$ (e) $4x^3 + x + 2$ (g) $4t^4 - 2t^3 - t + 3$
   (i) $3x^2 + 9x - 10$
2. (a) x^{13} (b) x^6 (c) y^2 (d) m^8 (e) a^7 (f) x^3 (g) b^5 (h) u^9 (i) n^8 (j) t^8
3. (a) $x + 8$ (c) $x + 5$ (e) $x^2 + 5x - 3$ (g) $7y^2 - 19$ (i) $4x - 3$
4. (a) $x^2 + 7x + 8$, R: 21   (c) $x^2 + 4x - 10$, R: 26   (e) $x^2 + 2x + 17$, R: 87

5. (a) $x^2 + x + 2$ (c) $3x^3 - 2x^2 + 2x + 5 + \dfrac{2}{x + 1}$ (e) $x^2 - x + 1$ (g) $x^3 + x^2 + x + 1$

6. (a) $x + 5 + \dfrac{14}{x - 2}$ (c) $x^2 + 10x + 43 + \dfrac{239}{x - 5}$ (e) $5x^2 + x$ (g) $x^2 + 5x - 3 - \dfrac{5}{x - 2}$

CHAPTER 11 REVIEW EXERCISES (page 152)

1. (a) $-x$ (b) $-4x - 4y$ (c) $13x - 12$
2. (a) x^{17} (c) x^7 (e) $7x^{13}$ (f) $3u^{14}$
3. (a) $-x - 17$ (b) $7 + 9x$ (c) $6x + 11$ (d) $16 - 47x$
4. (a) $6x^5 + 21x^2$ (b) $4x^6 + 36x^2$ (c) $m^2 + 10m + 21$ (e) $x^2 + 20x + 100$
   (g) $9y^2 - 18y + 5$ (i) $x^3 + 5x^2 - 33x + 27$ (j) $10x^3 + 29x^2 - 23x - 7$
5. (a) $2x^2 - 3x - 1$

6. (a) $x - 10$ (b) $3x^2 + 2x + 5 - \dfrac{3}{x - 4}$ (c) $x^2 + 2x + 4$ (e) $x^2 + 2x - 2 + \dfrac{12}{x + 3}$

CUMULATIVE REVIEW EXERCISES (CHAPTERS 10–11) (page 153)

1. (a) -11 (b) -30 (c) $+25$ (d) -16 (e) -24 (f) $+5$
2. (a) 20 (b) 50 (c) 10
3. (a) $-x$ (b) $9xy + 4x + 2y$
4. (a) -7 (b) 1 (c) $2/7$
5. (a) $12x^3$ (b) $30x^{16}$ (c) cannot simplify   (d) $3x^5y^2$
6. (a) $12x + 3$ (b) $27x - 20$
7. (a) $2y^2 - 11y - 6$ (b) $x^2 - 18x + 81$
8. (a) x^3 (b) y^{13}
9. $x^2 + x - 6$, R: -5

EXERCISES 12.2 (page 156)

1. (a) 5 (b) 3 (c) 9 (d) 17 (e) -3 (f) -7 (g) -9 (h) 160 (i) 0 (j) 2
   (k) -6 (l) -6 (m) 9 (n) -26 (o) 0 (p) 5
2. (a) $3/8$ (b) $1/12$ (c) 5.6 (d) 1.6 (e) $15/8$ (g) 0 (i) -1.2 (k) -14.1

EXERCISES 12.3 (page 158)

1. (a) 3 (b) 6 (c) 48 (d) 21 (e) $17/4$ (f) $3/8$ (g) $2/5$ (h) $19/3$ (i) -5
   (j) -14 (k) -12 (l) -6 (m) $-17/5$ (n) $-5/8$ (o) 10 (p) 92 (q) $9/5$
   (r) $1/6$ (s) -19 (t) 7 (u) 0 (v) 0 (w) 20 (x) 22.6 (y) -500 (z) 6.5
2. (a) $65/4$ (c) -16 (e) $-63/8$ (g) $17/65$ (i) 0
3. (a) -20 (c) $-1/4$ (e) $9/2$ (g) 0 (i) $37/6$

EXERCISES 12.4 (page 161)

1. (a) 14 (b) $28/3$ (c) 15 (d) -24 (e) $-9/10$ (f) $8/5$ (g) -28 (h) $99/10$
   (i) $4/15$ (j) $11/91$ (k) $77/12$ (l) $6/35$
2. (a) $45/2$ (c) 14 (e) $-13/42$ (g) $-35/12$ (i) $6/5$
3. (a) $-3/7$ (c) -54 (e) $-27/44$ (g) $-6/5$ (i) $-27/7$

EXERCISES 12.5 (page 164)

1. (a) 10/3 (b) 2 (c) −15/7 (d) −1 (e) −3/7 (f) 0 (g) −1 (h) −44/7 (i) 6
(j) −75 (k) 6 (l) 5/7 (m) 4 (n) 5 (o) −35 (p) 75/2
2. (a) 25/3 (b) 20 (c) 12/7 (d) 2/3 (e) 1/15 (f) 0 (g) −3/10 (h) −4/15
(i) −2
3. (a) 39 (c) −6 (e) −4 (g) 1/12 (i) 4 (k) 2

EXERCISES 12.6 (page 168)

1. (a) 5/3 (b) −11/3 (c) −3/2 (d) 4 (e) 7/3 (f) −8 (g) 19/18 (h) −2 (i) 5/7
(j) −14/11 (k) 11/12 (l) −9 (m) 29/11 (n) −2 (o) 21/13 (p) −3 (q) −5/2
(r) 4/5
2. (a) 1/5 (c) −5/3 (e) −6 (g) 17 (i) 7/2 (k) 1/5 (m) −8 (o) 5/3 (q) 25/7
(s) 3/2 (u) 19/25 (w) 1/10
3. (a) −1 (b) −32 (c) 72/17 (d) 1 (e) 1.25 (f) 2

EXERCISES 12.7 (page 171)

1. (a) yes (b) no (c) no (d) yes (e) yes (f) yes (g) no (h) no (i) yes
(j) yes
3. (a) 4 (b) 3 (c) 0 (d) 12/5 (e) 3 (f) 1 (i) −1/7

EXERCISES 12.8 (page 172)

1. (a) none (b) infinite (c) one (e) none (g) infinite

CHAPTER 12 REVIEW EXERCISES (page 172)

1. (a) 43 (b) 17/5 (c) −19/6 (d) 4/7 (e) −32 (g) −11/9 (i) −11/4 (k) 7
(m) 35/4 (n) −9/5 (o) −1/4 (q) 12/7 (s) −5/3 (u) 4

CUMULATIVE REVIEW EXERCISES (CHAPTERS 10–12) (page 173)

1. (a) +3 (b) −36 (c) 9 (d) 16
2. 33
3. (a) $12x^2 + 4x$ (b) $2 - 8x$
4. (a) $20x^{10}$ (b) cannot simplify
5. (a) $6x - 7$ (b) $13x - 20$
6. (a) $x^2 - 15x + 56$ (b) $x^2 - 25$
7. $x^2 + x + 9$
8. (a) 12 (b) −12/7 (c) 4/5 (d) 11/2 (e) −45 (f) −2/9 (g) 3/8 (h) 1

EXERCISES 13.2 (page 177)

1. (a) $y + 4$ (b) $w - 17$ (c) $5x + 10$ (d) $\frac{1}{2}m$ or $\dfrac{m}{2}$ (e) $2t$ (f) $2x + 3$ (g) $3y - 5$

(h) $n + m$ (i) $y - x$ (j) xy (k) $7 \cdot 6c$, which simplifies to $42c$ (l) $.23x$ or $\dfrac{23}{100}x$

(m) $\dfrac{k}{100}$ or $.01k$ (n) $\dfrac{k}{100}x$ or $.01kx$ or $\dfrac{kx}{100}$

2. (a) $5t$ (c) $v/100$ (e) $5x + 50y$ (g) $150x + 300y$
3. (a) $2p$ (c) $60H$ (e) $T/24$ (g) $15y$ (i) $p/12$ (k) mx (m) $ax + by$
   (o) $x - y$ (not $y - x$)   (p) $\$5000 - x$

4. (a) $.06d$ or $6d/100$ or $\dfrac{6}{100}d$ (b) $2dp/100$ or $.02dp$ (e) $.01mnw$ or $mnw/100$

(f) $x + .23x$ or $1.23x$ or $123x/100$
5. (a) $1000x$ (b) $y/1000$ (c) $300x$ (d) $2000y$ (e) $m/1000$ (f) $1000x$ (g) $y/1000$
6. (a) $x + y + z$ yards   (b) $2x + 2y$ or $2(x + y)$ (c) xy (d) $4x$ (f) $(x + 2)7$ or $7x + 14$
   (h) $x \cdot 2x$, which simplifies to $2x^2$ (j) $2x + 2(x + 4)$, which simplifies to $4x + 8$ (l) $a + b + c$

EXERCISES 13.3 (page 184)

1. 37, 38, 39   3. 114, 228   5. 73, 80   7. $7\frac{1}{2}$, $9\frac{1}{2}$ meters   9. 225   11. 260   13. $55
15. $43.75   17. $2375   19. $280   21. $1080   22. $17,200   24. $325   26. 4
28. $15.40   30. length 312 meters, width 104 meters   32. 91 centimeters   34. 41°, 57°, 82°
36. 61 inches each   38. 18°, 72°   40. 38°, 142°   42. 13 inches   44. 8.5 centimeters
45. 4.5 hours   47. 5.5 hours   49. 3.75 mph   51. 15   53. 60 miles

EXERCISES 13.4 (page 192)

1. 78 dimes, 39 nickels   3. 11 quarters, 33 dimes   4. 23   5. 19 25¢ stamps, 23 15¢ stamps
7. 75 $4 tickets, 27 $5 tickets   8. 72 general public tickets, 144 student tickets
10. 5 quarters, 10 dimes, 14 nickels   12. 8 dimes, 7 nickels   13. Charles is 32, Liz is 16
15. Ginger is 23, Yung is 17   17. 51 years ago   19. 26   21. $2200 at 5%, $2300 at 6%
23. $7200 at 9%, $3600 at 6%   25. $2600 at 8%, $2800 at 5%   27. 85 mph, 115 mph
29. Jack 40 mph, Wilbur 55 mph   31. 385 mph   32. 3.25 hours   33. train 85 mph, car 51 mph

CHAPTER 13 REVIEW EXERCISES (page 195)

1. (a) $ab - 15$ (b) $10x + 100y$ (c) $17x/3$ (d) $2x + 2(x - 4)$, which simplifies to $4x - 8$
   (e) $213cx/100$ or $2.13cx$ (f) $1.37y$ or $137y/100$
2. (a) 42 and 126   (b) 105
3. (a) 17 nickels, 15 quarters   (b) $6500 at 9%, $3500 at 8%

CUMULATIVE REVIEW EXERCISES (CHAPTERS 10–13) (page 195)

1. (a) -13 (b) 196   (c) -5 (d) 5
2. (a) $-11x$ (b) $8 - 5y$
3. (a) $15m^9$ (b) $-20x^3y^{12}$ (c) $-7x - 6$ (d) $-y^2 + 10y - 9$

4. (a) $x^2 - x - 30$ (b) $u^2 + 10u + 25$

5. $x^2 + 3x - 9$, R: 26

6. (a) 16/9 (b) 4/5 (c) $-7/5$ (d) $-1/2$

7. (a) $6t$ (b) $.47y$ or $47y/100$ (c) $6x + 16$

8. (a) \$24.60 (b) 48 inches (c) 20

EXERCISES 14.1 (page 200)

2. (a) $(1, 1)$ (b) $(1, -3)$ (c) $(-5, 3)$ (d) $(6, 4)$ (e) $(-3, -1)$ (f) $(-1, 3)$ (g) $(7, -1)$
(h) $(2, 6)$ (i) $(-6, -2)$ (j) $(-2, -3)$ (k) $(-4, 0)$ (l) $(0, 6)$ (m) $(0, 0)$ (n) $(6, 0)$
(o) $(0, -1)$ (p) $(4, -2)$

EXERCISES 14.2 (page 210)

1. (a)   (c)   (e)   (g)

(i)   (k)   (m)   (o)

(q)   (s)   (u)   (w)

2. (a)   (c)   (e)   (g)

3. (a)

(c)

(e)

(g)

(i)

(k)

(m)

4. (a)

(c)

5. (a)

(c)

6. (a)

(b)

(c)

(e)

(g)

(i)

(k)

(m)

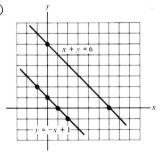

EXERCISES 14.3 (page 215)

1. (a) 1/2   (c) 3   (d) −1   (e) 5   (g) −4   (h) 5/4   (i) −1   (k) −7/9

2. (a)   (c)   (e)

(g)   (i)   (k)

(m)

EXERCISES 14.4 (page 219)

1. (a) 9   (c) −6   (e) 1   (g) 0
2. (a) slope = 5, y-intercept = 3   (c) −9, 4   (e) −1, 7   (f) 1, −8   (g) 12, −7   (h) 6, 0
3. (a) $y = 5x + 2$ (b) $y = 4x + 7$ (c) $y = -2x + 1$ (d) $y = 6x - 1$ (e) $y = 3x$
   (f) $y = -3x$ (g) $y = -x + 4$ (i) $y = x$
4. (a) $y = 3x - 11$ (c) $y = x + 4$ (e) $y = -4x + 2$ (g) $y = -2x - 3$ (i) $y = 12x + 9$
5. (a) $y = 2x - 1$ (c) $y = 4x - 1$ (e) $y = -3x + 17$ (g) $y = 2x + 8$ (i) $y = -3x + 5$
6. (a) parallel   (c) intersecting   (d) parallel   (e) intersecting   (f) intersecting
   (g) intersecting
7. (a) $y = 2x + 7$ (c) $y = -3x - 1$ (d) $y = -4x - 12$ (e) $y = -x - 5$ (g) $y = -7x + 14$

CHAPTER 14 REVIEW EXERCISES (page 220)

1. (a)   (b)   (c)

(d)   (e)   (f)

(g)   (h)   (i)

(j)

2. (a) 6   (b) −3
3. (a) slope = −7, y-intercept = 9   (b) slope = 6, y-intercept = −2
4. (a) $y = 4x + 3$ (b) $y = -2x + 10$ (c) $y = 4x - 10$ (d) $y = 5x + 5$

CUMULATIVE REVIEW EXERCISES (CHAPTERS 10−14) (page 220)

1. (a) 4   (b) −79   (c) −49   (d) +8
2. (a) $6x^7y^8$ (b) $-24m^8n^8$
3. (a) $15x^2 + 2x - 8$ (b) $t^2 + 8t + 16$
4. $x^2 - 5x + 21$, R: −85
5. (a) 19/9   (b) −6   (c) 5/13   (d) 8/3
6. (a) 85, 86, 87, 88   (b) $5700   (c) 30°, 60°   (d) $3200 at 7%, $1300 at 8%
7. (a)   (b)

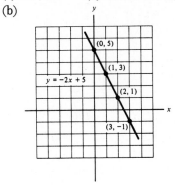

8. 2
9. $y = 4x - 11$

EXERCISES 15.2 (page 226)

1. (a) $x = 6, y = 7$ (c) $x = 4, y = 1$ (d) $x = -2, y = -1$ (e) $x = 2, y = 1$
   (f) $x = 1, y = -3$ (g) $x = -2, y = 3$ (i) $x = -1, y = 2$ (k) $m = 5, n = 0$
2. (a) $x = 2, y = 1$ (b) $x = 2, y = 3$ (c) $x = 3, y = 2$ (e) $x = 4, y = -3$ (g) $m = 5, n = 3$
   (i) $x = -2, y = 3$ (j) $a = -3, b = 5$ (k) $x = 4, y = 7$ (m) $x = 1/2, y = 3$
3. (a) $x = 1, y = 1$ (c) $x = -1, y = 10$
4. (a) hot dogs 75¢, milkshakes 60¢   (c) 14 and 27

EXERCISES 15.3 (page 229)

1. (a) $x = 2, y = 3$ (c) $x = 4, y = 2$ (e) $x = -1, y = -2$ (g) $x = 4, y = 4$
   (i) $x = 7/19, y = 54/19$ (k) $x = 3, y = 2$ (m) $a = -2, b = 3$ (o) $x = 11, y = 11$
2. (a) $x = 0, y = -1$ (c) $x = -2, y = 1/2$ (e) $x = -7, y = 8$
3. (a) $x = 2, y = 1$ (c) $x = 4, y = -3$
4. (a) width 115 meters, length 230 meters   (c) 24, 96   (e) 10°, 80°

EXERCISES 15.4 (page 231)

1. (a) $x = -1, y = 4$ (c) $x = 5, y = 3$ (e) $x = 11, y = 7$

EXERCISES 15.5 (page 232)

1. (a) $x = 4, y = 1$ (b) $x = 3, y = 1/2$ (c) Infinite number of solutions; both lines are the same.
   (d) $x = 6, y = 2$ (e) No solution; parallel lines.   (f) No solution; parallel lines.
   (g) $x = 3, y = 3$ (h) Infinite number of solutions; both lines are the same.

EXERCISES 15.6 (page 233)

1. (a) $x = 1, y = 2, z = 3$ (b) $x = 1, y = -1, z = 1$ (c) $x = 2, y = 0, z = -3$
   (d) $x = 5, y = 1, z = 0$ (e) $x = 1, y = -2, z = 4$ (f) $x = 4, y = 1, z = -2$
   (g) $r = 2, s = -3, t = 4$ (h) $m = 0, n = -1, p = 2$

CHAPTER 15 REVIEW EXERCISES (page 234)

1. (a) $x = -4, y = -6$ (b) $a = 4, b = -1$
2. (a) $x = 3, y = -1$ (b) $x = 4, y = 5$
3. (a) $m = -1, n = 1$ (b) $x = 1/4, y = -2$

CUMULATIVE REVIEW EXERCISES (CHAPTERS 10–15) (page 234)

1. (a) 1   (b) 22   **2.** (a) $15x^7 + 35x^9$ (b) $x^2 + x - 56$
3. $x^2 + 5x + 3$ **4.** $-2m$ **5.** (a) $-4/7$ (b) $7/22$
6. (a) width 48 feet, length 99 feet   (b) 60 mph, 85 mph
7.

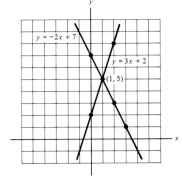

8. $y = -2x + 14$
9. $y = 4x + 15$
10. $x = -3, y = 5$
11. $x = 6, y = -4$

EXERCISES 16.1 (page 238)

1. (a) $(3, 8)$ (b) $[5, 6]$ (c) $[0, 5)$ (d) $(1, 19]$ (e) $[10, 100]$ (f) $[-6, 0)$ (g) $(-4, -1]$
   (h) $(-10, 10)$
2. (a) $2 < x < 5$ (b) $2 \le x \le 5$ (c) $1 < x \le 7$ (d) $3 \le x < 11$ (e) $-3 \le x \le 0$
   (f) $-2 < x \le 4$ (g) $-7 \le x \le -3$ (h) $0 < x < 7$
3. (a) ⟶ x   (b) ⟶ y

(c) 0   1

(d) −5   3   t

(e) 4   12   x

(f) 5   8   x

(g) 9   10   x

(h) −4   1   x

(i) 4   x

(j) −1   x

(k) −2   2   x

(l) −3   2   t

EXERCISES 16.2 (page 241)

1. (a) $x < 5$ (b) $t > -24$ (c) $x > 6/5$ (d) $x \le -4/3$ (e) $y < -7/4$ (f) $n > -9/14$
(g) $x < -1$ (h) $x \ge -13/7$ (i) $x > 11/3$ (j) $x \le -1/4$
2. (a) $y > 6$ (c) $x > -3/2$ (e) $x \ge 6$ (g) $t > -16/3$ (i) $x \ge 9/2$ (k) $x < 1/4$
3. (a) at least 62 inches   (b) at least 64   (c) 66 feet or less   (e) all numbers greater than 24
(f) all numbers less than 168

EXERCISES 16.3 (page 245)

1. (a)

(c)

(e)

(g)

(i)

(k)

$2x + 3y < 12$

(m)

$x \geq 5$

(o)

$y \geq 0$

2. (a)

(c)

(e)

(g)

(i)

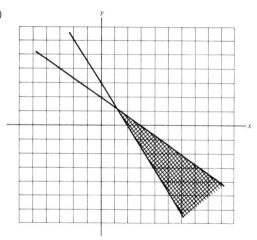

3. (a) $x \geq 40$ (b) $x < 100$ (e) $x > y$ (f) $x \geq 40, x \leq 60$ or $40 \leq x \leq 60$
(h) $x \geq 20, y \geq 40, x + y \leq 75$

CHAPTER 16 REVIEW EXERCISES (page 246)

1. (a) $(-4, 6)$ (b) $[-5, 0)$ (c) $[78, 100]$ (d) $(-9, 7]$
2. (a) $-3 \leq x \leq 10$ (b) $2 < x \leq 11$
3. (a) $x < 5$ (b) $x < -13/2$ (c) $x \leq 3/5$ (d) $x > 7/8$ (e) $x > -13/28$ (f) $x \geq -2$
4. (a) (b) (c)

(d)

5. (a) (b)

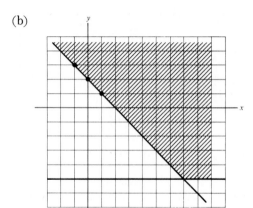

CUMULATIVE REVIEW EXERCISES (CHAPTERS 10–16) (page 246)

1. (a) 6   (b) 27   **2.** 22   **3.** $6y - 11x$ **4.** $-72x^9y^9$
5. $32x + 9$ **6.** $4x^2 + 51x - 135$
7. (a) 16/5   (b) $-11/2$ (c) $-7/10$ (d) -10
8. (a) $.21hk$ or $21hk/100$ (b) $85x$ (c) $(x + 3y)/2$
9. (a) 21 centimeters   (b) 32 25¢ stamps and 15 40¢ stamps
10.

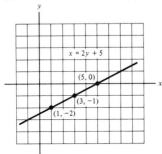

11. $y = 5x + 14$
12. (a) $x = -5, y = 2$
       (b) $x = 1, y = -4$
13. (a) $x > -19/2$
       (b) $x \geq -2/3$

EXERCISES 17.1 (page 250)

1. (a) $(7 + n)x$ (b) $(m + n)x$ (c) $(p + q)r$ (d) $(3 - t)y$ (e) $5w(x + 2y)$
     (f) $3x(y + 4z)$ (g) $6y(2x - 3z)$ (h) $5mn(3 + 4p)$ (i) $x(x + 6)$ (j) $y(y - 8)$
     (k) $a(a + 1)$ (l) $y(y - 1)$ (m) $m(5 - m)$ (n) $x(3x - 1)$ (o) $5mn(4m + 3n)$
     (p) $6xy^2(3x - 4)$ (q) $x(6 + 7x)$ (r) $t(17t - 6)$ (s) $3(x + 1)$ (t) $10(1 - y)$
2. (a) $5(2x + 1)$ (c) $8(1 + 5x^2)$ (e) $x^3(x^2 + 1)$ (g) $ab(a - 2c^4)$ (i) cannot be factored
     (k) $(p + q + r)c$ (m) $m(m^2 + m + 1)$ (o) $(x + 3 + w)yz$ (q) $10m^2(10n + 5n^2 + 2)$
     (r) $ab(5b + 10a + ab)$ (s) $3u(5v^2 - 2u + 4v)$ (t) cannot be factored   (u) $mx(m + x + 1)$
     (v) cannot be factored

EXERCISES 17.2 (page 256)

1. (a) $x^2 + 5x + 6$ (b) $x^2 + 14x + 45$ (c) $m^2 + 10m + 24$ (d) $m^2 + 2m + 1$
     (e) $x^2 - 7x + 10$ (g) $m^2 - 10m + 24$ (i) $y^2 + 2y - 15$ (k) $x^2 - 49$ (l) $x^2 - 25$
     (m) $m^2 - 5m - 24$ (o) $y^2 + 5y - 24$ (q) $x^2 - 7x - 18$ (s) $m^2 - 12m + 20$
2. (a) $(x + 3)(x + 2)$ (b) $(x + 5)(x + 1)$ (c) $(x + 2)(x + 2)$ (d) $(x - 5)(x - 1)$
     (e) $(t - 7)(t + 1)$ (f) $(x - 10)(x - 2)$ (g) $(m - 6)(m + 1)$ (h) $x(x + 3)$
     (i) $(x + 4)(x - 3)$ (j) $a(a - 9)$ (k) $(y - 6)(y + 2)$ (l) $(m + 6)(m - 5)$
     (m) $(x + 2)(x - 1)$ (n) $(y + 8)(y - 3)$ (o) $(x - 3)(x - 5)$ (p) $(x - 7)(x - 5)$
     (q) $(y - 11)(y + 1)$ (r) $(x + 13)(x - 1)$ (s) $(x)(x - 8)$ (t) $(x - 3)(x - 1)$
     (u) $(n - 3)(n - 7)$ (v) $(n - 3)(n + 2)$ (w) $(x + 7)(x - 3)$ (x) $(x + 4)(x - 1)$
     (y) $(m)(m + 10)$ (z) $(x - 2)(x - 5)$
3. (a) $(x - 4)(x - 5)$ (c) cannot be factored   (e) $(x - 1)(x - 1)$ (g) $(x + 6)(x - 6)$
     (i) cannot be factored   (k) cannot be factored   (m) $(x + 5)(x + 5)$ (o) $(x - 5)(x + 3)$
     (q) $(t + 5)(t - 4)$ (s) $x(x - 9)$ (u) $(x - 5)(x - 9)$ (w) $m(m - 8)$ (y) $x(x + 9)$

4. (a) $(x + m)(x - m)$ (c) $(x + 9y)(x - 9y)$ (e) cannot be factored
    (g) $(3x + 5y)(3x - 5y)$ (i) $(10y + 7x)(10y - 7x)$ (k) cannot be factored
    (m) $(5 + 2t)(5 - 2t)$ (o) $(4 + 3x)(4 - 3x)$ (q) $(1 + n)(1 - n)$ (s) $(3 + x)(3 - x)$
    (u) $(8 + y)(8 - y)$

EXERCISES 17.3 (page 260)

1. (a) $3(m + 3)(m + 2)$ (c) $2(x - 10)(x + 1)$ (e) $3(x + 1)(x - 1)$ (g) $5(x - 3)(x - 3)$
    (i) $7x(x - 3)$ (k) $2(t - 4)(t - 1)$ (m) $4(u + 14)(u - 1)$ (o) $5(2 + x)(2 - x)$
    (q) $3y(y + 1)$
2. (a) $(2x + 3)(x + 5)$ (b) $(3x + 2)(x + 1)$ (c) $(3x + 1)(x - 2)$ (d) $(5y - 2)(y - 3)$
    (e) $(5n - 2)(n + 3)$ (g) $(3x + 2)(2x - 3)$ (i) $(6x + 1)(x + 10)$ (k) $(2x - 3)(2x + 5)$
    (m) $(5t + 2)(t + 4)$ (o) $(3y - 7)(y + 2)$ (q) $(2x - 3)(x + 5)$
3. (a) $(5y + 2)(y + 4)$ (b) $(2m - 3)(m + 4)$ (c) $(6x + 1)(x + 10)$ (d) cannot be factored
    (e) $2(y - 3)(y - 3)$ (g) $3(m + 5)(m - 5)$ (i) $2(3x - 5)(x - 1)$ (k) $(4x - 9)(x - 1)$
    (m) $2(x^2 + 1)$ (o) $2(x + 1)(x - 1)$ (q) cannot be factored   (s) $(3x + 5)(x - 2)$
    (u) $(5x - 6)(x - 2)$ (w) $(4t - 3)(t + 1)$

EXERCISES 17.4 (page 265)

1. (a) $-1, -6$ (b) $-1, -3$ (c) $1, 5$ (d) $2, 5$ (e) $-2, 3$ (f) $-8, 1$ (g) $-1, -1$
    (h) $-1, 3$ (i) $-4, 4$ (j) $-7, 7$ (k) $0, -7$ (l) $0, 8$ (m) $3, 8$ (n) $-5, 6$
    (o) $-2, 6$ (p) $-3, -3$
2. (a) $-4, 5$ (b) $0, 1$ (c) $-1, 1$ (d) cannot be factored   (e) cannot be factored   (g) $1, 8$
    (i) $-5, -1/2$ (k) $-4, 2/3$
3. (a) $-1, 3/5$ (b) $-2/3, -2$ (c) $-5/2, 1/2$ (e) $-3, 3$ (g) $4, 6$ (i) $-4/3, -3/2$
    (k) cannot be factored   (m) $-5, 3$ (n) $0, 8$ (o) $-2, 7$ (q) $-5/2, 3$
    (r) cannot be factored
4. (a) cannot be factored   (c) $1, 7$ (e) $-2, 5/3$ (f) $-8, -4$ (g) $-1/2, 7$ (i) $3/4, 2$
    (k) $-1, -2/3$ (m) $-3, 3$ (o) $-2, 9$
5. (a) at $t = 16$ sec.   (c) at $t = 30$ sec.   (e) at 10 seconds and at 20 seconds
6. (a) length 21 meters, width 7 meters   (c) length 12 feet, width 8 feet   (e) 3 or -12
    (f) -9 or 9 (g) -20 or 0

CHAPTER 17 REVIEW EXERCISES (page 267)

1. (a) $(m + n + p)x$ (b) $ab^2c(a + c)$ (c) $13b(2a + 3c)$ (d) $x(x - 10y)$ (e) $10(1 - 3y)$
    (f) cannot be factored   (g) $(x + 5)(x - 2)$ (h) $(x - 13)(x - 1)$ (i) $x(x - 21)$
    (j) $m(m + 30)$ (k) $(n + 10)(n - 10)$ (l) $(y + 1)(y - 1)$ (m) $(2x + 3)(2x + 1)$
    (n) $2(x - 7)(x + 2)$ (o) $(4x - 9)(x + 1)$ (p) $3(x + 4)(x - 4)$
2. (a) $-12, -1$ (b) $1, 1$ (c) $5, 5$ (d) $-2, 7$ (e) cannot be factored   (f) $-8, -2$
    (g) $-12, -2$ (h) $-2, -7$ (i) $-2, 3$ (j) cannot be factored   (k) $3, 5$ (l) $-2, 7/2$
    (m) $-3, 1/5$ (n) $-1, 5$ (o) $-7, 7$ (p) $0, 12$

CUMULATIVE REVIEW EXERCISES (CHAPTERS 10–17) (page 267)

1. (a) 74 (b) -50 (c) -33 (d) -38 (e) $+1$ **2.** (a) $-x$ (b) $6 - 7x$
3. (a) $-11/3$ (b) $75/2$ (c) $-27/2$ (d) $7/39$ (e) $9/4$

4. (a) x^{14} (b) y^{38} (c) m^{21} (d) x^4 (e) y^8 (f) m^6
5. (a) $x^2 + x + 7$, R: 3   (b) $x^2 - 6x + 3$
6. (a) $z + 10$ (b) $5x - 9$ (c) $7x/12$ (d) $.07qr$ or $7qr/100$
7. (a)   (b)   (c)

(d)

8. $-3/2$
9. (a) $x = 3$, $y = -1$ (b) $x = -1$, $y = 1$
10. (a) $x < -13$ (b) $x \le -6/5$
11. (a) $x(x + 1)$ (b) $(7 + n)y$ (c) cannot be factored
  (d) $(x + 12)(x - 2)$ (e) $(x + y)(x - y)$ (f) $5x(x + 2)$
  (g) $(3x - 4)(x + 7)$ (h) $(2x - 5)(x + 4)$
12. (a) 6, 6 (b) $-18, 2$ (c) $-3, 5$ (d) $-5, -3$ (e) 2, 9
  (f) 0, 1 (g) $-7, 7$ (h) 2, 5

EXERCISES 18.1 (page 270)

1. (a) $\dfrac{10}{21}$ (b) $\dfrac{8}{45}$ (c) $\dfrac{xy}{15}$ (d) $\dfrac{7b}{6t}$ (e) $\dfrac{5x + 5}{14}$ (f) $\dfrac{2x - 8}{5x}$ (g) $\dfrac{4x + 20}{x^2}$

  (h) $\dfrac{x^2 + 5x + 4}{x^2}$ (i) $\dfrac{3m^2n^2}{5t}$ (j) $\dfrac{x^2y^2}{16w^2}$ (k) $\dfrac{x^2 - 2x}{21}$ (l) $\dfrac{3w^2}{2x}$

2. (a) $\dfrac{6x^2y}{35}$ (c) $\dfrac{10r^3w^2}{3x^2}$ (e) $\dfrac{35x}{y^5}$ (g) $-\dfrac{15x^2y^2}{14z}$ (i) $-\dfrac{12tu^5}{7v^3}$

3. (a) $-\dfrac{17x^4 + 17}{12x^2}$ (c) $\dfrac{x^2 + 4x + 3}{(x - 4)^2}$ (e) $\dfrac{5x^2 + 32x - 21}{3x^2}$

EXERCISES 18.2 (page 275)

1. (a) $\dfrac{1}{5}$ (b) $\dfrac{1}{50}$ (c) $\dfrac{3}{4}$ (d) $\dfrac{4}{5}$ (e) $\dfrac{x}{4}$ (f) $\dfrac{5}{2x^2}$ (g) $\dfrac{3}{x-1}$ (h) $\dfrac{1}{2(1-m)}$ (i) -1

(j) cannot be reduced (k) -1 (l) -1 (m) $\dfrac{1}{2}$ (n) $\dfrac{1}{x+1}$ (o) $\dfrac{1}{x-2}$

(p) $\dfrac{1}{x-3}$ (q) $x-4$ (r) $x+4$

2. (a) $\dfrac{x+3}{x+1}$ (c) cannot be reduced (e) $\dfrac{x+y}{4}$ (g) -2 (i) cannot be reduced (k) $\dfrac{10}{3}$

(m) -3 (o) $\dfrac{5}{7}$ (q) $-\dfrac{x+5}{2}$ (r) $\dfrac{3x+2}{3x-5}$ (t) $\dfrac{x+1}{2}$

EXERCISES 18.3 (page 277)

1. (a) 6 (b) $\dfrac{xy}{4}$ (c) $\dfrac{3}{2}$ (d) $\dfrac{10x}{a^2b^2c^2}$ (e) $\dfrac{x-5}{x+1}$ (f) $\dfrac{20}{(x+3)7}$ (g) $\dfrac{1}{2x^2}$ (h) $\dfrac{2n-8}{n+4}$

(i) $\dfrac{5}{x-3}$ (j) $\dfrac{(1-x)(x-3)}{x}$

2. (a) $x+3$ (c) $-\dfrac{y}{2x}$ (e) $\dfrac{x+4}{x}$ (g) $2x$ (i) $-\dfrac{4x}{3}$ (k) $-\dfrac{2}{7}$ (m) $\dfrac{(x-1)(5-x)}{x(x+1)}$

EXERCISES 18.4 (page 280)

1. (a) $\dfrac{3x+15}{7}$ (b) $\dfrac{1}{4}$ (c) $\dfrac{15xy}{x+3}$ (d) $\dfrac{x+1}{3x(x-1)}$ (e) $-\dfrac{1}{2}$ (f) $\dfrac{x+3}{3x}$ (g) $\dfrac{m-4}{3m}$

2. (a) $\dfrac{15}{x}$ (c) $\dfrac{x}{3}$ (e) $\dfrac{2x^2+2x+24}{5(x+3)(x+4)}$ (g) $-\dfrac{2x}{x+4}$ (i) $\dfrac{2(x+y)(x+4)^2}{xy(x+8)(x-8)}$

EXERCISES 18.5 (page 284)

1. (a) $\dfrac{t+3}{x}$ (b) $\dfrac{5+x}{y}$ (c) $\dfrac{7y+5x}{xy}$ (d) $\dfrac{4x+1}{x}$ (e) $\dfrac{3m+5}{m}$ (f) $\dfrac{2-9c}{c}$ (g) $\dfrac{11}{2x}$

(i) $\dfrac{15w+2x^3}{5w^2x^2}$ (k) $\dfrac{32cx+9y}{60abc}$ (l) $\dfrac{3bc+4ac+5ab}{abc}$ (n) $\dfrac{6adf+4bcf-3bde}{12bdf}$

(o) $\dfrac{10y+21x}{30x^2y^2}$ (q) $\dfrac{2bx+3ay}{36a^3b^2}$ (r) $\dfrac{5xy+4xz+3yz}{x^2y^2z^2}$ (s) $\dfrac{3+4abc-5ac}{a^2bc^2}$

2. (a) $\dfrac{26x+14}{(3x+2)x}$ (c) $\dfrac{2x^2+2x+7}{x+1}$ (e) $\dfrac{2x-1}{x-2}$ (g) $\dfrac{8x+13}{(x+1)(x+2)}$ (i) $\dfrac{38-6x}{(x-5)(x+3)}$

(k) $\dfrac{9x-1}{6x(x-1)}$ (m) $\dfrac{8x+15}{2x(x+3)}$ (o) $\dfrac{x^2+9x-4}{8x(x+1)}$ (q) $\dfrac{2x^2+7x+7}{6x(x+1)}$

3. (a) $\dfrac{2x+12}{x^2-4}$ (c) $\dfrac{10x+21}{(x+1)(x+2)}$ (d) $\dfrac{5x^2-3x+6}{x(x-7)(x-2)}$ (e) $\dfrac{2x-6}{(x+4)(x+2)(x+2)}$

(g) $\dfrac{-21x^2+34x+64}{2x^2(x+4)}$

EXERCISES 18.6 (page 288)

1. (a) $\dfrac{33}{20}$ (c) $\dfrac{ad}{bc}$ (e) $\dfrac{3}{7x}$ (g) $\dfrac{xy + 3y}{xy + 2x}$ (i) $x^2 - 4x + 3$

2. (a) $\dfrac{x + 1}{x - 1}$ (c) $\dfrac{b + a}{b - a}$ (e) $\dfrac{3 + xy}{2y + 3x}$ (g) $\dfrac{3xy - 42y}{48x + 2xy}$ (i) $\dfrac{mxy + x^2}{y^2 - mxy}$ (k) $\dfrac{9b}{a + 4b}$

EXERCISES 18.7 (page 293)

1. (a) 4/45   (c) 5   (d) −2/9   (e) 8   (f) 31/10   (g) −5   (i) −5
2. (a) 3   (c) −68/9   (e) −4
3. (a) −16   (b) 5/4   (c) 39/8   (d) −13/5
4. (a) 88   (c) 102 (impossible)
5. (a) $2\frac{2}{9}$ hours   (c) $2\frac{2}{11}$ hours
6. (a) 6 and 10   (c) 5/7

EXERCISES 18.8 (page 297)

1. (a) $x = 16$ (c) $x = 10$ (e) $m = 12\frac{1}{2}$ (g) $x = 12/7$ (i) $x = 75/14$
2. (a) 80   (b) 200   (c) 120   (d) 60   (e) $17\frac{1}{2}$
3. (a) 70 feet and 30 feet   (c) 25,000   (e) $1.45   (f) 1050 square feet   (g) 4 pounds
   (h) 8 gallons   (i) 20 inches   (k) 5 teaspoons

EXERCISES 18.9 (page 300)

1. (a) 14   (b) 7   (c) 133   (d) 2   (e) 10/7
2. (a) 10   (b) 3   (c) 30   (d) 1/2   (e) 60
3. (a) 22   (c) 126
4. (a) 112   (c) 8
5. (a) $c = 6$; $192   (b) 17 hours   (d) $c = 2800$; 175; 350

CHAPTER 18 REVIEW EXERCISES (page 301)

1. (a) $\dfrac{x}{2y}$ (b) $-4x$ (c) $\dfrac{3x^2}{2}$ (d) $\dfrac{2(x - 2)}{x + 3}$ (e) $\dfrac{wz + xy}{xz}$ (f) $\dfrac{29}{35x}$ (g) $\dfrac{9x + xy + y}{(x + 1)x}$

   (h) $\dfrac{25z + 45y - 21x}{15xyz}$ (i) $\dfrac{x + 5}{6x}$ (j) $-\dfrac{x + 1}{2}$ (k) $12bxy^2$ (l) $\dfrac{bce + ade + acf}{ace}$

   (m) $\dfrac{2t^2 - 2tu}{5u}$ (n) 2   (o) $\dfrac{10x + 114y + 3x^2}{6x^2y}$ (p) $\dfrac{25y + 24x}{90xy}$ (q) −1   (r) $\dfrac{15 + 4x}{5x(x - 1)}$

2. (a) $\dfrac{x^2 + x}{xy - y}$ (b) $\dfrac{3m - 1}{2 + m}$

3. (a) 17   (b) 5

4. (a) $2x - 2$ (c) $\dfrac{x - 3}{x + 2}$ (e) $\dfrac{x + 2}{2(x - 2)}$

CUMULATIVE REVIEW EXERCISES (CHAPTERS 10–18) (page 302)

1. (a) 4   (b) $13x - 5$ (c) $-10x^3z^{11}$ (d) y^{16}
2. (a) $15x - 20$ (b) $r^2 + 8r + 16$
3. (a) $-3/2$ (b) $-11/7$
4. (a) 325   (b) 118, 162   (c) Tammy is 12, Kim is 4   (d) 84 mph
5. $y = -3x + 19$
6. $x = 4, y = 0$
7. $x < -3/5$
8. (a) $(x + 6)(x + 6)$ (b) $(3y - 5)(y + 4)$
9. (a) $-1, 4$ (b) $-8, 8$ (c) $-1, 9/2$ (d) $0, 17$
10. (a) $\dfrac{x^2 + x - 3}{x(x - 3)}$ (b) $\dfrac{5x - 5}{2y}$

EXERCISES 19.1 (page 307)

1. (a) $r = \dfrac{d}{2}$ (b) $r = \dfrac{d}{t}$ (c) $d = \dfrac{C}{\pi}$ (d) $r = \dfrac{C}{2\pi}$ (e) $l = \dfrac{A}{w}$ (f) $h = \dfrac{V}{lw}$ (g) $r = \dfrac{i}{pt}$

   (h) $h = \dfrac{V}{\pi r^2}$

2. (a) $x = \dfrac{y}{6}$ (b) $x = -\dfrac{y}{5}$ (c) $x = \dfrac{y - b}{a}$ (e) $a = \dfrac{y - b}{x}$ (g) $w = \dfrac{P - 2l}{2}$

   (i) $y = \dfrac{x + 2}{7}$ (k) $x = \dfrac{5y + 2}{4}$ (m) $x = \dfrac{8 - 7y}{5}$ (o) $b = -ax$ (p) $x = -\dfrac{b}{a}$

   (q) $x = \dfrac{-by - c}{a}$ (s) $p = \dfrac{A - m}{m}$ (u) $d = \dfrac{l - a}{n - 1}$ (v) $p = \dfrac{R}{1 + rt}$

EXERCISES 19.2 (page 310)

1. (a) $h = \dfrac{2A}{b}$ (c) $m = \dfrac{2E}{v^2}$ (e) $E = IR$ (f) $R = \dfrac{E}{I}$ (g) $s = tv$

   (i) $m_1 = \dfrac{Fd^2}{m_2}$ (j) $h = \dfrac{2A}{b_1 + b_2}$ (k) $c = \dfrac{1}{2\pi fR}$ (l) $l = \dfrac{2S - na}{n}$

2. (a) $E = \dfrac{C(R + nr)}{n}$ (b) $R = \dfrac{nE - Cnr}{C}$ (c) $r = \dfrac{nE - CR}{Cn}$ (d) $n = \dfrac{CR}{E - Cr}$

   (e) $b_2 = \dfrac{2A - hb_1}{h}$ (g) $p = \dfrac{fq}{q - f}$ (i) $l = \dfrac{Sr - S + a}{r}$ (j) $r = \dfrac{S - a}{S - l}$

   (k) $s = \dfrac{Rg}{g - R}$ (m) $x = \dfrac{by}{y - a}$

CHAPTER 19 REVIEW EXERCISES (page 311)

1. (a) $e = \dfrac{p - 3d}{4}$ (b) $w = \dfrac{t - uv}{u}$ (c) $a = cm - b$ (d) $b = 2a - c$

2. (a) $v = \dfrac{T - 4n}{nr}$ (b) $a = \dfrac{cx + dx + 2b}{3}$ (c) $x = \dfrac{wy}{y - w}$ (d) $c = \dfrac{a + b - dx}{x}$

CUMULATIVE REVIEW EXERCISES (CHAPTERS 10–19) (page 311)

1. (a) -18 (b) $-x - 3$

2. $21x^2 - 11x - 2$

3. (a) $-5/8$ (b) $7/3$

4. (a) $\dfrac{17mn}{3}$ (b) $5x + 3$ (c) $12x + 6$ (d) $5x^2 + 3x$ (e) $2n - 2$

5. (a)   (b)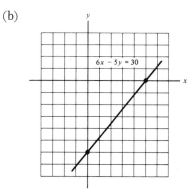

6. $y = -x + 6$

7. $x = 12, y = -3$

8. $x > -3/19$

9. (a) $-9, 4$ (b) $5/2, 3$

10. (a) $\dfrac{10x^2 + x + 1}{2x(x + 1)}$ (b) $\dfrac{x^2}{9}$

11. (a) $x = \dfrac{y - a}{ab}$ (b) $x = \dfrac{a - bm}{m - 1}$

EXERCISES 20.1 (page 316)

1. (a) x^5 (c) d^5 (d) 2^{25} (e) x^9 (g) m^{25} (h) x^{589} (i) 5^{615} (j) cannot simplify
(k) a^{250} (l) cannot simplify   (m) cannot simplify   (o) $40x^{11}$ (p) $12x^9y^{11}$
(q) $-56x^{18}y^{13}$ (r) $-6x^{13}$ (s) $63a^7b^7$ (u) $-90m^9n^{11}$

2. (a) x^{12} (c) r^{28} (e) $r^{14}s^{14}$ (g) $x^{50}y^{50}$ (h) $p^{17}q^{17}r^{17}$ (i) $x^{20}y^{30}$ (k) a^8c^{14}
(l) $125x^9y^{24}$ (m) cannot simplify   (o) $64x^2y^{12}$ (p) $-27x^{15}y^{30}$ (q) $16m^{12}n^4p^8$
(s) $-125a^9b^{21}$

3. (a) $75m^9n^{16}$ (b) $648x^{31}y^{42}$ (c) $-8p^{52}q^{13}$ (e) cannot simplify   (g) $-288b^{11}c^{33}d^8$
(h) cannot simplify   (i) $\dfrac{x^9}{y^9}$ (k) $\dfrac{32m^5}{n^5}$ (m) $\dfrac{x^8y^{10}}{z^6}$ (n) $\dfrac{b^6c^{21}}{a^{12}d^{15}}$ (o) $\dfrac{m^{28}n^{32}p^8}{625t^8}$

4. (a) 8   (b) 64   (c) 625   (d) 243   (e) 18   (f) 48   (g) 256   (h) 729   (i) 16/81
(j) 27/64   (k) 625   (l) 125

5. (a) 2^{10}

EXERCISES 20.2 (page 321)

1. (a) 1/8   (c) 1/27   (e) 1/6   (g) 3/5   (i) 7/8   (k) 1/36

2. (a) $\dfrac{1}{x^5}$ (c) $\dfrac{4}{m^6}$ (e) $-\dfrac{7}{t^4}$ (g) $\dfrac{1}{9x^2}$ (i) $\dfrac{1}{4n^7}$ (k) $\dfrac{2}{7m^4}$ (m) $\dfrac{5}{x^7y^4}$ (o) $\dfrac{7n^7}{m^6}$

   (q) $\dfrac{a^2c^4}{b^3d^5}$ (s) $-\dfrac{8n^2}{m^7p^9}$

3. (a) $\dfrac{3y}{x^4z^5}$ (c) $\dfrac{5}{2a^4b^6c}$ (e) $\dfrac{c^8}{a^4b^3}$ (g) $\dfrac{7b^4y^3}{a^6c^2x^5}$ (i) $-\dfrac{5v^3z}{w^2x^4y^5}$

4. (a) $\dfrac{1}{x^9}$ (c) $\dfrac{6}{t^{13}}$ (e) $x^{12}y^{20}$ (g) $\dfrac{x^{18}y^{15}}{8}$ (i) $\dfrac{a^6b^{14}c^{10}}{25}$

EXERCISES 20.3 (page 323)

1. (a) x^{30} (c) y^3 (e) $\dfrac{1}{m^3}$ (g) x^4y^3 (i) x^4y^3 (k) $\dfrac{1}{a^4b^{11}}$ (m) $\dfrac{1}{p^5q^3r}$

2. (a) $\dfrac{4x^7}{y^7}$ (c) $\dfrac{4x^8}{5yz}$ (e) $\dfrac{4x^7}{5w^7}$ (g) $\dfrac{2x^7z^4}{y^4}$ (i) $\dfrac{m^2}{np^5}$ (k) $\dfrac{1}{7w^4x^6y^9z^5}$

EXERCISES 20.4 (page 325)

1. (a) 1   (b) 3   (c) 1   (e) 5   (g) 6   (h) 2   (i) 7   (k) -3 (m) 1
   (n) $5x - 1$ (o) 0

EXERCISES 20.5 (page 329)

1. (a) 5   (c) 10   (e) 2   (g) 3   (i) 5   (j) 10   (k) 63   (m) 15   (o) 1
2. (a) 3/5   (c) 2/3   (e) 6
3. (a) 8   (c) 64   (e) 25   (g) 20   (h) 133   (i) 10   (j) 10   (k) 270   (l) 264
4. (a) 1/7   (c) 1/3   (e) 1/32   (g) $2\frac{1}{2}$ (i) 28
5. (a) 32   (c) 243   (e) 1   (g) 1/27   (i) 1
6. (a) x^4y^2 (c) $7ab^3$ (d) $-2x$ (g) x^2 (i) $25r^2s^6$ (k) $1331x^{15}y^{24}$ (l) $5a^3y^5$
   (n) $2xy^2z^3$ (o) $12a^5b^3$ (q) $6m$ (r) $-4xy^2$
7. (a) $2x^2$ (b) $-2y$ (e) $3a^3b^7$
8. (a) $x^{3/4}$ (c) x (e) $x^{5/6}$
9. (a) $x^{\frac{3}{2}}$ (c) $x^{\frac{5}{3}}$ (e) $y^{\frac{7}{4}}$
10. (a) $\sqrt{x^5}$ (c) $\sqrt[3]{n^7}$ (e) $\sqrt[5]{t^4}$

EXERCISES 20.6 (page 333)

1. (a) 4.7×10^6 (c) 3.5×10^{11} (e) 4.53×10^{17} (g) 2.3×10^{-4} (i) 7.0×10^{-7}
   (k) 1.3×10^{-12} (m) 1.2×10^{13}
2. (a) 340,000,000   (c) 942,000,000,000   (e) .0000000013   (g) .00000000050   (i) 75.63
3. (a) 4.5×10^9 (b) 1.2×10^{19} (c) 6.2×10^{22} (d) $7\,31 \times 10^{12}$ (e) 4.73×10^{-10}
   (f) 8.41×10^{-14} (g) 6.21×10^{18} (h) 4.56×10^{-22} (i) 3.0×10^{24} (j) 5.2×10^{-22}
4. (a) 5.0×10^9 (b) 2.0×10^3 (c) 2.0×10^{-11} (d) 10^{11}

CHAPTER 20 REVIEW EXERCISES (page 333)

1. (a) x^{12} (c) x^{48} (e) $15x^{16}$ (g) $8m^{12}y^{15}$ (i) $45x^{15}y^{23}$ (k) $\dfrac{81x^8}{y^{20}}$ (m) t^{21}

(o) $-32x^{10}y^5$ (q) $-35a^4b^{11}c^4$

2. (a) $\dfrac{1}{x^7}$ (c) $\dfrac{4}{m^2}$ (e) $-\dfrac{7}{x^3}$ (g) $\dfrac{1}{3y^6}$

3. (a) x^3y^7 (c) $\dfrac{y}{x^3}$ (e) $\dfrac{1}{16x^2}$ (g) $\dfrac{5y^4}{x^3z}$ (i) $-\dfrac{3}{x^7y^8}$ (k) $\dfrac{4s^{12}u^5}{r^6t^3}$ (m) $\dfrac{1}{x^{10}}$ (o) $\dfrac{3}{x^9}$

(q) $b^{12}c^{15}$ (s) $\dfrac{m^{13}}{n^{10}p^4q^3}$

4. (a) 24   (b) 1   (c) 8   (d) -2 (e) 10   (f) 2   (g) 2   (h) 7/8   (i) 12   (j) 32
(k) 9   (l) 24   (m) 1/16   (n) 5/9   (o) 1   (p) 128

5. (a) $7m^5n^{12}$ (c) $16x^2y^8z^{20}$

CUMULATIVE REVIEW EXERCISES (CHAPTERS 10–20) (page 334)

1. 10
2. $x^2 + 3x + 29$, R: 108
3. $-51/10$
4. 42.5°, 65.5°, 72°
5.

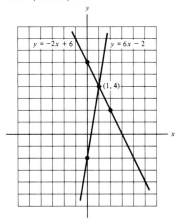

6. (a) $x = 5, y = 2$ (b) $x = 4, y = -9$
7. (a) $(y - 21)(y + 3)$ (b) cannot be factored

8. $\dfrac{20x + 3}{2 - 10x}$

9. $-17/42$

10. (a) $y = \dfrac{x - w + 1}{3}$ (b) $y = \dfrac{bx}{tx - a}$

11. (a) $\dfrac{y^4}{x^5}$ (b) 1/7   (c) -67

EXERCISES 21.2 (page 340)

1. (a) $2\sqrt{2}$ (b) $2\sqrt{3}$ (c) $3\sqrt{3}$ (d) 6   (e) $5\sqrt{2}$ (f) $2\sqrt{15}$ (g) $2\sqrt{7}$ (h) $2\sqrt{11}$
   (i) $3\sqrt{5}$ (j) 10   (k) $5\sqrt{5}$ (l) $3\sqrt{7}$ (m) $4\sqrt{2}$ (n) $4\sqrt{3}$ (o) $2\sqrt{13}$ (p) $2\sqrt{14}$
   (q) $10\sqrt{2}$ (r) 30
2. (a) $2\sqrt{17}$ (c) $3\sqrt{10}$ (e) cannot be simplified   (g) $4\sqrt{7}$ (i) $8\sqrt{2}$ (k) 50
3. (a) 7   (c) $5\sqrt{5}$ (e) 9   (g) $2\sqrt{5}$ (i) 10
4. (a) $10\sqrt{2}$ (b) 15   (e) $20\sqrt{17}$ (g) $6\sqrt{5}$ (i) $50\sqrt{3}$
5. (a) $\sqrt{2}$ (c) $\sqrt{5}/6$ (e) $\sqrt{5}$
6. (a) $1 + \sqrt{5}$ (c) $2 + \sqrt{7}$ (e) $\dfrac{1 + \sqrt{6}}{3}$
7. (a) $3\sqrt{13}$ (c) $7\sqrt{2}$ (e) $2\sqrt{26}$ (g) $9\sqrt{2}$ (i) cannot be simplified   (k) $7\sqrt{3}$
   (m) $3\sqrt{17}$ (o) cannot be simplified   (q) $3\sqrt{31}$

EXERCISES 21.3 (page 342)

1. (a) $2\sqrt{7}$ (c) $14\sqrt{2}$ (e) $11\sqrt{2}$ (f) 6   (g) $10\sqrt{2}$ (i) $5\sqrt{2}$ (k) cannot combine
   (l) 104   (m) $32\sqrt{3}$ (o) $11\sqrt{5}$ (p) cannot combine   (q) $5\sqrt{10}$ (r) $12\sqrt{3}$
   (s) $3\sqrt{5} + \sqrt{10}$

EXERCISES 21.4 (page 345)

1. (a) $\dfrac{3\sqrt{2}}{2}$ (c) $\dfrac{2\sqrt{5}}{5}$ (e) $2\sqrt{3}$ (g) $2\sqrt{5}$

2. (a) $\sqrt{5}$ (c) 15   (e) $\dfrac{2\sqrt{3}}{5}$ (g) $\dfrac{3\sqrt{10}}{2}$

3. (a) $\dfrac{2}{5\sqrt{2}}$ (c) $\dfrac{4}{3\sqrt{2}}$ (e) $\dfrac{1}{\sqrt{6}}$ (g) $\dfrac{1}{\sqrt{2}}$

4. (a) $\dfrac{\sqrt{3}}{3}$ (c) $\dfrac{2\sqrt{14}}{7}$ (e) $\dfrac{1}{3}$ (g) $\dfrac{2\sqrt{14}}{7}$

5. (a) $\sqrt{30}$ (c) $3 + 2\sqrt{2}$ (e) 18   (g) $48 - 6\sqrt{3} - 8\sqrt{13} + \sqrt{39}$
6. (a) $\sqrt{5} - 2$ (b) $3\sqrt{2} - 3$ (c) $2\sqrt{7} + 4$ (d) $7\sqrt{10} - 21$

EXERCISES 21.5 (page 349)

1. (a) 81   (c) 49   (e) 17   (g) 80/7   (i) 31/9   (k) 17/20   (m) 4   (o) 10/11
2. (a) 7   (c) no solution   (e) 23/2   (g) 3/4   (h) 11   (i) −1   (j) 447/7

CHAPTER 21 REVIEW EXERCISES (page 350)

1. (a) $2\sqrt{6}$ (c) $10\sqrt{5}$ (e) $3\sqrt{6}$ (f) $2\sqrt{19}$ (g) $4\sqrt{10}$ (i) cannot be simplified
   (j) $3\sqrt{19}$ (k) $9\sqrt{3}$
2. (a) $26\sqrt{2}$
3. (a) $7\sqrt{5}/5$
4. (a) $\sqrt{11}/11$
5. (a) 119/3

CUMULATIVE REVIEW EXERCISES (CHAPTERS 10–21) (page 350)

1. $6x^2 + 7x - 20$
2. $-1/2$
3. (a) $12r$ (b) $4x + 3$ (c) $5y - 3x$
4. $y = -3x + 19$
5. $m = 6, n = 3$
6. (a) $5x(x - 2)$ (b) $2(x + 9)(x - 9)$ (c) $3(x + 5)(x - 4)$ (d) $(4x - 3)(x + 4)$
7. (a) $\dfrac{2x + 21}{6(x - 2)}$ (b) $\dfrac{4np^2}{3m^2t^7}$
8. $x = \dfrac{t + y}{m + n}$
9. (a) $\dfrac{b^8}{a^{11}}$ (b) $\dfrac{x^{17}}{y^3}$ (c) 3
10. (a) $2\sqrt{30}$ (b) $3\sqrt{10}$

EXERCISES 22.1 (page 353)

1. (a) ± 8 (b) ± 1 (c) ± 6 (d) ± 10 (e) $\pm\sqrt{7}$ (f) $\pm\sqrt{11}$ (g) $\pm\sqrt{15}$ (h) $\pm\sqrt{47}$
2. (a) $-1 \pm \sqrt{11}$ (c) $3 \pm \sqrt{19}$ (e) $4 \pm 2\sqrt{2}$ (g) $-1 \pm 2\sqrt{7}$
3. (a) $2 \pm \sqrt{3}$ (c) $-1 \pm \sqrt{7}$

EXERCISES 22.2 (page 357)

1. (a) $-4 \pm \sqrt{15}$ (c) $-3 \pm \sqrt{2}$ (e) $-6 \pm \sqrt{21}$ (g) $4 \pm 2\sqrt{2}$ (i) $5 \pm 2\sqrt{6}$
   (k) $6 \pm \sqrt{39}$ (m) $-5 \pm \sqrt{29}$ (o) $-2 \pm \sqrt{5}$
2. (a) $\dfrac{3 \pm \sqrt{5}}{2}$ (c) $\dfrac{-5 \pm \sqrt{21}}{2}$ (e) $\dfrac{-3 \pm \sqrt{41}}{2}$ (g) $\dfrac{-3 \pm \sqrt{11}}{2}$ (i) $\dfrac{-3 \pm \sqrt{29}}{10}$

EXERCISES 22.3 (page 361)

1. (a) $\dfrac{-3 \pm \sqrt{5}}{2}$ (b) $\dfrac{-5 \pm \sqrt{17}}{2}$ (c) $\dfrac{5 \pm \sqrt{37}}{2}$ (d) $\dfrac{3 \pm \sqrt{5}}{2}$ (e) $\dfrac{-5 \pm \sqrt{97}}{6}$
   (f) $\dfrac{-3 \pm \sqrt{33}}{4}$ (g) $\dfrac{7 \pm \sqrt{41}}{2}$ (h) $\dfrac{11 \pm \sqrt{113}}{2}$ (i) $\dfrac{9 \pm \sqrt{89}}{4}$ (j) $\dfrac{7 \pm \sqrt{97}}{6}$
2. (a) $2 \pm \sqrt{3}$ (c) $-3 \pm 2\sqrt{2}$ (e) $1 \pm \sqrt{2}$ (g) $-4 \pm 2\sqrt{5}$ (i) $5 \pm \sqrt{22}$ (k) $\dfrac{1 \pm \sqrt{22}}{3}$
   (m) $\dfrac{2 \pm \sqrt{3}}{2}$ (o) $\dfrac{-3 \pm \sqrt{11}}{2}$
3. (a) $\dfrac{-7 \pm \sqrt{69}}{2}$ (c) $\dfrac{3 \pm \sqrt{41}}{2}$ (e) $\dfrac{-9 \pm \sqrt{89}}{2}$ (g) $\dfrac{1 \pm \sqrt{37}}{6}$ (i) $\dfrac{-1 \pm \sqrt{41}}{2}$
4. (a) $-3/5, 2$ (c) $-6, 0$ (e) $\pm\sqrt{3}$ (g) $-1/2, 1$ (i) $-2/3, 1$

EXERCISES 22.5 (page 365)

1. (a) $4i$ (c) $i\sqrt{3}$ (e) $2i\sqrt{6}$ (g) $3i\sqrt{3}$ (i) $4i\sqrt{5}$

2. (a) $\pm 2i$ (c) $1 \pm 2i$ (d) $\dfrac{-1 \pm i\sqrt{3}}{2}$ (e) $\dfrac{-1 \pm i\sqrt{7}}{2}$ (g) $\dfrac{3 \pm i\sqrt{23}}{4}$ (i) $1 \pm 2i\sqrt{2}$

EXERCISES 22.6 (page 366)

1. (a) real and unequal   (c) imaginary and unequal   (e) real and equal   (g) real and unequal
(h) imaginary and unequal   (i) real and unequal   (l) real and unequal   (m) real and unequal
(n) real and unequal

EXERCISES 22.7 (page 369)

1. (a)

(b)

(c)

(d)

(e)

(f)

2. (a)

(c)

(e)

(g)

(i)

(k)

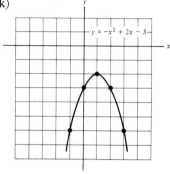

CHAPTER 22 REVIEW EXERCISES (page 369)

1. (a) $-1/3, 1$ (b) $\dfrac{3 \pm \sqrt{3}}{3}$ (c) $\dfrac{-9 \pm \sqrt{73}}{4}$ (d) $-7/2, -1/2$

2. (a) $-5 \pm 2\sqrt{6}$

3. (a) imaginary and unequal (b) real and unequal (c) real and equal

4.

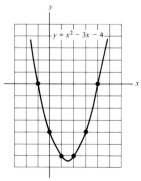

CUMULATIVE REVIEW EXERCISES (CHAPTERS 10–22) (page 369)

1. (a) 48   (b) -33 (c) $4x - 11$ (d) $13 - 29x$
2. $x^2 - 2x + 1$
3. 7/12
4. (a) length 285 meters, width 142 meters   (b) Marcy 53, Janice 46
5. $y = 2x + 3$
6. (a) $x = 10, y = -1$ (b) $x = -4, y = 2$
7. $x \geq -29/16$

8. (a) 20/7   (b) $\dfrac{5}{3x(x - 6)}$ (c) $\dfrac{20 + 9y^2}{24xy}$ (d) $\dfrac{2x + 19}{(x - 3)(x + 2)}$

9. (a) $-3/8$ (b) $\dfrac{-7 \pm 3\sqrt{5}}{2}$

10. (a) $64x^{30}y^{18}$ (b) $-54m^{26}$ (c) $-\dfrac{5y^6}{x^{19}}$ (d) $15x^2y^{50}z^7$

11. (a) 3/2   (b) 1   (c) 27   (d) 8

12. (a) $m = \dfrac{y - b}{x}$ (b) $y = \dfrac{-5x - 7}{9}$ (c) $a = \dfrac{cx + dx}{b}$ (d) $d = \dfrac{ab - cx}{x}$ (e) $d = \dfrac{a - b + ce}{c}$

13. $\dfrac{\sqrt{13}}{5}$

14. -31

15. (a) 3/2, 7   (b) $-2 \pm i$ (c) 1, 6   (d) $\dfrac{9 \pm \sqrt{105}}{4}$

16. $3 \pm \sqrt{10}$

EXERCISES 23.1 (page 372)

1. (a) $\{2, 4, 6, 8, 10, 12, 14, 16, 18, 20, 22, 24\}$ (c) $\{7, 8, 9, 10, 11, 12, 13, 14, 15\}$
   (e) {March, May}
2. (a) $5 \in \{1, 3, 5, 7, 9, 11, 13, 15\}$ (b) $7 \notin \{18, 16, 14, 12, \ldots\}$ (c) $x \notin \{a, e, i, o, u\}$
3. (a) true   (b) false   (c) false   (d) false   (e) false   (f) false   (g) false

EXERCISES 23.2 (page 374)

1. (a) $\{x | x > 13\}$ (b) $\{x | x < 1/2\}$ (d) $\{x | x \leq 2\}$ (e) $\{x | x \geq 0\}$ (g) $\{x | -5 \leq x \leq 16\}$
   (i) $\{x | x \text{ even}, 3 < x < 95\}$ (k) $\{x | -25 \leq x \leq 25, x \neq 0\}$
2. (a) all the positive numbers   (b) all numbers except zero   (d) 6, 8, 10, 12, 14, . . .
   (f) 0, 1, 2, 3, 4, 5, . . .   (i) all numbers between 3 and 9 exclusive
   (k) all numbers between 3 and 9, including 3 but excluding 9

EXERCISES 23.3 (page 375)

1. (a) $\{5, 6, 8, 12, 19, 35\}$ (c) $\{3, 4, 7, 8, 12, 14\}$ (e) $\{1, 2, 3, 4, 5, 6\}$
2. (a) $\{19\}$ (c) $\varnothing$ (e) $\{1, 3, 4\}$
3. (a) $\{x, y, z\}$ (c) $\{a, b, c, d, e, f\}$

EXERCISES 23.4 (page 376)

1. (a) false   (b) false   (d) true   (f) true   (i) true   (j) true
2. (a) $\{a, b\}, \{a\}, \{b\}, \{\ \}$ (b) $\{x, y, z\}, \{x, y\}, \{x, z\}, \{y, z\}, \{x\}, \{y\}, \{z\}, \{\ \}$

EXERCISES 23.5 (page 377)

1. (a) $\{7, 9, 11, 13, 15, 17\}$ (b) $\{1, 3, 5, 9, 11, 13, 17\}$ (c) $\mathcal{U}$ (d) $\{\ \}$
(e) $\{1, 3, 5, 7, 9, 11, 15, 17\}$

EXERCISES 23.6 (page 382)

1. (a) 1, 2   (b) 2, 3   (c) 1, 2, 3   (d) 2   (e) 1, 2, 3, 4   (f) 3, 4   (g) 1, 4   (h) 4
(i) 1, 3, 4
2. (a) 1, 2, 3, 4   (b) 3   (f) 5, 6   (g) 1, 2, 4, 5, 6
3. (a) 1, 2, 4, 5   (d) 1, 2, 3, 4, 5, 6, 7, 8   (e) 2, 5   (h) 5   (i) 1, 2, 3, 4, 5, 6
(l) 1, 2, 3, 4, 5, 6, 7
4. (a)

(c)

(e)

(g)

(j)

EXERCISES 23.7 (page 386)

1. (a) yes   (b) yes   (c) no   (d) yes   (e) yes   (f) no
2. (a) yes   (b) yes   (c) no   (d) yes   (e) yes   (f) yes   (g) no   (h) no

EXERCISES 23.8 (page 388)

1. (a) -2 (b) 1   (c) -5 (d) -20 (e) $3a - 5$ (f) -4 (g) -6 (h) $9m - 5$
(i) $3x - 2$ (j) $3x - 20$ (k) $3t - 8$ (l) $3x + 3h - 5$
2. (a) 19   (b) 32   (c) 179   (d) -1 (e) -8 (f) -16 (g) $m^2 + 8m - 1$ (h) 13/4
(i) $-47/16$ (j) $x^2 + 12x + 19$ (k) $x^2 + 6x - 8$ (l) $x^2 + 2xh + h^2 + 8x + 8h - 1$

EXERCISES 23.9 (page 393)

1. (a) 8/10 or 4/5   (b) 6/10 or 3/5   (c) 8/6 or 4/3

2. (a) $\dfrac{3}{\sqrt{13}}$ (b) $\dfrac{2}{\sqrt{13}}$ (c) $\dfrac{3}{2}$ (d) $\dfrac{2}{\sqrt{13}}$ (e) $\dfrac{3}{\sqrt{13}}$ (f) $\dfrac{2}{3}$

3. (a) $\dfrac{1}{\sqrt{2}}$ (b) $\dfrac{1}{\sqrt{2}}$ (c) 1   (d) $\dfrac{1}{\sqrt{2}}$ (e) $\dfrac{1}{\sqrt{2}}$ (f) 1

5. 86 feet   6. 848 feet   9. 70 feet

CHAPTER 23 REVIEW EXERCISES (page 395)

1. All numbers between 4 and 9, including 4 but excluding 9.
2. (a) {1, 2, 3, 4, 5, 6}   (b) Ø   (c) {1, 2, 3, 5, 7}   (d) {5}   (e) {1, 2, 3, 7}   (g) Ø
    (h) Yes, all elements of A are also elements of the universe.
4. (a) no   (b) yes
5. (a) 33   (b) 5
6. (a) $\sqrt{113}$ (b) $\dfrac{7}{\sqrt{113}}$ or $\dfrac{7\sqrt{113}}{113}$ (c) $\dfrac{8}{\sqrt{113}}$ or $\dfrac{8\sqrt{113}}{113}$ (d) $\dfrac{7}{8}$

INDEX